LADY MARGARET HALL LIBRARY
OXFORD

Characterisation of Catalysts

Characterisation of Catalysts

Edited by

J.M. Thomas and R.M. Lambert
Department of Physical Chemistry
University of Cambridge

A Wiley–Interscience Publication

JOHN WILEY & SONS
Chichester · New York · Brisbane · Toronto

British Library Cataloguing in Publication Data:

Characterisation of catalysts.
1. Catalysts—Congresses
2. Heterogeneous catalysis—Congresses
I. Thomas, John Meurig
II. Lambert, R.M.
660.2'9'95

ISBN 0 471 27874 2

Printed in Great Britain

CONTENTS

ACKNOWLEDGMENTS

We would like to thank the following for permission to reproduce some of the illustrations and tables used in this volume:

Academic Press Inc

American Chemical Society

American Institute of Physics

The Chemical Society

N M Fisher

McGraw-Hill, Inc

North-Holland Publishing Company

Pergamon Press Ltd

Taylor & Francis Ltd

PREFACE

Unlike the situation that prevailed only a few years ago, several techniques are now available for carrying out in situ, dynamic studies of catalysts. Until recently, essentially all the methods used for catalyst characterisation could be classified as either post-mortem or pre-natal, in the sense that tests were carried out either on the expired, poisoned, or partly consumed catalyst or, alternatively, on the newly prepared, preactivated or 'simulated' solid. Great progress was achieved in this way, a fact borne out by the virility of the chemical industry in which heterogeneous catalysts continue to play a crucial rôle.

Fortunately, however, significant advances have been made very recently in extending traditional methods, such as those based on infrared and Raman spectroscopy and radioisotope exchange. But even greater progress has been registered in the development of novel techniques such as those employing neutron beams, intense X-ray sources and other methods. There are now good reasons for believing that dynamic studies of catalysts and catalysis, under the actual conditions of industrial processes, will soon become more or less routinely feasible.

It was an awareness of these facts that led us to organize, in the summer of 1979 in King's College, Cambridge, a short appreciation course on catalyst characterisation. Encouraged by the response of the participants and many others, we undertook to collate most of the presentations given at that course. They are reproduced here in expanded form, along with other contributions which together cover all the important viable techniques with which the catalyst manufacturer, user or researcher must nowadays be familiar.

Considerable prominence is given here to newer methods, particularly neutron scattering, EXAFS, Mössbauer and other types of spectroscopy, electron microscopy and cyclic voltammetry. But the foundations and scope of older techniques such as radioisotope exchange, surface area and pore volume determinations, are also assessed.

The book is aimed at all those interested in the theory and practice of heterogeneous catalysis, and no knowledge beyond that normally covered in undergraduate chemistry courses is presupposed. All the techniques that are covered are applicable to, or relevant in, the characterisation of commercial catalysts. They are critically examined and discussed by acknowledged experts, and many specific examples of their uses are given. It is hoped that this will enable the research worker to evaluate their scope and relative merits, and, if necessary, to adapt and extend them to suit his particular needs.

We express our appreciation to Mrs. N.F. Hurley for her invaluable work in preparing this text, and to Messrs E.L. Smith, N.F. Cray and I. Cannell for their help with the illustrations. Not least we acknowledge with gratitude the friendly co-operation of our colleagues for whose work we take editorial responsibility.

J.M. Thomas

R.M. Lambert

King's College and
Department of Physical Chemistry,
Cambridge, 1980.

LIST OF CONTRIBUTORS

ACRES, Dr. G.J.K. BIRD, Dr. A.J. JENKINS, Dr. J.W. KING, Dr. F.	Johnson Matthey Research Centre, Blount's Court, Sonning Common, Reading RG4 9NH.
CAIRNS, Dr. J.A.	Metallurgy Division, AERE Harwell, Oxfordshire OX11 0RA.
COX, Mr. A.D.	Department of Physics, University of Warwick, Coventry, Warwickshire CV4 7AL.
EDMONDS, Dr. T.	The British Petroleum Co. Ltd. BP Research Centre, Chertsey Road, Sunbury-on-Thames, Middlesex TW16 7LN.
HOWIE, Dr. A.	Department of Physics, University of Cambridge, Cavendish Laboratory, Madingley Road, Cambridge CB3 0HE.
JONES, Dr. W.	Department of Physical Chemistry, University of Cambridge, Lensfield Road, Cambridge CB2 1EP.
JOYNER, Dr. R.W.	School of Studies in Chemistry, University of Bradford, Bradford, West Yorkshire BD7 1DP.
KENNEY, Dr. C.N.	Department of Chemical Engineering, University of Cambridge, Pembroke Street, Cambridge CB2 3RA.
LAMBERT, Dr. R.M.	Department of Physical Chemistry, University of Cambridge, Lensfield Road, Cambridge CB2 1EP.
McNICOL, Dr. B.D.	Shell International Petroleum Co.Ltd PNTL Division, Shell Centre, London SE1 7NA.
MURRAY, Dr. R.T.	Imperial Chemical Industries Ltd. Corporate Laboratory, P.O. Box 11, The Heath, Runcorn, Cheshire WA7 4QE.

cont.

LIST OF CONTRIBUTORS (cont)

PETTIFER, Dr. R.F. — Department of Physics, University of Warwick, Coventry, Warwickshire CV4 7AL.

SAMPSON, Dr. R.J. — Imperial Chemical Industries Ltd. Research & Development Department, Petrochemical Division H.Q. PO Box 90, Wilton, Middlesbrough, Cleveland TS6 8JE.

SING, Prof. K.S.W. — Department of Applied Chemistry, Brunel University, Uxbridge, Middlesex UB8 3PH.

THOMAS, Prof. J.M. — Department of Physical Chemistry, University of Cambridge, Lensfield Road, Cambridge CB2 1EP.

THOMSON, Prof. S.J. — Department of Chemistry, University of Glasgow, Glasgow G12 8QQ.

WRIGHT, Dr. C.J. — Materials Physics Division, AERE Harwell, Oxfordshire OX11 0RA.

I

RECENT TRENDS IN SURFACE SCIENCE AND THEIR IMPACT ON CATALYST CHARACTERISATION

By

J.M. Thomas

1. INTRODUCTION

In the real world of the industrial scientist there is generally much respect, even admiration, for the significant advances that have recently been made by academically oriented investigators in the study of solid surfaces. During the past decade several ingenious ways of establishing the compositions, crystallographic structures and electronic properties of the last few layers of adsorbents and catalysts - or sub-monolayer amounts of adsorbed species - have been devised thanks largely to the arrival of new techniques such as photoelectron spectroscopy, Auger and electron-energy-loss spectroscopy, ion scattering procedures, low energy electron diffraction (LEED), and the extension of more traditional ones such as infrared and Raman spectroscopy.

It is, however, undeniable that a certain sense of disappointment - if not frustration which, at times, borders on cynicism - pervades the catalyst community when it reflects on the paucity of techniques that are capable of being utilized to study 'live' catalysts, under actual operating conditions. Some individuals, whose task it is to design new catalysts or to improve existing ones, have become inured to the waves of enthusiasm that frequently overtake (or overwhelm) the purist surface chemist or chemical physicist when the discovery of another potent, all-conquering technique is announced. They greet with scepticism descriptions of identification of a new state of a surface-bound molecule or a new 'electronic surface state', that exist under conditions (typically 10^{-10} to 10^{-4} torr) widely removed from those relevant to commercial catalytic reactors (e.g. a few hundred atmospheres pressure and temperatures of several hundred kelvin).

No one denies the great importance of an academic concern for fundamentals: indeed the dictum that 'the more closely we enquire into the nature of things the greater is our reward' is universally valid. Without paying due attention to fundamentals it ceases to become possible to confirm or reject plausible models of adsorption and catalysis, irrespective of whether or not the conditions under which the model apply may be far removed from a given, desired set. And it often follows that, with due allowance for error, an effect or feature identified under one set of well-defined conditions may indeed be equally valid under another. Thus, if it is incontrovertibly established (by EXAFS for example) that rafts of metal atoms, rather than three-dimensional arrays (clusters) of active metal catalysts are distributed in a given manner or an appropriate oxide support, it is not improbable though not impossible that these rafts also exist under the actual real-life conditions of the 'live' catalyst. Clearly the principal point here is to know whether the facts pertaining to the existence of the rafts have indeed been incontrovertibly established.

To take another, rather extreme situation, few catalyst experts are likely to argue that because LEED, for example, signifies the existence of a given kind of surface reconstruction under rather clinical conditions - ultra-pure metal, fully outgassed and well-annealed, with pressures in the vicinity of 10^{-4} torr of an ambient gas - it will follow that the same state of surface reorganization will persist to elevated (say 100 atm) pressures of the reactant gas. It is not so much the facts that we already know which cause problems when we extrapolate to other conditions: rather it is those that we have not yet uncovered, and which may apply uniquely to the actual, catalytic operating conditions, that arouse doubts. Because of this and other reasons, the quest for methods of studying catalysts *in situ* at high temperatures and pressures is currently pursued with considerable vigour. What bonding states are there at catalyst surfaces when the reactant gases are at several hundred atmospheres pressure ? And what relevant or crucial surface phases exist under a combination of elevated pressure, temperature and reaction mixture ?

2. ORGANISATION OF SUBSEQUENT CHAPTERS

All the techniques discussed in this monograph are, in greater or lesser measure, applicable to, or relevant in, the characterisation of commercial catalysts. Some are classical, well-established and widely used. Surface area and pore-volume measurements, for example, play so basic a rôle in the assessment and specification of catalysts, that it is prudent to reconsider the essentials of current practice, and, more particularly, also to examine the principles upon which area and volume measurements are based. The opening chapter by Sing, a notable pioneer in this field, places this topic into appropriate perspective.

The two succeeding chapters survey several aspects of catalyst characterisation from the viewpoint of the industrial user (Sampson) and the manufacturer (Acres). These chapters exemplify the need to employ a multiplicity of techniques rather than relying on the results of a few convenient ones. The combined use of several key procedures, elaborated in subsequent chapters, is obviously a necessity; and Acres' account highlights revealingly which collection of tools serve the manufacturer best.

Electron microscopy has burgeoned in the past decade. Not only have accelerating voltages been increased and lens design and performance greatly improved in conventional transmission electron microscopes (TEM), but a basically different kind of electron-optical instrument - the scanning transmission electron microscope (STEM), where a fine pencil of focussed electrons, of

5 Å or less cross-section, is scanned across the surface to be studied under ultra-high resolution conditions - has emerged. Apart from structural information, extracted either from diffraction patterns, or from direct imaging, extremely localised microanalysis is possible. Using either the X-ray emission (electron beam stimulated) or the electron-energy loss signals, amounts of as little as 10^{-20}g (of say a metal such as Cu or Ru on a silica support) are detectable. The wealth of information pertaining to catalysts that electron microscopy has uncovered is considered in detail by Howie. And a brief account of a typical industrial application in polyolefin catalysts is given by Murray.

One of the most important attributes of XPS (X-ray induced photoelectron spectroscopy or ESCA, electron spectroscopy for chemical analysis) is its extreme sensitivity. It can, along with other techniques involving ion beams rather than photons, detect the presence of minute fractions of monolayers of adsorbed entities even on single crystal surfaces. But much more than mere detection may be achieved. Often the valence state, the nature of the electron transfer to or from the surface, as well as the existence of new phases may be deduced from photoelectron spectroscopic studies. Such techniques which, by the very limitations inherent in electron mean free paths, are unlikely ever to be used under in situ catalytic conditions, have contributed much to our understanding of catalyst behaviour, and especially in tracing the delicate compositional changes (including poisoning) which catalytic material exhibit during activation. Such applications of XPS/ESCA are surveyed by Edmonds, whose experience in industrial environments serves him well in this task. Even though ESCA is a demonstrably post-morten approach to catalyst characterisation, it offers considerable insight into questions associated with promoter action and catalyst selectivity.

It may at first seem strange that Mössbauer spectroscopy appears as a viable tool for the study of catalysts. The Mössbauer effect, as is well-known, involves resonant emission-absorption of energetic γ-photons, and has, ever since it was first used to elucidate solid-state phenomena, been regarded as a bulk technique. Less well-known is the fact that some excited Mössbauer nuclei (and ^{57}Fe is one of these) relax by a process which entails a substantial degree of internal conversion. This means that electrons (from the extranuclear core orbitals) are liberated on resonance. Bursts of electrons are therefore emitted only when nuclei in the Doppler modulated specimen obey the resonance condition. What Jones describes in his chapter is how, by detecting these conversion electrons (sometimes designated back-scattered electrons), a surface sensitive tool becomes available for the study of catalysts which contain iron, tin and other elements. He also discusses how, for high-area support catalysts, the bulk (γ-photon) Mössbauer effect can be put to good use for surface structural investigations. Much more is likely to be heard of this dual approach, especially in the study of highly porous, including zeolitic, catalysts.

Also of value and greater range of applicability for porous

or any high-area catalytic solid are the techniques which rely on temperature programming (i.e. on thermally stimulated desorption and/or reduction) or on voltage programming (i.e. cyclic voltammetry). These two complementary approaches have proved nicely applicable in the assessment of new catalysts for fuel-cell development, especially in the hands of McNicol, our reviewer here.

3. IN SITU CHARACTERISATION

A. Neutrons and Other Projectiles; Radioisotopes; EXAFS; and Kinetic Measurements

In varying degrees of effectiveness all the remaining techniques that are described in this monograph are, in principle, capable of being used for *in situ* i.e. dynamic studies of catalysts under live conditions.

Perhaps the most promising of all these is that which is based on neutron scattering. The advantages, potential and scope of neutron scattering studies are analysed by Wright (Chapter X). Suffice it to record here that (i) elastic neutron scattering yields diffraction data and hence crystallographic and structural information; (ii) inelastic neutron scattering reveals insights into diffusive motions of sub-monolayer quantities of adsorbate as well as direct information pertaining to the vibrational and other frequencies of bound precursors or reaction complexes; and (iii) neutrons can penetrate stainless steel and several other contaminent vessels which are frequently necessary for catalytic reactors that operate at elevated temperatures and pressures.

Not quite as advantageous for *in situ* studies, but of considerable general value nevertheless, are methods based on beams of protons or other projectiles. Cairns considers the merit for proton-induced X-ray emission studies. It transpires that the quality of the X-ray spectrum thus generated surpasses those produced by more conventional (i.e. fluorescence or electron-beam) methods. If certain other highly energetic projectiles are fired at catalyst surfaces, advantage can be taken of the subnuclear particles liberated as a consequence of the nuclear reactions that ensue above a threshold impinging energy. In this way 'carbon profiles' and other relevant items of information relating to the fouling of catalysts by carbon, or the accumulation of poisons (such as sulphur) may be obtained. As a bonus, projectile-induced X-ray emission simultaneously identifies the culprit impurities at which fouling or poisoning takes place preferentially.

Radioisotope techniques have long served as valuable tools for elucidating mechanisms of chemical conversion at solid surfaces. It was one of the earliest _in situ_ techniques to be invoked in the domain of catalyst characterisation. And Thomson, who summarizes (Chapter XII) radioisotope and related methods, has been one of the principal pioneers of this approach.

One of the newest of tools, though albeit of ancient lineage, for it was first identified by Kronig in the 1930's, is extended X-ray absorption fine structure (EXAFS) which, intrinsically, possesses the great advantage of being able to yield structural information even though the material in question may not possess long-range order. Again, this is principally a bulk technique, but, in ways that are outlined later, it may be adapted so as to give surface structural information (coordination numbers, interatomic distances and mean-square displacements of atoms and ions). EXAFS is best carried out using a synchrotron. With such a centralised facility, photon fluxes far exceeding those that are generated even by rotating anode devices, are available.

In view of its potential importance as a powerful tool for catalyst research, we here devote three chapters to EXAFS. Apart from its merits as a post-mortem method, it can, like the neutron-scattering approach, readily be adapted for _in situ_ studies. The contribution by Joyner reviews the general prospects of the technique, whereas that by Cox concentrates on a single case history. Pettifer's chapter underlines some of the problems that still remain on the interpretive front. Doubtless we shall hear a great deal about EXAFS in the future.

Up until a decade or so ago, kinetics and operational kinetic parameters such as orders of reaction and magnitudes of activation energy were used extensively for assessing the performance of catalysts and interpreting the nature of catalytic activity. Although this approach is still adopted occasionally, there has, of late, been a tendency to underutilize kinetics as an instrument for catalyst research. The pendulum has, as it were, swung too much in the other direction: this is a point that is touched upon, along with other relevant ones, by Kenney, who briefly analyses the present scene in catalyst characterisation from the viewpoint of the chemical engineer. Prospects and prognoses are also summarized, but from the viewpoint of the physical chemist, by Lambert who _inter alia_ faces up to some of the practical (including cost) factors that inevitably supervene in real-life situations.

B. Other Viable _In Situ_ Techniques

In addition to the use of neutrons, synchrotron radiation or radioisotopes, all of which are discussed critically in this book, there are other ways of achieving _in situ_, dynamic studies of catalysts under reactor conditions. Two of these techniques, infrared and Raman spectroscopy, have been available for some time, but only in very recent years has it become possible to employ them for _in situ_ studies. Raman methods are especially attractive because they are usable even when water is present in large amounts (contrast the infrared technique). A third technique, nuclear magnetic resonance, although rather more restricted in its potential field of application, may also be adapted for in situ studies. As none of these three is discussed at length in later chapters, summarizing details on the present position relating to them are given in this introductory chapter.

(a) Raman Spectroscopy

Two technical developments have been responsible for the renaissance of interest and application of Raman spectroscopy generally: (i) the discovery of the laser and its use as a convenient source; and (ii) the computerisation of the necessary instrumentation, and in particular the data processing that is nowadays associated with a Raman spectrometer. Thus fluorescence, which has long been a serious problem in the Raman study of solids, may now be minimized by mathematical manipulation (Fourier transform) of the baseline slopes. Moreover, sensitivity to trace components in absorbate or adsorbent mixtures, or to poor Raman scatterers, may be enhanced by signal averaging and scaling for spectral display. Molecular interactions may be detected through subtle frequency shifts in subtracted spectra; and it is now possible readily to record accurate peak positions, depolarization ratios and peak intensities. Full experimental details including the kind of lasers used, for example the 180° 'viewing platforms', focussed radiation (permitting spatial resolution of ca 1 μm) and heated (in situ) cells may be gleaned from the work of Grasselli, Hazle and Wolfram[1], Kiefer and Bernstein[2], Payen et al.[3], Fleischmann et al.[4], Girlando et al.[5], and of Delhaye and Dhamelincourt[6]. (The last-named workers describe what has now become known as the "MOLE" laser microprobe which in a sense offers a photonic complement to other microprobe procedures such as scanning Auger and the conventional electron microprobe. In principle the in situ molecular dynamics of a catalytic reactor could be traced at picosecond time resolution using such a laser microprobe. By taking advantage of non-linear Raman effects, and other phenomena, for example the so-called giant Raman effect, it should prove feasible to observe the Raman spectrum of essentially any adsorbed molecule or fragment.)

One area of catalyst characterisation in which the Raman spectroscopic study has proved illuminating is molybdate catalysts, which are important in reactions such as the selective oxidation of methanol to formaldehyde, the selective ammoxidation of propylene to acrylonitrile and the oxidative dehydrogenation of 1-butene to butadiene. Bismuth molybdate, Bi_2MoO_6, is known to exist in three crystallographic modifications, the α, β and γ phases. The γ phase is metastable with a structure thought to be similar to the mineral koechlinite[7]. It has been alleged (see Gates, Katzer and Schuit[8]) that the γ phase, upon heating in air to 660°C undergoes an irreversible transition to the γ' phase. Grasselli, Hazle, Mooney and Mehicic[9], using their *in situ* Raman technique, with a heated cell, demonstrated that a surface γ' phase forms at *ca* 600°C, a fact which has important practical repercussions in the industrial conversion of 1-butene to butadiene. These workers have also completed *in situ* Raman studies of the promotional effect of iron in bismuth molybdate catalysts, and have explored the question of surface *versus* bulk structure in the Sb-Sn-O catalyst system that plays a prominent part in olefin oxidation. It transpired that combined Raman and XPS studies afforded much insight into the nature of segregated phases in this system.

(b) Infrared Spectroscopy

The rôle of infrared studies in surface chemistry of solids and catalysis has been recognized for a very long time (see Rodebush et al.[10] and Terenin[11]) and its value as a tool recognized ever since Eischens[12] and Sheppard[13] adapted the technique for studying supported catalysts. Many reviews dealing with the contribution of infrared spectroscopy to catalyst characterisation are available (see Thomas[14], Pritchard[15], Rochester[16], Sheppard and Nguyen[17], Roberts and McKee[18] for further details of work carried out up until *ca* 1978). We note here some of the triumphs of this technique: that when benzene is adsorbed on certain catalyst surfaces, loss of aromatic character ensues; when H_2 is adsorbed on zinc oxide surfaces, dissociation leads to the production of Zn-H and O-H bonds; when formaldehyde contacts a rutile surface, no CH_2 only CH links are detected; when formic acid is brought into contact with a wide range of metal surfaces, a monomolecular layer of formate is formed just as when formaldehyde is adsorbed on rutile; and when gaseous NH_3 is introduced to a dry silica-alumina cracking catalyst, the Lewis acid sites on the surface are attacked, whereas in the presence of moisture protonic acidity predominates and the adsorbate molecules are converted to NH_4^+ species. These examples illustrate well the immediate relevance, and power of the infrared technique. It must be borne in mind, however, that as extinction coefficients do not vary much from one surface bond to the next, the species which is most tenaciously bound to the surface is that which tends to dominate in the resulting spectrum.

As with Raman spectroscopy, the advent of the laser and the computer has profoundly influenced the way in which infrared spectra can nowadays be recorded. Fourier transform methods (FTIR), based as they are on interferometry, are capable of coping with intensely coloured solids (e.g. transition-metal based catalysts) and a whole host of other kinds of solids. Display systems can be used either to record the results in the frequency- or in the time-domain; and minute changes in spectral intensity accompanying various subtle surface processes can be relatively routinely monitored.

Quite apart from the advent of the laser and the computer, however, what makes the infrared approach particularly relevant to catalyst characterisation is that robust heatable cells, capable of withstanding pressures of several tens of atmospheres have recently been constructed (using, for example, polycrystalline CaF_2 windows and a heavy-walled stainless-steel cylinder) and shown[19,20] to be suitable for in situ studies of many kinds of catalytic reactions involving silica and alumina-supported metal catalysts. Thus King[19] has traced the synthesis of hydrocarbons from $CO + H_2$ mixtures over Ru/SiO_2, Ru/Al_2O_3 and Fe/SiO_2 catalysts under reaction conditions (typically at 500 K and 150 psig CO/H_2). He found that, other than CO, the surface species which are observed during reaction contain only carbon and hydrogen. Strong CH_2 bands, arising from long-chain saturated hydrocarbons, were observed at relatively low temperatures with supported Ru but not with supported Fe. King found[19] no evidence for the existence of hydroxy carbene or formyl species during reaction, but some positive evidence that hydrocarbon syntheses over Ru and Fe proceed via initial formation of surface carbides.

(c) Magnetic Resonance

Electron spin resonance (esr) measurements, particularly in favourable circumstances, can reveal both the source and the destiny of transferred electrons in a heterogeneous (solid-gas or solid-liquid) situation. Whenever both the adsorbate and catalyst contain, either initially or finally, an unpaired electron, analysis by esr is, in principle, possible. Naccacha's work[21] on perylene adsorbed into a MoO_3-γAl_2O_3 catalyst is a nice example since it illustrates the ease of detecting the loss of an electron by the aromatic hydrocarbon and its ultimate transfer to the Mo, thereby yielding a d^1 transition-metal ion. Many other useful characterisation tasks have been completed using esr, particularly when the catalyst contains Cu^{2+} (d^9) ions (see Delgass et al.[22]).

Until quite recently, nuclear magnetic resonance (nmr) was rather scantily used in the chemical characterisation of catalysts and adsorbents, but the situation has changed dramatically in the past few years. Whenever solids of high surface area (and not too rich in paramagnetic ions) are employed, conventional nmr spectrometers can yield quite useful information, provided the adsorbed entities are mobile enough to give rise to the relatively sharp lines associated with the translationally free molecules of the liquid state. Water, for example, sorbed inside the pores of various adsorbents such as silica gel, will, under the appropriate conditions of temperature and pressure, exhibit quite sharp proton spectra, from which chemical shift data may be extracted.

Cirillo, Dereppe and Hall[23] recently studied the low-temperature adsorption of hydrogen on alumina and molybdena-alumina catalysts by nmr. They detected, by means of an _in situ_ procedure, at least two, and on extensively reduced molybdena-alumina catalysts, three, types of adsorbed hydrogen species. A single, averaged absorption line was observed originating from fast exchange between a small amount of strongly chemisorbed gas and the majority species which was molecular and much more weakly held. On alumina and unreduced molybdena-alumina the strongly chemisorbed species is also thought to be molecular, but evidence was obtained for dissociative adsorption on extensively reduced catalysts at low coverage.

It is less widely known that ordinary commercial ^{13}C, Fourier Transform nmr spectrometers can be used as very revealing tools for the identification of various bound hydro-carbons in catalyst channels, cages, slits or pores. To illustrate this point the nmr spectrum xylene intercalated in the interlamellar spaces of a synthetic hectorite is shown in Fig. 1. This illustration is taken from the work of Lyerla, Fyfe and Thomas[24], who have also shown that _in situ_ keto-enol equilibria of appropriate adsorbates, and the composition of certain adsorbate mixtures (for example, ethyl-benzene and p-xylene) may be monitored by this technique. Such results augur well for dynamic studies, using ^{13}C nmr, of catalytic conversions involving benzene, toluene, and xylene (the so-called 'BTX' problem).

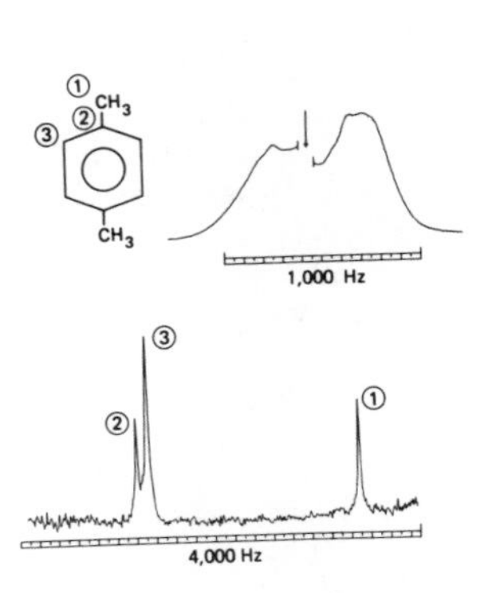

Fig. 1

A. The ^{1}H nmr spectrum of the p-xylene intercalate at 80 MHz and 30°C (50 pulses). The vertical arrow indicates the position of a small, sharp peak due to the HOD impurity in the D_2O lock reference that has been deleted for clarity.

B. The ^{13}C nmr spectrum of the same material recorded at 20 MHz and 30°C (16,000 pulses). Note how easily the three distinct kinds of carbon atoms in the xylene are readily distinguished.

Of even greater significance, possibly, is the use of magic angle spinning and cross-polarization techniques which serve to sharpen the wide-line (dipolar broadened) spectra normally obtained from solids. Studies of the rigid condensed state have been profoundly influenced by the arrival of magic-angle and related techniques. It is not yet clear where the greatest impact in surface chemistry and catalysis will be made by these powerful new variants of nmr, but recognising that there are many other nuclei such as Aℓ, Si, F, Li etc which possess favourable magnetogyric ratios and which also figure quite eminently in many catalyst preparations, it is certain that dynamic nmr studies offer considerable scope in the general area of catalyst characterisation.

REFERENCES

1. J.G. Grasselli, M.A.S. Hazle and L.E. Wolfram in "Molecular Spectroscopy", A.R. West, Ed. Heyden & Son Ltd., London (1977), p.200.
2. W. Kiefer and H.J. Bernstein, Appl. Spec., 25, 609 (1971).
3. E. Payen, J. Barbillat, J. Grimblot, J.P. Bonnelle, Spectros. Letters, 11, 997 (1978).
4. M. Fleischmann, P.J. Hendra, A.J. McQuillan, R.L. Paul and E.S. Reid, J. Raman Spec., 4, 269 (1976).
5. A. Girlando, J.G. Gordon II, D. Heitmann, M.R. Philpott, H. Seki and J.S. Swalen, Surface Sci., in press.
6. M. Delhaye and P. Dhamelincourt, J. Raman Spec., 3, 33 (1975).
7. K. Aykan, J. Catalysis, 12, 281 (1968).
8. B.C. Gates, J.R. Katzer, G.C.A. Schuit, "Chemistry of Catalytic Processes", McGraw-Hill Book Co., New York (1979).
9. J.G. Grasselli, M.A.S. Hazle, J.R. Mooney and M. Mehicic, Sohio Report No. 5705, Aug. 1979.
10. W.H. Rodebush, A.M. Buswell and V. Diltz, J. Chem. Phys., 5, 501 (1931).
11. A. Terenin, Proc. Roy. Soc. A108, 105 (1949).
12. R.P. Eischens, Accounts Chem. Res., 5, 74 (1972).
13. N. Sheppard, Proc. Roy. Soc., A259, 242 (1960).
14. J.M. Thomas, Progr. in Surface and Membrane Science, Ed. D.A. Cadenhead, J.F. Danielli and M.D. Rosenberg, Academic Press, Vol. 8, p.49 (1974).
15. J. Pritchard, Surface Sci., 79, 231 (1979) : (see also J. Pritchard and M.L. Sims, Trans. Faraday Soc., 66, 427 (1970)).
16. C.H. Rochester, D.A. Trebilco, J.C.S. Faraday Trans. I, 75, 2211 (1979).
17. N. Sheppard and T.T. Nguyen in "Advances in Infrared and Raman Spectroscopy", Vol. 5, Ed. R.J.H. Clark and R.E. Hester, Heyden & Sons Ltd., London 1978, p. 67.
18. M.W. Roberts and C.S. McKee, 'Chemistry of the Metal-Gas Interface', Oxford, 1978.
19. D.L. King, J. Catalysis, 61, 77 (1980).
20. J.M.L. Penninger, J. Catalysis, 56, 287 (1979).
21. C. Naccacha, J. Catalysis, 25, 334 (1972).

22. W.N. Delgass, G.L. Haller, R. Kellerman and J.H. Lunsford, 'Spectroscopy in Heterogeneous Catalysis', Academic Press, New York, 1979.
23. A.C. Cirillo, J.M. Dereppe and W.K. Hall, J. Catalysis, 61, 170 (1980).
24. J.R. Lyerla, C.A. Fyfe and J.M. Thomas, in preparation.

II

THE USE OF PHYSISORPTION FOR THE DETERMINATION OF SURFACE AREA AND PORE SIZE DISTRIBUTION

By

K.S.W. Sing

1. INTRODUCTION

Physisorption measurements are widely used for the determination of the surface area and pore size distribution of catalysts. The interpretation of physisorption data is not always straightforward, however, and a considerable amount of research has shown that the computation of the surface areas and pore sizes is justified only if certain conditions are fulfilled. To establish the scope and limitations of physisorption measurements, it is therefore necessary to identify the different mechanisms of physisorption and ascertain their dependency on porosity and other factors.

Physisorption (or physical adsorption) occurs whenever a gas (the adsorptive) is brought into contact with an outgassed solid (the adsorbent). The phenomenon is thus a general one - unlike chemisorption - and is brought about by the same balance of intermolecular attractive and repulsive forces that are responsible for the condensation of vapours and the deviations from ideality of real gases. Dispersion forces always provide the non-specific source of attraction between the adsorbate (i.e. the adsorbed species) and the adsorbent, but there are various types of specific adsorbent-adsorbate interactions which may contribute to the interaction energies for the adsorption of polar molecules. These specific physisorption interactions have been characterised by Barrer[1] and Kiselev[2].

It has been found expedient to classify pores according to their effective width (IUPAC, 1972, 1976 - see refs 3 and 4). The narrowest pores, of width not exceeding about 2.0 nm (20 Å), are called micropores; the widest pores of width greater than about 50 nm (0.05 μm or 500 Å) are called macropores; and pores of intermediate width, which were for a time termed intermediate or transitional pores, are now called mesopores. Although these limits are necessarily somewhat arbitrary because of the complexity of real gas-solid systems, a good deal of evidence suggests that we may picture the whole of the accessible volume in micropores as adsorption space. The process of micropore filling thus occurs as an alternative to monolayer-multilayer adsorption on an open surface. Because of its dependence on the overlap of adsorbent-adsorbate interaction in very narrow pores, micropore filling may be regarded as a primary physisorption process. On the other hand, capillary condensation in mesopores is always preceded by the formation of an adsorbed layer on the pore walls and is consequently a secondary process. Capillary condensation in macropores does not occur until the vapour pressure is very close to the saturation value and a macroporous solid therefore exhibits very similar adsorptive properties to those of a non-porous solid. For this reason, it has been necessary to develop other methods (e.g. mercury porosimetry) for the study of macro-pore structures.

2. EXPERIMENTAL PROCEDURES

The solid surface must be cleaned - at least to the extent of removing physisorbed material - before any adsorption measurements are undertaken. Previously adsorbed gases are usually removed by outgassing the adsorbent with the aid of a combination of vacuum pumps (rotary and diffusion) and pumping through a cold trap. The exact outgassing conditions (temperature of adsorbent, time of pumping and residual pressure) required to attain reproducible physisorption isotherms depend to a large extent on the nature of the adsorption system and the purpose of the investigation. For the determination of surface area and pore size distribution by nitrogen adsorption, a vacuum of the order of 10^{-4} Torr is normally regarded as adequate.

The rate of desorption is strongly temperature-dependent and to minimise the time required for outgassing the temperature should be the maximum consistent with avoidance of changes in the adsorbent structure or surface composition. It is usually considered undesirable to remove chemisorbed species (especially if surface dissociation is involved): for this reason inorganic oxides and hydroxides are generally outgassed at a temperature of about 150°C. Microporous solids require higher outgassing temperatures to remove physisorbed material (e.g. molecular sieve zeolites and microporous carbons are outgassed at temperatures of 350° - 400°C).

Volumetric methods are generally used for the determination of nitrogen adsorption isotherms at temperatures $\sim$ 77 K (Gregg and Sing[5]; British Standard[6]; American National Standard[7]). A known quantity of gas is admitted to a confined volume containing the adsorbent and the volume of gas adsorbed at the equilibrium pressure is the difference between the volume of gas admitted and that required to fill the dead space at the equilibrium pressure. The adsorption isotherm is constructed point-by-point by the admission of successive charges of gas, allowing sufficient time for equilibration at each point. Helium is often used for the dead space calibration, but Everett[8] has pointed out that some microporous solids adsorb appreciable amounts of helium over a range of temperature.

There are a number of potential sources of experimental error in the application of volumetric methods for the determination of nitrogen adsorption isotherms[9]. Particular care is required to control and check the temperature of the adsorbent. The adsorbent bulb should be immersed to a depth of at least 5 cm below the liquid nitrogen surface and the level of liquid nitrogen in the cryostat bath kept constant to within a few millimetres.

The adsorbent temperature should be monitored - preferably by using a nitrogen vapour pressure thermometer (to ± 0.1 °C). If possible, the whole apparatus should be erected in an air thermostat.

A continuous volumetric method has been described recently (see Grillet et al.[10]) for the determination of argon and nitrogen isotherms at 77 K. The method is employed in conjunction with adsorption calorimetry and is capable of giving excellent agreement with the corresponding isotherms determined using a conventional volumetric apparatus.

Gravimetric techniques have been in use for many years for adsorption measurements and catalyst characterisation. Considerable progress has been made[11,12] in the design and application of recording microbalances, and a number of versatile microbalances are commercially available to meet the demand for adsorption measurements in the submicrogram range. It is now possible to obtain a continuous record of the thermal decomposition and change in texture of a crystalline solid (see Alario Franco and Sing[13]) or the deposition of carbon during catalyst poisoning[14].

One problem associated with the gravimetric measurement of adsorption[15] at low temperature is the difference in the temperature of the adsorbent and that of the surrounding cryostat bath. Special precautions may be taken to minimise this temperature difference (Cutting[16]; Partyka and Rouquerol[17]), and in some types of investigation it can be allowed for by the use of reference adsorbents[18].

Other sources of error, which are often overlooked in the use of microbalances, are thermal transpiration and buoyancy effects; the former is significant at low pressure and the latter becomes more important as the pressure is increased (see Gregg and Sing[5], and Thomas and Williams[19]).

Gas chromatographic methods provide an alternative to the 'static' volumetric and gravimetric methods for the determination of adsorption isotherms. This approach is especially useful if the adsorption measurements are to be conducted at above ambient temperature or if the adsorptive is of low volatility[12]. By taking advantage of the high sensitivity of modern g.c. detectors and by applying the principles of mass balance, one can obtain[20] a rapid and accurate measurement of an adsorption isotherm over a wide range of relative pressure. In favourable cases, g.c. techniques can be used for the determination of small surface areas and isosteric enthalpies of adsorption at low surface coverage[12].

Many adsorption calorimetric studies are described in the literature, but the results are amenable to rigorous thermodynamic analysis in relatively few cases. Thus, 'heat of adsorption' data cannot be interpreted in an exact manner unless the conditions of measurement of the adsorption energy are specified and the initial and final thermodynamic states properly defined (see Everett[8] and Létoquart et al.[21]). Various types of adsorption calorimeter have been compared by Rouquerol[22], who concludes that the Tian-Calvet microcalorimeter has a number of advantages. This microcalorimeter employs two thermopiles as detectors - each containing 1000 thermocouples - and measures the heat flux under nearly isothermal conditions.

Recent measurements using this technique have provided convincing evidence for bidimensional phase changes taking place in the monolayers of nitrogen and argon on graphitised carbon black[23], and the method has also been used to characterise the various adsorption sites on rutile and silica[24].

3. CLASSIFICATION OF ADSORPTION ISOTHERMS

The majority of physisorption isotherms may be grouped into the six types shown in Fig. 1. The isotherm Types I-V are essentially those given in the original classification of Brunauer, Deming, Deming and Teller[26a] - often referred to as the Brunauer classification[26b].

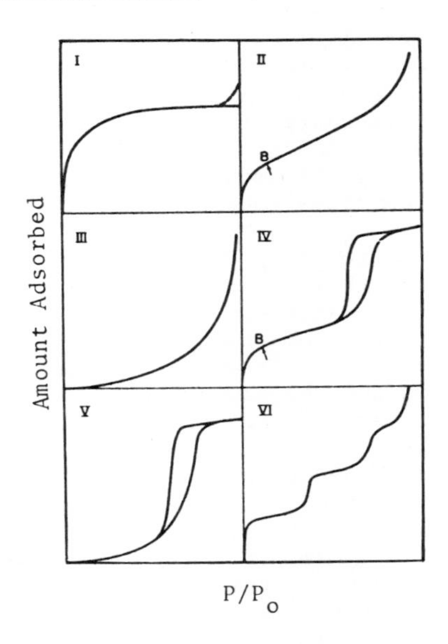

Fig. 1

Types of physisorption isotherms

Physisorption isotherms are often found to exhibit hysteresis (i.e. the adsorption and desorption curves follow different paths), which in some cases extends to very low pressures. The three most characteristic shapes of hysteresis loop, which were included in the 1958 classification of de Boer[25], are given in Fig. 2. Loops in the form of A or E are usually associated with Type IV isotherms and those similar to B are found with certain Type II isotherms.

In the most straightforward case, the Type I isotherm is reversible and concave to the relative pressure, p/p_o, axis; the adsorption, n, approaches the limiting value, n_s, as $p/p_o \rightarrow 1$. With some systems, however, the value of n begins to increase at $p/p_o < 1$ and the isotherm is then likely to exhibit a hysteresis loop (of flattened B shape).

For many years it was thought that n_s corresponded to monolayer coverage ($\theta = 1$) in accordance with the Langmuir theory (see Brunauer[26b]). The more widely accepted view at present is that Type I isotherms result from the adsorption taking place in narrow pores (i.e. by the process of micropore filling) and that the uptake n_s is controlled by the pore volume. Type I isotherms

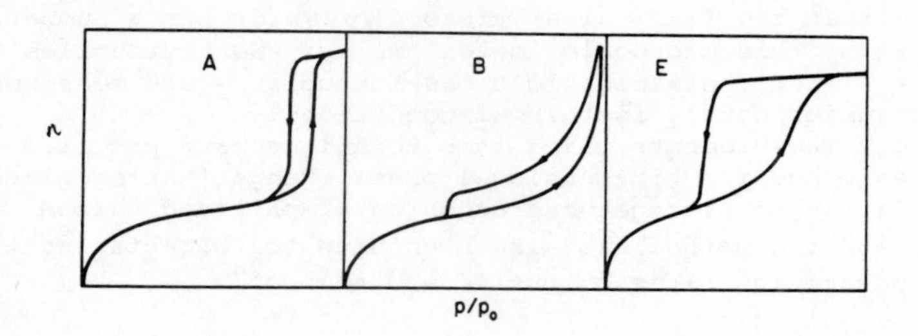

Fig. 2

Types of hysteresis loops

are obtained with many porous carbons and molecular sieve zeolites; the latter are generally very steep at low pressure (indicating high adsorption affinity) with a long nearly horizontal plateau at high p/p_o (see Barrer[27]).

Reversible Type II isotherms are characteristic of unrestricted monolayer-multilayer adsorption or non-porous or macroporous solids. If the knee of the isotherm is sharp, the uptake at Point B - the beginning of the middle linear section- provides a measure of the 'monolayer capacity', i.e. the amount of adsorbate required to cover the surface with a complete monomolecular layer.

True Type III isotherms are convex to the p/p_o axis over their entire range and therefore do not exhibit a Point B. Isotherms of this type are not common; the best-known examples are found with water vapour on carbons or other non-polar materials. However, there are a number of systems (e.g. nitrogen on various polymers) which give isotherms of such gradual curvature that there is no identifiable Point B. In such cases, the adsorbent-adsorbate interaction is relatively weak.

Type IV isotherms are obtained with mesoporous solids. The hysteresis loop is associated with the secondary process of capillary condensation, which results in complete filling of the mesopores at $p/p_o < 1$. In the simplest case, the initial part of a Type IV isotherm follows the same path as that of a Type II isotherm obtained with the given adsorptive on the adsorbent in a non-porous form. If Point B is well-defined it again indicates monolayer coverage. Type IV isotherms are obtained with many industrial catalysts.

Type V isotherms are uncommon and are difficult to interpret. They are related to Type III isotherms in that the adsorbent-adsorbate interaction is weak, but pore filling also takes place - leading to a limited uptake at high p/p_o.

Type VI isotherms occur as a result of stepwise multilayer adsorption on uniform surfaces. The adsorption of each layer takes place within a limited range of p/p_o and lateral interactions between the adsorbed molecules contribute to the layer-by-layer process. Stepwise isotherms are comparatively rare because most adsorbent surfaces of practical importance are to some extent heterogeneous. Some of the best examples have been reported[28] for krypton adsorption on graphitised carbon and on certain clean metal surfaces.

4. SURFACE AREA DETERMINATION

It is an indication of the success of the BET method[29] that it has been adopted by various official bodies as a standard procedure for surface area determination (e.g. British Standard[6]; Deutsche Normen[30]; Norme Francaise[31]; American National Standard[7]). Despite its popularity, the BET method is open to criticism on a number of grounds, and there is a general awareness of the shortcomings of the underlying theory.

By the introduction of a number of simplifying assumptions, the BET theory extends the Langmuir model to multilayer adsorption. Thus, adsorption in the first layer is assumed to take place on an array of surface sites of uniform energy. Molecules in the first layer act as sites for multilayer adsorption, which in the simplest case approaches infinite thickness as $p \rightarrow p_o$. It is further assumed that the evaporation-condensation characteristics are identical for all layers except the first and that the heats of adsorption for the second and higher layers are equal to the heat of condensation of the adsorptive. Summation of the amount adsorbed in all layers then gives the isotherm equation

$$\frac{n}{n_m} = \frac{C\ p/p_o}{(1-p/p_o)\ [1 + (C-1)\ p/p_o]}\ , \tag{1}$$

where n is the amount adsorbed at the equilibrium relative pressure p/p_o, n_m is the monolayer capacity and C is a constant which, according to the original theory, is related exponentially to the 'first layer' heat of adsorption. More strictly, C may be regarded as a free energy term[32].

For convenience the BET equation is usually expressed in the form

$$\frac{p}{n(p_o - p)} = \frac{1}{n_m C} + \frac{(C - 1)}{n_m C} \cdot \frac{p}{p_o}\ , \tag{2}$$

which demands a linear relation between $p/n(p_o - p)$ and p/p_o (i.e. the BET plot). With all known experimental isotherms, the range of linearity of the BET plot is severely restricted - usually to within the p/p_o range of 0.05 - 0.30. With some systems (e.g. N_2 on graphitised carbon at 77 K), the linear plot may not even extend above $p/p_o = 0.1$.

Many attempts have been made to modify the BET model to improve the range of fit of the theoretical isotherm with experimental data (see Gregg and Sing[5]; Sing[28]). Unfortunately, most of these treatments achieve little more than to provide a set of adjustable parameters which are evaluated by curve fitting. It is not surprising that most _users_ of the BET method have preferred to employ the original two-parameter equation on an empirical basis and to define the p/p_o range of application for a given system and temperature[9].

It has been noted above that if Point B is well-defined it can be identified with the completion of the monolayer. In support of their theory, Brunaúer, Emmett and Teller[29] claimed that the close agreement often obtained between the value of n_m and the amount adsorbed at Point B helps to confirm the validity of n_m. The exact location of Point B is difficult to ascertain if C is less than about 80, and indeed it becomes impossible to identify Point B if C is very low (say, $C < 20$). Under these circumstances the isotherm begins to take on Type III character (through weak adsorbent-adsorbate interaction) and the validity of n_m is questionable[5]. The origin of these limitations in the application of the BET method have been revealed[28] by physisorption studies conducted before and after the modification of well-defined non-porous surfaces: such results indicate that BET monolayer capacities from isotherms with low C values may be _ca_ 30% in error, even on a comparative basis. These results support the view that it is Point B which represents the true monolayer point and that the role of the BET method is to provide a graphical procedure for locating this characteristic point[28]. High C values pose a different problem and will be discussed later.

The second stage in the application of the BET method is the calculation of the surface area, A_{BET}, from the value of n_m. This step requires a knowledge of the average area, a_m, occupied by the adsorbate molecule in the completed monolayer since

$$A_{BET} = n_m N_A a_m , \tag{3}$$

where N_A is Avogadro's constant. Such a procedure is based on the assumption that, for a given adsorptive and temperature, the value of a_m remains constant and independent of the nature of the adsorbent.

In their early work, Emmett and Brunauer[33] assumed that the packing of adsorbate molecules in the completed monolayer is the same as in the bulk condensed phase (usually taken as the liquid). For nitrogen at 77 K and assuming liquid-packing, this assumption leads to $a_m = 0.162\ nm^2$, a value which has been adopted by most workers[5].

In view of the various types of adsorbent-adsorbate interactions in physisorption[12,28,34], one might expect an appreciable variation in the value of a_m for a given adsorbate on different surfaces. For many years little was known about the exact mode of packing of adsorbate molecules in the monolayer, but the study of physisorbed structures on certain surfaces can now be undertaken with the aid of Auger spectroscopy, LEED, THEED and neutron scattering techniques (see, for example, ref. 34). It is too early to come to any firm conclusions about these studies, but it would appear from the results already obtained that the close-packed monolayer is a fairly good approximation for the adsorption of Kr and Xe on certain metals, although differences in packing have been observed, e.g. for Xe on Pt and Ni.

Two-dimensional compression of the monolayer from a localised epitaxial state to a close-packed state is the most likely explanation for the origin of the sub-steps detected in stepwise isotherms of Kr and Xe on graphitised carbon (see Price and Venables[35]). The sub-step obtained with Ar at 77 K on graphitised carbon has been attributed to the transformation from a liquid-like state to the close-packed 'solid'[23a,b], whilst the sub-step given under the same conditions by N_2 has been explained as a change from a liquid-like state to a localised state[23b]. Adsorption calorimetry was used to establish that the transition is just complete at Point B, and Rouquerol et al.[23b] suggest that one nitrogen molecule then occupies a site over three carbon hexagons, i.e. a surface coverage of 0.1575 nm^2.

It has been noted above that the value of a_m for N_2 at 77 K is usually taken as 0.162 nm^2. A good deal of experimental evidence suggests that this value is reasonably satisfactory for surfaces (e.g. ungraphitised carbons, hydroxylated silicas) which give BET C values of around 100 (see ref. 5 for further details). In these cases it seems likely that the specific quadrupolar interaction between adsorbed N_2 and surface OH groups (or other polar species) provides a significant contribution to the adsorbent-adsorbate interaction energy and thus helps to minimise the overlap between monolayer and multilayer formation (i.e. sharpening Point B). It has also been suggested by Mikhail and Brunauer[36] that specific lateral interactions between adsorbed N_2 molecules help to stabilise the close-packed liquid-like monolayer on most surfaces.

With other adsorptives, arbitrary adjustment of a_m values has been required to bring the BET areas into agreement with the corresponding nitrogen values. A case in point is Ar at 77 K for which recommended a_m values cover the range 0.125 - 0.182 nm^2 (see Nicholson and Sing[34]). Another example is Kr (at 77 K), which apparently has an effective value of a_m in the region of 0.20 nm^2 instead of 0.152 nm^2, the close-packed liquid-like value.

The adsorption of water vapour must be treated as a special case because of the relatively large scale of the specific interactions (usually hydrogen bonds) between adsorbed H_2O and surface functional groups or ions. Removal of this specific interaction (e.g. by dehydroxylation of silica) has the direct effect of considerably reducing the amount of water adsorption per unit area. These findings demonstrate that water vapour cannot be used as an adsorptive for surface area determination unless the chemical structure of the surface is unchanged. On the other hand, the adsorption of water is extremely useful for characterising the hydrophilic (or hydrophobic) properties of certain surfaces.

High BET C values are likely to be associated either with localised monolayer adsorption or with micropore filling. If the adsorbent is microporous and also has an appreciable external area, the nitrogen isotherm will have Type II appearance even though the uptake at Point B includes a micropore filling contribution. The magnitude of C may provide a useful indication of the complexity of the adsorption process, but the BET area should not be accepted as a real surface area.

In the application of the BET method it is strongly recommended that the range of lirearity of the BET plot and the value of C are recorded along with the value assumed for the molecular area in the completed monolayer. Caution should be exercised in the interpretation of the BET area - especially if the nitrogen C values are outside the approximate range 80-120.

A number of empirical procedures, summarized elsewhere[37], have been devised to establish the mechanism of adsorption and analyse the isotherm shape. These methods make use of standard isotherm data obtained with suitable non-porous reference solids. In favourable cases this approach may be used to separate surface coverage and micropore filling contributions and enable an assessment to be made of the external surface area[38].

5. ASSESSMENT OF MICROPOROSITY

If the area of the external surface is very small, the amount of gas adsorbed, n_s, at the plateau of a Type I isotherm can be taken as the amount of adsorptive required to fill the micropores. An estimate of the total micropore volume, V_p, can then be made by expressing n_s as the volume of liquid adsorptive. It is remarkable that this simple procedure, when used with a number of adsorptives - even at widely different temperatures - on a given adsorbent, usually gives[5] values of V_p in good agreement (generally to within ± 5%). This conformity to the 'Gurvich rule' obviously fails if the pores are too narrow to accommodate the larger adsorptive molecules; in that event, the values of V_p will be dependent on the molecular dimensions as well as on the pore size; in extreme cases (e.g. molecular sieve carbons and zeolites), V_p will fall to a very low value at a critical molecular diameter (see Breck[39]).

A number of attempts have been made to evaluate microporosity from physisorption isotherms. In their pioneering studies of micropore filling, Dubinin and his co-workers (see refs 40-43) have utilized the 'characteristic curve' principle of the Polanyi potential theory. The fractional filling, n/n_p, of the micropore volume is expressed in the general form

$$n/n_p = \exp\left[-(A/E)^m\right] \quad (4)$$

where m is a small integer, E is a characteristic free energy of adsorption for the given system, and A, originally termed the 'adsorption potential', is the change in differential free energy of adsorption, viz

$$A = -RT \ln(p/p_o) . \tag{5}$$

In the special case of m = 2, the adsorption isotherm equation in linear form becomes

$$\log n = \log n_p - D (\log p_o/p)^2 , \tag{6}$$

where D is a constant related to E. Equation (6) is generally known as the Dubinin-Radushkevich (DR) equation.

According to Equation (6), a linear relationship should be obtained between log n and $(\log p_o/p)^2$ with the intercept equal to $\log n_p$. In principle, therefore, the DR equation provides a simple means for the assessment of the micropore volume (again taken as the liquid volume equivalent to n_p), provided that the characteristic curve for micropore filling has the $(A/E)^2$ form.

With a number of microporous solids, the DR plots are indeed linear over a wide range of relative pressure, which may extend from 10^{-5} to close to saturation, whilst in other cases there is either no linear region or a very limited range of linearity[34]. Dubinin and his co-workers[40-43] have found that some zeolites and active carbons with very narrow micropores give isotherms which obey Equation (4) with m = 3 and therefore do not give linear DR plots.

Unfortunately, the extent of linearity of the DR plot is not in itself a sufficient test that the physisorption process is confined to micropore filling (i.e. that mesopores are absent) and it is essential that other features of the isotherm are taken into account. Attempts made to verify the micropore volumes obtained by extrapolation of DR plots have produced rather disappointing results (see Nicholson and Sing[34]).

Stoeckli[44] has recently proposed a different form of generalised Dubinin relation based on the principle that the simple DR equation is valid when it is applied to a nearly homogeneous system (i.e. narrow size distribution) of micropores and that physisorption by heterogeneous micropore systems is better described by summation of the individual contributions,

$$n = \sum_j n_{pj} \exp\left[- D (\log p_o/p)^2\right] \tag{7}$$

By adopting this approach, Stoeckli has attempted to establish the physical meaning of the variables which appear in the generalised equation. Clearly its success must depend on calibration of the DR parameters with the aid of a number of nearly homogeneous micropore systems. Some progress has been made in this direction, by Stoeckli and co-workers[45], with the aid of electron microscopy and low-angle scattering of X-rays.

A method for the analysis of the micropore size distribution (the MP method), which was put forward by Mikhail et al.[46], is essentially an extension of the t-method of Lippens and de Boer[47]. In the t-method, the amount of nitrogen adsorbed by a given solid is replotted against t, the corresponding multilayer thickness for the adsorption of nitrogen on a non-porous reference solid. Any deviation in shape of the experimental isotherm from the standard multilayer thickness curve is therefore detected as a departure of the 't-plot' from linearity. In the MP-method, the downward deviations of the t-plot from linearity are used to evaluate the micropore volume and surface distribution (by drawing tangents to obtain the surface areas of groups of pores).

Brunauer[32] has stressed that it is essential that the correct t-curve is used in the application of the MP method. He argues that this is the one giving approximately the same heats of adsorption as the microporous system and that it may be selected by matching the BET C values. The logic of Brunauer's approach has been challenged by other workers (Dubinin[41] and Sing[28,38]) who point out that the overlap of the adsorbent-adsorbate interactions across the micropore produces an enhanced adsorption energy so that surface areas calculated from the low p/p_o region of the t-plot are anomalously high for microporous solids. According to this view, if the t-method (or any similar empirical procedure) is to be used, it is essential that standard isotherms (t-curves) are obtained on non-porous reference solids of known structure.

A novel method for the evaluation of microporosity was introduced by Gregg and Langford in 1969[48]. In this method, the outgassed adsorbent is exposed at 77 K to n-nonane vapour and then pumped at successively higher temperatures (from ambient to 350°C). In the case of a microporous carbon black it was found that pumping at ambient temperature resulted in the removal of the nonane from the external surface but not from the micropores and the resultant BET-nitrogen area agreed closely with the geometric area determined by electron microscopy. In the multilayer range, the initial nitrogen isotherm ran parallel to that obtained after nonane pre-treatment and the vertical separation was attributed to the micropore filling contribution[49].

It must be concluded that no reliable method has been developed for the determination of the micropore size distribution. At present, the most promising approach appears to be that of pre-adsorption linked with the use of various probe molecules of known size and shape. To make further progress it will be necessary to employ well-defined microporous adsorbents and have available non-porous reference solids of the same surface structure.

6. ASSESSMENT OF MESOPORE SIZE DISTRIBUTION

Mesopore filling occurs by the process of capillary condensation, which follows monolayer-multilayer coverage on the pore walls. The mesopore size is usually calculated with the aid of the Kelvin equation in the form

$$\ln\left(\frac{p}{p_o}\right) = -\frac{2\gamma v_L}{r_k RT}, \tag{8}$$

which relates p/p_o, the relative pressure at which condensation takes place, to r_k, the radius of a hemispherical meniscus. Here, γ and v_L are the surface tension and molar volume, respectively, of the condensed liquid.

Equation (8) may be derived quite simply from the Young-Laplace equation,

$$\Delta P = \gamma\left(\frac{1}{r_1} + \frac{1}{r_2}\right), \tag{9}$$

which gives the difference in pressure on the two sides of a meniscus (or interface) as a function of its principal radii of curvature, r_1 and r_2. To obtain Equation (8), either $r_1 = r_2$ or r_k is taken as the mean radius of curvature. A refined treatment of the nature of Kelvin and Laplace equilibria is given by Everett[50], who draws attention to the problems involved in establishing a sound basis for the thermodynamic description of capillary condensation.

If the Kelvin equation is to be applied to porous solids, it is necessary to assume that r_k is dependent on the dimensions of the pore. A relatively simple pore shape is cylindrical: then, if the pore radius is r_p and a correction made for the multilayer thickness, t, we have

$$r_p = r_k + t. \tag{10}$$

In this case it is assumed that the meniscus shape is hemispherical and thus the pore radius is given by the equation

$$r_p = \frac{2\gamma v_L}{RT \ln (p_o/p)} + t. \tag{11}$$

Another simple model is that of a cylindrical meniscus in a slit-shaped pore. If the slit is parallel-sided, its width, d_p, is given by

$$d_p = \frac{2\gamma v_L}{RT \ln (p_o/p)} + 2t. \tag{12}$$

Other pore shapes may be postulated (see refs 49 and 51), but it must be kept in mind that it is strictly the meniscus shape and curvature which determines the p/p_o at which the capillary condensate is in thermodynamic equilibrium with the vapour.

Various computational procedures have been used[49] to calculate the mesopore size distribution with the aid of Equations (11) or (12). The calculation is complicated by the fact that capillary condensation and multilayer thickening are taking place over the same range of p/p_o; this leads to a progressive change in the dimensions of the internal regions (sometimes called the 'cores') in which capillary condensation takes place. A further complication is that the exposed area of the multilayer changes as the adsorption progresses.

Values of the multilayer thickness, required for insertion in Equations (11) or (12), are either interpolated from experimental t-curves or calculated[52] using the Halsey equation, which for nitrogen takes the form

$$t/\text{nm} = 0.354 \left[5/\ln(p_o/p)\right]^{1/3} \qquad (13)$$

where the value 0.354 nm is adopted as the <u>average</u> thickness of a molecular layer in the nitrogen multilayer.

The derived pore size distribution is usually presented in the graphical form, $\Delta V_p/\Delta r_p$ vs. r_p. Obviously the form of the distribution curve will depend on whether the <u>ad</u>sorption or <u>de</u>sorption branch of the hysteresis loop is used in the calculations. With a Type E hysteresis loop the difference is especially large. For many years, there appeared to be a slight preference for the desorption curve, on the grounds that it was more likely to represent conditions closer to thermodynamic equilibrium[5]. On the other hand, it has been recognised for many years that in the case of a wide-bodied pore with a narrow neck (an 'ink bottle' pore), liquid is likely to remain trapped on desorption until the p/p_o is reduced sufficiently to allow evaporation from the neck. The general principles of blocking effects in pore networks have been discussed recently by Everett[50] and by Doe and Haynes[53], who conclude that the distribution calculated from the desorption curve is likely to be strongly biassed towards smaller pore sizes. In such cases, the <u>adsorption</u> branch is to be preferred for pore size calculations.

It has been argued (e.g. by Broekhoff[54]) that since both the adsorption and desorption branches of the loop are probably irreversible, neither branch can be used with confidence for the calculation of the pore size distribution. Everett[50] points out that in fact the pressure at which the condensate loses its stability is governed by the curvature of the meniscus, irrespective of the lack of reversibility of the ensuing process. Detailed examination of the shape and scanning characteristics of the hysteresis loop should therefore provide a basis for further work, but the success of this type of approach will depend on the availability of model mesoporous systems.

Harris[55] first drew attention to the fact that many nitrogen hysteresis loops (at 77 K) have a very steep desorption branch which meets the adsorption branch at $p/p_o \sim 0.42$ and that with N_2 at 77 K the lower closure point of the loop never occurs at

$p/p_o < 0.42$. The most likely explanation for the existence of this limiting p/p_o is that it corresponds to the limit of stability of the capillary condensate, which is under increasing tension as p/p_o is decreased (i.e. ΔP is increased). The phenomenon is a general one, the value of the limiting p/p_o being dependent on the properties of the liquid adsorptive and the temperature[56,57].

There seems little doubt[5,50] that the Kelvin equation cannot hold in very narrow pores. In decreasing the pore width, there must come a point at which the chemical potential of the 'liquid' adsorbate is influenced by interaction with the force field of the solid as well as by the curvature of the meniscus. This effect may be regarded as an overlap of capillary condensation and micropore filling and probably occurs in the pre-hysteresis part of the isotherm. Indeed, there appears to be many porous solids with pore sizes extending from the micropore into the mesopore range. With such materials, the interpretation of physisorption isotherms is likely to be especially difficult because of a compensation effect in the multilayer range (see Bhambhani et al.[58]).

Brunauer et al.[59] have attempted to avoid the problem of pore shape in their so-called 'model-less method'. They make use of an equation, attributed to Kiselev[60],

$$\gamma dA = -\Delta\mu\, dn \tag{14}$$

where dA is the multilayer area removed when mesopores are filled with dn moles of condensed liquid at the p/p_o corresponding to the change in chemical potential $\Delta\mu$ (i.e. $RT\, \ln p/p_o$). The process involves the filling of the 'core', i.e. the region remaining when pore walls are already covered by multilayer, and A should therefore be identified with the 'core area'.

Integration of Equation (14) over the limits of capillary condensation gives the total core area

$$A = -\frac{RT}{\gamma} \int \ln p/p_o \, dn \, . \tag{15}$$

For the calculation of the core size distribution, Brunauer et al. replace the integration by summation applied to desorption by a number of small finite steps - starting at saturation pressure when all the mesopores are full of liquid. The method makes use of the concept of hydraulic radius which for a group of pores (or cores) may be defined as

$$r_{h,i} = \frac{2\, V_i}{A_i}$$

where V_i is the volume of the group of pores (or cores) and A_i the area of the pore (or core) walls.

For each desorption step, the change in area accompanying the decrease in adsorbed volume is calculated, taking into account the thinning of the multilayer on pores already emptied. The magnitude of the multilayer correction increases as p/p_o is decreased until the 'correction' becomes equal to the amount desorbed: at this stage the mesopores are assumed to be empty of condensate.

The procedure is said to be model-less because it gives the core size distribution, $\Delta V_c/\Delta r_h$ versus r_h, which is claimed[32] to be adequate for many purposes. If required, the core volumes can be converted into pore sizes by the application of the appropriate t-curve and pore model. Although the Brunauer method has been applied by some workers, it has been criticised by others (see Everett and Haynes[57]) on the grounds that Equation (14) is valid only for reversible processes and that it must in principle be inaccurate when applied to the hysteresis region. Another serious limitation of Equation (14) is that it is strictly applicable only to pores of uniform cross section which are completely wetted by the adsorbate ($\cos\theta = 1$): to this extent, therefore, a model pore system has been tacitly assumed.

7. GENERAL CONCLUSIONS

It should be evident from the above discussion that there are a number of formidable problems involved in the determination of the pore size by physisorption. Recently, Broekhoff[54] has provided a set of instructions for analysing nitrogen isotherms, based on his experience in the Unilever Research Laboratories at Vlaardingen. The following recommended procedure departs in some respects from Broekhoff's treatment, but it can be applied more generally for the analysis of routine physisorption measurements on catalysts.

Stage 1. Inspect nitrogen isotherm: identify isotherm Type, nature of hysteresis and any peculiarities of isotherm shape. If hysteresis is given at $p/p_o < 0.4$, the experimental procedure should be checked. Reproducible low-pressure hysteresis is likely to be associated with a non-rigid pore system (e.g. in a weakly bonded aggregate structure) or an 'activated entry' effect[49].

Stage 2. With Type II, IV or VI isotherm. Construct BET plot: ascertain range of linearity, calculate n_m, A_{BET} and C value.

Stage 3. With Type II, III or IV isotherm. Apply empirical method of isotherm analysis (t-method or α_s-method): ascertain range of linearity and hence check validity of A_{BET}.

With Type I or Composite I and II isotherm. Apply α_s-method to obtain external area and V_p (treatment of Sing[38]).

Stage 4. Examine independent evidence for pore shape, e.g. the existence of slit-shaped pores, as revealed by electron microscopy.

Stage 5. With Type IV isotherm. Select and apply procedure for computation of mesopore size distribution. Compare cumulative area and pore volume with A_{BET} and V_p, respectively.

Stage 6. Critically assess the significance of the calculated values of surface area and pore size distribution. Particular care is required in the interpretation of an isotherm with a large hysteresis loop (E shape) with a lower point of closure at $p/p_o = 0.42$. Such features are indicative of a wide distribution of pore size which extends from the micropore into the mesopore range. If this is the case, a unique solution for the pore size distribution cannot be obtained from the nitrogen isotherm at 77 K. Consideration should then be given to the use of other adsorptives or the pre-adsorption method.

REFERENCES

1. R.M. Barrer, J. Colloid Interface Sci., 21, 415 (1966).
2. A.V. Kiselev, Disc. Faraday Soc. No. 40, 205 (1965).
3. IUPAC. "Definitions, Terminology and Symbols in Colloid and Surface Chemistry", Part I of Appendix II, (D.H. Everett) Pure Appl. Chem. 31, 579 (1972).
4. IUPAC. "Terminology in Heterogeneous Catalysis", Part II of Appendix II, (R.L. Burwell, Jr.) Pure Appl. Chem., 45, 71 (1976).
5. S.J. Gregg and K.S.W. Sing, "Adsorption, Surface Area and Porosity", Academic Press, London and New York, (1967).
6. British Standard. 4359: Part 1 (1969).
7. American National Standard. ASTM D 3663-78: (1978).
8. D.H. Everett, In "Thermochimie" (Ed. M. Lafitte) (Colloques Internationaux du Centre National de la Recherche Scientifique, No. 201, Marseille), p.54, C.N.R.S., Paris, (1972).
9. D.H. Everett, G.D. Parfitt, K.S.W. Sing and R. Wilson, J. Appl. Chem. Biotechnol. 24, 199 (1974).
10. Y. Grillet, F. Rouquerol and J. Rouquerol, J. Chim. Phys. 74, 199, 778 (1977).
11. D.A. Cadenhead and N.J. Wagner, In "Experimental Methods in Catalytic Research", Vol. II (Eds R.B. Anderson and P.T. Dawson) p.223, Academic Press, London (1976).
12. N.D. Parkyns and K.S.W. Sing, Specialist Periodical Report, "Colloid Science" (Ed. D.H. Everett) 2, p.1, The Chemical Society, London (1975).
13. M.A. Alario Franco and K.S.W. Sing, Anales de Quim. 71, 296 (1975).
14. D. Trimm, Laboratory Practice, 132 (1974).
15. A.K. Galwey, Chemistry of Solids, Chapter 4, Chapman and Hall, London (1967).

16. P.A. Cutting, In "Vacuum Microbalance Techniques" Vol.7, (Eds C.H. Massen and H.J. van Beckum) p.71, Plenum Press, New York (1970).
17. S. Partyka and J. Rouquerol, In "Progress in Vacuum Microbalance Techniques" (Ed. C. Eyraud) Vol.3, p.83, Heyden & Son, London (1975).
18. M.A. Alario Franco, F.S. Baker and K.S.W. Sing, In "Progress in Vacuum Microbalance Techniques" (Eds S.C. Bevan, S.J. Gregg and N.D. Parkyns), 2, p.51, Heyden, London (1973).
19. J.M. Thomas and B.R. Williams, Quart. Rev. Chem. Soc., 19, 231 (1965).
20. A.V. Kiselev and Ya. I. Yashin, "La Chromatographie gaz-solide", Masson et Cie, Paris (1969).
21. C. Létoquart, F. Rouquerol and J. Rouquerol, J. Chim. Phys., 70, 559 (1973).
22. J. Rouquerol, In "Thermochimie" (Ed. M. Lafitte) (Colloques Int. du Centre National de la Recherche Sci., No.201, Marseille), p.537, C.N.R.S. Paris (1972).
23a. J. Rouquerol, S. Partyka and F. Rouquerol, Compt. ren., 282, 1057 (1976).
23b. J. Rouquerol, S. Partyka and F. Rouquerol, J.C.S. Faraday I, 73, 306 (1977).
24. D.N. Furlong, F. Rouquerol, J. Rouquerol and K.S.W. Sing, J.C.S. Faraday I (1980) (in press).
25. J.H. de Boer, In "The Structure and Properties of Porous Materials" (Eds D.H. Everett and F.S. Stone), J. Colloid Interface Sci., 21, 405 (1958).
26a. S. Brunauer, L.S. Deming, W.E. Deming and E. Teller, J. Amer. Chem. Soc., 62, 1723 (1940).
26b. S. Brunauer, "The Adsorption of Gases and Vapours", Oxford University Press (1944).
27. R.M. Barrer, In "Zeolites and Clay Minerals as Sorbents and Molecular Sieves", P.174, Academic Press, London (1978).
28. K.S.W. Sing, Spec.Per. Report, "Colloid Science" (Ed. D.H. Everett) 1, p.1, The Chem. Soc., London (1973).
29. S. Brunauer, P.H. Emmett and E. Teller, J. Amer. Chem.Soc., 60, 309 (1938).
30. Deutsche Normen. DIN 66131, Bestimmung der spezifischen Oberflache von Feststoffen durch Gasadsorption nach Brunauer, Emmett and Teller (BET) (1973).
31. Norme Francaise. Determination de l'aire massique (surface specifique) des poudres par adsorption de gaz. 11-621.(1975).
32. S. Brunauer, In "Proc. of Intern. Symp. Surface Area Determination" (Eds D.H. Everett and R.H. Ottewill) p.63, Butterworths, London (1970).
33. P.H. Emmett and S. Brunauer, J. Amer. Chem. Soc., 59, 1553 (1937).
34. D. Nicholson and K.S.W. Sing, Spec. Per. Report, "Colloid Science", 3, p.1, The Chem. Soc., London (1979).
35. G.L. Price and J.A. Venables, Surface Sci., 59, 509 (1976).
36. R.Sh. Mikhail and S. Brunauer, J. Colloid Interface Sci., 52, 572 (1975).

37. K.S.W. Sing, In "Characterisation of Powder Surfaces" (Eds G.D. Parfitt and K.S.W. Sing) p.1, Academic Press, London (1976).
38. K.S.W. Sing, In "Proc. of the Intern. Symp. on Surface Area Determination" (Eds D.H. Everett and R.H. Ottewill), p.25, Butterworths, London (1970).
39. D.W. Breck, "Zeolite Molecular Sieves", New York (1974).
40. M.M. Dubinin, In "Chemistry and Physics of Carbon", (Ed. P.L. Walker), 2, 51, Arnold, London and New York (1966).
41. M.M. Dubinin, In "Proc. of Intern. Symp. on Surface Area Determination (Eds D.H. Everett and R.H. Ottewill) p.123, Butterworths, London (1970).
42. M.M. Dubinin, In "Progress in Surface and Membrane Science", (Eds J.F. Danielli, M.D. Rosenberg and D.A. Cadenhead) Vol. 9, p.1, Academic Press, New York (1975).
43. M.M. Dubinin, In "Characterisation of Porous Solids" (Eds S.J. Gregg, K.S.W. Sing and H.F. Stoeckli), p.1, Soc. of Chem. Ind., London (1979).
44. H.F. Stoeckli, J. Colloid Interface Sci., 59, 184 (1977).
45. H.F. Stoeckli, J. Ph. Houriet, A. Perret and U. Huber, In "Characterisation of Porous Solids", (Eds S.J. Gregg, K.S.W. Sing and H.F. Stoeckli) p.31, Soc. of Chem. Ind., London (1979).
46. R. Sh. Mikhail, S. Brunauer and E.E. Bodor, J. Colloid Interface Sci., 26, 45 (1968).
47. B.C. Lippens and J.H. de Boer, J. Catalysis, 4, 319 (1965).
48. S.J. Gregg and J.F. Langford, Trans. Faraday Soc., 65, 1394 (1969).
49. S.J. Gregg and K.S.W. Sing, In "Surface and Colloid Science" (Ed. E. Matijevic) 9, 231, Wiley, New York (1976).
50. D.H. Everett, In "Characterisation of Porous Solids" (Eds S.J. Gregg, K.S.W. Sing and H.F. Stoeckli), p.229, Soc. of Chem. Ind., London (1979).
51. A.P. Karnaukhov, In "Pore Structure and Properties of Materials" (Ed. S. Modrý) Pt. 1, A-3, Academia, Prague (1973).
52. D. Dollimore and G.R. Heal, J. Applied Chem., 14, 109 (1964).
53. P.H. Doe and J.M. Haynes, In "Characterisation of Porous Solids (Eds S.J. Gregg, K.S.W. Sing and H.F. Stoeckli) p.253 Soc. of Chem. Ind., London (1979).
54. J.C.P. Broekhoff, In "Preparation of Catalysts II" (Eds B. Delmon, P. Grange, P. Jacobs and G. Poncelet) p.663, Elsevier, Amsterdam (1979).
55. M.R. Harris, Chem. and Ind., 268 (1965).
56. C.G.V. Burgess and D.H. Everett, J. Colloid Interface Sci., 33, 611 (1970).
57. D.H. Everett and J.M. Haynes, Spec. Per. Report "Colloid Science", 1, p.123, The Chem. Soc., London (1973).
58. M.R. Bhambhani, P.A. Cutting, K.S.W. Sing and D.H. Turk, J. Colloid and Interface Sci., 38, 109 (1972).
59. S. Brunauer, R.Sh. Mikhail and E.E. Bodor, J. Colloid Interface Sci., 24, 451 (1967).
60. A.V.Kiselev,in "The Structure & Properties of Porous Materials"(Eds D.H.Everett & F.S.Stone)p128,Butterworths (1958).

THE CHARACTERISATION OF INDUSTRIAL CATALYSTS WITH ESCA

By

T. Edmonds

1. INTRODUCTION

Despite the efforts of many chemists and physicists using a range of techniques, the forecasting of catalytic behaviour and the preparation of the catalysts themselves remains an essentially empirical art. In part this is because techniques for studying the chemistry of the active site have not been available, but chiefly it is an indication of the difficulty and complexity of the problem. This is apparent even from an outline discussion. A viable industrial catalyst needs to exhibit high selectivity to the desired product and high rate of production of this product. Moreover it should maintain these two attributes over a long period of operation. The key factors which influence these major properties are the nature of the active catalyst phase, the support material and the presence of promoter(s). The rôles of the latter are, in their turn, influenced by several factors.

a) The exact method in which the catalyst components are combined and their loading. Thus the character of a catalyst is different depending on whether it is prepared, for example, by ion exchange, co-precipitation or impregnation and the pH during this step; on whether the promoter is added before, after, or with the main component; on the temperature and rate of drying and calcining. All these and subsequent steps affect the dispersion (particle size) and distribution of the components through the catalyst pellet. Indeed they also determine whether the active catalyst is in itself a unique phase or just a physical mixture of a known bulk phase formed between the added components and the support.

b) The method in which the catalyst is activated. Simple factors during activation such as rate of temperature rise and final temperature, the dilutent phase and its pressure, the chemical form in which the activator is introduced, its concentration and exposure time are among the factors which influence and control the final chemical state of the catalyst when it is first exposed to the feedstock.

c) The conditions under which the catalyst is operated, e.g. the chemical composition of the feedstock and the nature of impurities therein; operating pressure, temperature and contact time; the care taken in the introduction of additives to aid performance.

d) Finally, many industrial catalysts are regenerated (and even rejuvenated). Control of conditions during the oxidation step where runaway exothermic reactions can easily occur is vital to the continuing satisfactory performance of any catalyst.

The major domain of the catalyst upon which all these influences impinge and which controls its behaviour is its surface. The characterisation of the surface chemical properties of a catalyst is the basis of this review.

2. ELECTRON SPECTROSCOPY FOR CHEMICAL ANALYSIS

Clearly no single method of analysis can monitor every step and all the factors listed above. However, Electron Spectroscopy for Chemical Analysis (ESCA), also known as X-ray Photoelectron Spectroscopy (XPS), is one technique which can reveal information about many more of the catalyst factors than other techniques. The features which make it so powerful are its ability to monitor the chemistry occurring on the outermost layers of a catalyst and, irrespective of the elements from which the catalyst is prepared (with the exception of hydrogen and helium) to detect changes of relative concentration of surface atoms. The principles are simple. A sample is bombarded with monoenergetic photons, normally $A\ell K_{\alpha}$ (1486.6 eV) or MgK_{α} (1253.6 eV), which eject electrons from core and valence shells in which the ionisation potential, or binding energy, is less than the primary photon energy. The kinetic energy of the ejected electrons is measured and is related to the binding energy by the relationship

$$\text{binding energy} + \text{kinetic energy} = \text{photon energy.}$$

The element from which the electron was ejected can be identified by comparing the measured binding energy with tables of X-ray absorption data. The method is not limited to chemical detection because the binding energy is a function of the chemical environment of the element, i.e. the exact value depends on the valency of the element and the ligands surrounding it. By utilising standard compounds the chemistry of the element can be readily characterised. The extreme surface sensitivity of ESCA arises because the kinetic energy of the ejected electron must be less than the photon energy and electrons with energies up to 1500 eV can escape only from within 20 Å of the surface. The relative concentration of surface atoms is revealed by the intensity of the specific signal from the atom under investigation. However, it is difficult to be quantitative. This is because of the complex range of factors which affect signal intensity such as photoelectron cross-section, atomic number and chemical state of the atom, escape depth, sample orientation with respect to the collector slit, surface roughness, surface geometry, distribution of species within the sampling depth, phase homogeneity and contamination. It is impossible to be specific about all these factors in differing samples, but where, in a group of catalysts, the same active phase is added to the same support, then the relative concentration of surface atoms can be compared following subsequent activation and use.

One of the limitations of ESCA is that the environment under which the catalyst is studied is never that under which the catalyst is prepared, activated or used. This means that prior to ESCA analysis, at pressures of approximately 10^{-8} torr, one attempts to fix the chemical state under real conditions and minimise chemical changes during transit. The chemical state determined must always be inferred from the spectra and not directly determined. Even sc-called _in situ_ preparation chambers do not completely overcome this limitation because reactor conditions are not reproduced. The second outcome from the conditions of study is that the nature of the adsorbed species, or adsorbed reaction intermediate, is not amenable to study. The effects of this major limitation are frequently not stated explicitly. Other limitations are :-

a) sample size. Analysis is from a sample area of approximately 3 mm by 15 mm allowing only macroscopic spatial resolution;

b) sensitivity. In a catalyst this is a function of surface area (and pore volume). The limit of detection of an element is about 1 percent of a monolayer, but to identify a chemical state a considerably higher concentration of a component is required;

c) non-quantitative analysis. Although mentioned above, this must be included in any list of limitations;

d) sample degradation in the photon beam. This is not a frequent occurrence but manifests itself as a change of valence state. It has been noted with chromium on silica[1], Ni_2O_3 on alumina[2], $PdC\ell_2$ on alumina[3] and Cu^{2+} in Y-zeolite[4];

e) signal overlap. This happens, unfortunately, in one of the most significant catalyst systems, Pt on $A\ell_2O_3$, where the major platinum signal Pt(4f) overlaps the major aluminium signal $A\ell$(2p). It is also found with ruthenium and carbon;

f) chemical changes with ion bombardment. Depth profiling via ion bombardment is known to cause chemical changes which could introduce chemical state ambiguity in the analysis of any catalyst so treated;

g) although ESCA can identify chemical states where there has been a major change of co-ordination environment, e.g. changes in anion, it cannot in general detect more subtle changes such as between octahedral and tetrahedral co-ordination as in spinels.

3. AIMS OF REVIEW

The aims of this review are to survey the application of ESCA to the identification of chemical states following the preparation and activation of catalyst systems of industrial relevance. Little information is available on such important features as the rôle of the promoter, changes in used catalysts, with the limited exception of the detection of catalyst poisons or modifications induced by regeneration or rejuvenation procedures. Consequently, at this time, these factors must be omitted. Some progress has been made in relating the activity or selectivity of a catalyst to a specific chemical state for a given reaction. Details of these systems will form the final section of the review leading to a short discussion on whether we are approaching a greater degree of understanding of catalysts or whether we are still in the age of empiricism.

4. INDUSTRIAL CATALYSTS OF IMPORTANCE

Discussions of the nature of industrial catalysts are hampered by lack of definition because of the commercial importance of these materials. Unless one is involved in both catalyst manufacture and use in industry one seldom knows all the chemical components of a catalyst for a given reaction, and one can never combine chemical composition with method of preparation and activation, details of feedstock and conditions of operation. Occasionally the veil is lifted, but it is never completely removed. In two recent articles[5,6] the chemical nature of catalysts used in industrial processes in the United States and their value was published. The information in these two articles is reproduced in a slightly abridged form in Tables 1 and 2 to provide the basis for the scope of catalyst systems which can be considered here. In the aspects discussed in subsequent sections, the major problem is to ensure realism with respect to industrial practice, particularly with regard to the loading of the active phases and the conditions for activation. In every case it is difficult to be exact, but knowledge from other sources will be used to ensure an approximation to reality.

Table 1

Use of Catalysts in Refining 1978

(After D.P. Burke, Chem. Week 28/3/79 p.42)

Process	Catalyst	Value x 10^6 \$ yr^{-1}
Catalytic cracking	Zeolites ; alumina-silica	144.3
Hydro-treating	Co/Mo/Al_2O_3	
	NiMo/Al_2O_3	
	WMo/Al_2O_3	40 - 48
Catalytic reforming	Pt/Al_2O_3	5.7
	bimetallic/Al_2O_3	16.8
	(metal costs)	4.9
Hydro-cracking	CoMo/Al_2O_3	
	Pd/zeolite	19.5
Alkylation	Sulphuric acid	115.6
	HF	13.2
		360 - 368

Table 2

Use of Catalysts in Chemicals Production 1978
(After D.P. Burke, Chem. Week 4/4/79 p.46)

Operation	Process	Catalyst	Value $\times 10^6$ \$ yr^{-1}
Steam Reforming	Reforming	Supported nickel (αAl_2O_3)	
	Shift	Cr promoted Fe, $CuO/ZnO/Al_2O_3$	
	Methanation	Supported nickel	23.00
Synthesis	Ammonia	Promoted iron (K_2O, CaO, Al_2O_3)	
	Methanol	Cu-chrome-zinc	6.10
Hydrogenation	Various	Ni	23.50
Oxidation	Ethylene oxide	Ag/support	19 - 23
	HNO_3 ex-NH_3	Pt-Rh gauze	17.10
	Sulphuric acid	V_2O_5/SiO_2	3.40
	Maleic anhydride	V_2O_5	25 - 30
	Phthalic anhydride	V_2O_5	3.16 - 3.26
Ammoxidation	Acrylonitrile ex-propylene	-	10.40
Oxychlorination	Ethylene dichloride	$CuCl$	1.26

(continued)

Table 2 continued

Operation	Process	Catalyst	Value $\times 10^6$ \$ yr^{-1}
Organic Synthesis	Various	Mainly noble metals Supported on $A\ell_2O_3$	89.83
Polymerisation	Ziegler-Natta	Aluminium alkyls Titanium trichloride	50 - 55
	Polyethylene (Phillips)	Cr_2O_3 on SiO_2	1.00
	Low density polyethylene	Peresters	17 - 18
	PVC	Percarbonates	17.50
	Polystyrene	Benzoyl peroxide	3.40
	Urethanes	Amines, organotins	21.00
		TOTAL	314.32

5. ESCA STUDIES OF THE PREPARATION OF CATALYSTS

Catalysts are prepared by precipitation or co-precipitation, by impregnation or ion-exchange depending on the active phase, the support and the reaction which the system aims to catalyse. This initial step is always accompanied by drying at a temperature just above 100°C and, where necessary, by calcination normally in flowing air around 550°C. In each method of preparation it is not possible to predict, *a priori*, whether surface or bulk compounds will be formed between the active phase and the support or, indeed, whether the components remain as a physical mixture. ESCA is particularly invaluable in resolving these possibilities particularly at high dispersions and relatively low loadings. The aim of this section is to illustrate the results which have been obtained on the chemical characterisation of catalyst phases during their preparation. There will be some overlap with the section on activation where reduction forms an integral part of catalyst preparation. Reduction will be discussed in this section only where it is necessary to achieve a complete picture of the chemical phases which are feasible within a given system.

Co-precipitation is usually employed to obtain an extremely uniform distribution of components throughout a catalyst pellet. This can lead to bulk compound formation characterisable by X-ray diffraction. But this is not always the case. For example, in the co-precipitation of nickel-alumina catalysts, for use in methanation, with nickel loadings between 10 and 40 wt percent, it was found that, after calcination, the Ni($2p_{3/2}$) binding energy had the value close to that for NiO and, from the magnitude of the signal intensity, it was deduced that the particle size of the unreduced nickel remains constant[7]. On the other hand, a nickel oxide on silica catalyst[8], prepared by the precipitation of $Ni(OH)_2$ with ammonia onto the silica for use in olefin dimerisation, formed nickel silicate at loadings up to 40 wt percent nickel if the catalyst was calcined. Yet when a catalyst of the same nominal composition was formed by impregnation only NiO was detected at the silica surface[8].

This was expected - impregnation is the normal method by which the active catalyst phase is prepared, initially as a physical mixture, in high dispersion at the surface or in the pores, but not in the bulk, of a support. Nevertheless, both bulk and surface compounds have been detected following impregnation. The detection of surface compounds is readily accomplished by means of ESCA. They are monitored either by a shift in binding energy or by a broadening of the signal from the impregnate. The shift in binding energy is always to higher value indicating a donation of electrons from the impregnate to the support. Major examples of this phenomenon are the systems $MoO_3/\gamma A\ell_2O_3$[9] and $WO_3/\gamma A\ell_2O_3$[10] where binding energy shifts between 0.4 and 1.0 eV are detected. Yet these phases do not exhibit the chemical properties of bulk aluminium molybdate or tungstate but have quite novel properties particularly in reduction (see below). In effect these are hardly much more than monolayer compounds, which cannot be associated with a bulk phase. Their formation is totally dependent on the nature of the support. They do not seem to be formed in the MoO_3/SiO_2[11], WO_3/SiO_2[10] or MoO_3/TiO_2[12] systems. A weaker kind of association is suggested when the line width of the impregnate signal is broadened without a shift in binding energy. An example of such a system is $Re_2O_7/\gamma A\ell_2O_3$[13] in which the reduction of the Re_2O_7 is retarded relative to Re_2O_7/SiO_2 but not modified. Bulk compounds can also be found via impregnation. Cobalt and nickel oxides form pseudo-aluminates with γ-$A\ell_2O_3$ even when impregnated in association with molybdena and tungsta[14,15]. However, the bulk compound formation is sensitive to loading and calcination temperature. This has been particularly well defined for the CuO/Al_2O_3 system. Freidman and his co-workers[16] report that, for less than 4 wt percent Cu per 100 m^2g^{-1} Al_2O_3, copper aluminate alone is formed. Above this loading free CuO also exists.

In the preparation step ion-exchanged catalysts must form a chemical compound with the support material, even if it is only a surface chemical compound at low exchange levels. Whether this compound remains intact depends on subsequent treatments. Thus, in the preparation of nickel alumina catalysts by ion exchange with $Ni(NO_3)_2$ or Ni_2SO_4, at low nickel concentrations the $Ni(2p_{3/2})$ binding energy has a value similar to that of $Ni(OH)_2$. The authors[2] argue that this indicates the incorporation of hydroxide (and oxide) anions around Ni^{2+} and not necessarily the formation of $Ni(OH)_2$ as an independent compound. At higher salt concentrations in the ion-exchange medium, both ion exchange and impregnation occur. With $NiSO_4$ the binding energy detected approaches the value for the salt itself, but $NiSO_4$ is readily removed by further washing.

The importance of further treatments is highlighted in studies of catalysts formed by ion exchange of palladium onto sodium Y zeolite[17]. Exchange was effected via the ammonia complex $Pd(NH_3)_4^{2+}$. Subsequent changes are listed in Table 3. With calcination at 500°C in flowing oxygen, the palladium is believed to be in cationic sites with a nominal valency of 2^+. In fact the binding energy determined is 2.9 eV greater than in PdO. Whether this is the formation of a surface chemical state, as seen above, or the palladium cations are co-ordinated with three lattice oxygens, i.e. to a larger extent to oxygen than in PdO, is a matter for conjecture. Following reduction at room temperature and increasing temperatures up to 300 - 400°C, the palladium state evolves through Pd^+ in cationic sites, then Pd^o atomically dispersed, to Pd^o in 20 Å particles (verified by XRD). A similar sequence of changes has been noted for NiNa Y zeolite[4] where the nickel metal segregation from the zeolite during the reduction stage can be reoxidised for certain catalytic applications (see below).

In considering the ion-exchanged palladium results given above, one must not forget the high palladium loading relative to conventional catalysts. Loadings at and below 1 wt percent are more normal. Intriguingly, a similar sequence of events has recently been reported[18] on a conventionally loaded palladium on alumina catalyst formed by impregnation. These trends are illustrated on the right-hand side of Table 3. In particular one notes :-

Table 3

Change occurring in the Preparation of Ion-exchanged and Impregnated Palladium Catalysts

Step	$Pd_{12.5}Na_{19.5}(NH_4)_{11.5}$ -Y zeolite (17) Author's assignments	Binding Energy $Pd(3d_{5/2})$		1 wt % $Pd/\eta-Al_2O_3$ impregnated via $[Pd(NH_3)_4](NO_3)_2$ (18) Author's assignments
After ion exchange	$Pd(NH_3)_4^{2+}$	337.6		
After calcination at 500°C	Pd^{2+} cationic sites	339.2	340.6 339.2	Pd^{4+} - O Pd^{2+} - O
After reduction at room temperature	Pd^{+} cationic sites Pd^{o} atomically dispersed	(339.2) 337.1 336.4		
Reduction at 160°C	Pd^{o} atomically dispersed	336.4		
Reduction at 350°C	Pd^{o} (20Å)	335.2	339.7 338.3	Pd^{2+} - O unreduced palladium compound
Reduction with hydrazine			339.3 337.4	Pd^{2+} - O Pd state reduced wrt Pd^{2+} - O
NB PdO_2 PdO PdO_{ads} Pd^{o} metal		337.9 336.3 335.6 335.0		

a) the exact correspondence in binding energy (BE) for states assigned to Pd^{2+}-O and Pd^{2+} in cationic sites;

b) if the Pd^{+} assignment is correct for the state with BE between 337.1 and 337.4 eV, and the detection of states with lower BE at higher palladium loading suggests this is reasonable, palladium is never reduced below a cationic state to a metallic state;

c) two additional states with binding energies of 340.6 and 338.3 eV are detected. The former state is possibly a Pd^{4+}-O state. The exact valency is unclear but the authors note the similarity in ΔBE between PdO_2 and PdO compared with their assignment for Pd^{4+}-O and Pd^{2+}-O. Their identification of the anion is less ambiguous since they have detected the same state with PdO impregnated onto alumina. This state is analogous with a platinum state on alumina, as discussed in the next paragraph.

The major trend revealed at lower loading is the incomplete reduction of the palladium even in the presence of as strong a reducing agent as hydrazine. Treatment with the latter enhances the state with binding energy 337.4 eV while at the same time the selectivity of styrene production from 4-vinyl cyclohexene-1[19] is increased.

With details of this significance revealed in Pd-zeolite and Pd-alumina systems, interest is stimulated in the more commercially important Pt-zeolite and Pt-alumina systems. Similar states do seem to be suggested even at the higher loadings necessary for study. Thus Vedrine et al.[17] with 6.5 wt percent platinum ion-exchanged onto Y-zeolite believe they have detected Pt^{2+} in cationic sites (ΔBE = 3.6 eV) and atomically dispersed Pt^{o} with a higher binding energy (ΔBE = 1.3 eV) than in 20 Å metal particles (ΔBE = 0.7 eV, all relative to ion-bombarded bulk metal). Larkins et al.[20] with 3 wt percent Pt ion-exchanged onto Pt (Na)/Y zeolite and Pt (La)/Y zeolite report three states following reduction at 350°C, Pt^{2+} with ΔBE = 2.4 eV, Pt^{+} with ΔBE = 1.2 eV and Pt^{o} (ΔBE = 0). With platinum supported on alumina the Pt ($4d_{5/2}$) signal is studied. Nevertheless the results are again similar. Bouwman et al.[21] for 3.8 wt percent Pt on γ-$A\ell_2O_3$ report a state with a binding energy 1.1 eV above the bulk value. Ushakov et al.[22] for a 7 wt percent platinum impregnated on alumina find Pt^{2+}-O following calcination (ΔBE = 2.5 eV), and a state with ΔBE = 0.8 eV following reduction. Finally, Escard et al.[23] reported a state whose binding energy was 1.9 eV above Pt^{2+}-O. This they assigned to

$$(> Al\text{-}O)_x\ PtCl_y \qquad \text{where } x + y = 4$$

i.e. the nominal valency of the platinum is 4+. There are thus very clear links between the Pd-zeolite/alumina and Pt zeolite/ alumina systems, and it is interesting to speculate on the feasibility of non-reducible platinum states in reforming catalysts when the loading of platinum is only 0.3 - 0.6 wt percent. Speculation on 'soluble platinum' and its rôle in reforming have been rife for many years[24].

Studies of the chemistry of catalyst preparation have far exceeded studies of physical parameters using ESCA. However, it is possible, in a macroscopic way, to study parameters such as catalyst distribution as a function of the method of preparation or the conditions of calcination. The distribution of a catalyst phase between the surface of a pellet or an extrudate and its interior can be measured by comparing the ratio of the active phase peak intensity to a support peak intensity for the same sample first as a pellet and then as a crushed specimen. Mitchell and Edmonds[25] have followed the change in distribution of MoO_3 at various loadings on alumina with calcination. Two trends emerge as can be seen in Table 4. Firstly, at all loadings, MoO_3 moves from the surface to the bulk of the pellet. This trend is confirmed with crushed samples. Secondly, as the surface loading decreases, the binding energy of MoO_3 increases from a value typical of bulk MoO_3 in the dried samples to a value some 0.7 eV higher, typical of surface interaction between MoO_3 and Al_2O_3. Indeed the existence of this interaction could provide the driving force for the movement of the molybdena.

6. ACTIVATION OF CATALYSTS

The activation stage of a catalyst is the final step before the catalyst is exposed to the feedstock. Consequently the identification of chemical states or modifications in the active phase of the catalyst during this treatment are crucial to any understanding of the performance of the catalyst. This step has been studied in detail for only a limited number of systems and usually not in association with subsequent behaviour. Consequently in this section it is planned to discuss changes which occur under activating conditions, such as reduction and sulphiding. Those systems where some attempt has been made to relate chemical state to catalytic behaviour will be discussed in the next section.

6.1 Activation by reduction to metal

In many systems activation by reduction leads to a metallic state for the major catalytic component in which the binding energy has the same value as that determined for the bulk metal.

Table 4

XPS Spectra of $MoO_3/A\ell_2O_3$ Catalysts

Sample No.	Physical form	Heat treat-ment ^oC	Mo wt-percent	B.E.[a] $Mo(3d_{5/2})$ eV	$\frac{I(Mo)}{I(A\ell)}$[b]
1	extrudate	120	2	232.0	0.56
2	powder	120	3	232.0	0.14
3	extrudate	120	4	232.0	0.92
4	extrudate	120	9	232.0	2.84
5	extrudate	120	15	232.1	5.81
6	extrudate	550	2	232.7	0.29
7	powder	550	3	233.0	0.16
8	extrudate	550	4	232.9	0.47
9	extrudate	550	9	232.7	6.35
10	extrudate	550	15	232.5	4.19
11	crushed extrudate	120	9	232.5	0.43
12	crushed extrudate	550	9	232.9	0.56

a Binding energy relative to $A\ell(2p)$ = 74.5 eV.

b Relative intensity of $Mo(3d_{5/2})$ and $A\ell(2p)$ peaks.

These systems will not be discussed here. Instead it is intended to focus on those systems in which a binding energy shift has been detected. They are listed in Table 5. The conclusions which may be drawn from this table are:-

a) interaction with the support has been detected with the metals Re, Ir, Pd, Pt. It is of interest to note that three of these four metals are significant components of catalytic reforming catalysts;

b) the major supports with which interaction occurs are γ- or δ-$A\ell_2O_3$ and Y-zeolite;

Table 5

Systems with Support Interaction in Reduced Catalyst

System	Reduction Conditions	ΔBE from bulk metal	Ref.
3.4 - 4.8 wt % $Re/\gamma Al_2O_3$	$500^oC/H_2/1$ atm	0.5 eV	(13)
$Re/\gamma Al_2O_3$	$200^oC/H_2 : C_6H_6=10/30$ atm	0.5 eV Re^o 2.6 eV Re^{4+}	
5 wt % $Ir/\delta Al_2O_3$	$400^oC/H_2$	1.8 eV	(26)
Ir/TiO_2	$400^oC/H_2$	1.2 eV	(26)
Ir/SiO_2	$400^oC/H_2$	0.9 eV	(26)
Ir/ZnO	$400^oC/H_2$	0.3 eV	(26)
12.5 wt % Pd/Y zeolite	$160^oC/H_2/$	1.4 eV (atomic ?)	(17)
	$350^oC/H_2/$	0.2 eV (10-20 Å)	
6.5 wt % Pt/Y zeolite	Unclear	With $Pt(4f_{7/2})$ 1.3 eV (atomic ?) 0.7 eV (10-20 Å)	(17)
3 wt % Pt/Y zeolite	$350^oC/H_2$	With $Pt(4f_{7/2})$ 2.4 eV Pt^{2+} 1.2 eV Pt^+ 0.0 eV Pt^o	(20)
7 wt % Pt/Al_2O_3	$550^oC/H_2/$	With $Pt(4d_{5/2})$ 0.8 eV	(22)
3.8 wt/$Pt/\gamma Al_2O_3$	$550/650^oC/H_2/1$ atm	With $Pt(4d_{5/2})$ 1.1 eV	(21)
[3.8 wt % Pt + 2.5 wt % $Ge/\gamma Al_2O_3$]	$550/650^oC/H_2/1$ atm	With $Pt(4d_{5/2})$ 1.6 eV	(21)

c) interaction with the support may be greater when the metal particles are atomically dispersed. In three systems, Re-$A\ell_2O_3$[13], Pd/Y-zeolite[17] and Pt/Y-zeolite[17], this is corroborated to some extent by XRD, i.e. the lowest BE state is measured at the limits of XRD detection which is 20 Å;

d) with platinum, binding energy shifts have been detected with both 4f and 4d signals;

e) in the Pt-Ge system the binding energy shift is increased by the presence of germanium.

It is normally argued that these binding energy shifts occur as a result of electron donation from the metal to the support. This is self-evident from the magnitude of the shift. Why it should occur particularly with alumina has never been explained. Maybe it is related in simple terms to the amphoteric nature of the support, i.e. its ability to form compounds with both acid and bases. Yet in both metal and inorganic systems the shift in binding energy which is induced is always so that electrons are donated to the support. The clue to this behaviour may, in fact, lie in the observation of a binding energy shift difference between atomically dispersed and 20 Å particles. It seems evident that alumina maintains metal and inorganic components at very high dispersions or small particle size under normal conditions of activation. At atomic dispersion individual atoms will reflect their environment, while, in larger particles, the support effect will be diluted across the total number of atoms in the particle. However, it has to be recognized that cause and effect may be difficult to distinguish here with the support interaction inducing the high dispersion. However inconclusive the outcome of this argument, it does seem that the unique character of alumina (and zeolites) as catalyst support materials is related to their ability to interact with active catalyst phases and moderate their behaviour.

Reduction causes a decrease in signal intensity, i.e. a probable increase in particle size. This has been reported for NiO/$A\ell_2O_3$[7], Pt/$A\ell_2O_3$[21,23], Rh/Charcoal[27] and Pd/Y zeolite[17]. One notes this effect takes place even in systems where there is strong interaction with the support.

Minachev et al.[4], on the other hand, have reported the opposite behaviour, at least in the initial reduction stages for a range of elements ion-exchanged into sodium faujasite. They argue that reduction is accompanied by migration of the metals onto the external surface of the zeolite causing the increase in relative intensity of the transition element lines in the spectra. The ability of reduced metals to migrate decreases in the series

$$Ag > Zn \gg Pd \gg Cu > Ni > Pt \geq Co.$$

From the migration kinetics measured by ESCA and the particle size determined by XRD, it has been calculated that the number of silver atoms, i.e. the most mobile element, migrating onto the external surface of the zeolite corresponds to 25 - 50 percent of the total number of Ag atoms in the zeolite. From this it is apparent that the metals are mostly retained in the zeolite cavities during reduction. Subsequent oxidation of zeolites containing reduced metals gives rise to positive shifts in the spectra. In this case the state of the transition element is different from the starting state and is similar to that of the cations in the corresponding oxides.

Finally, in two studies there is the suggestion that a greater degree of reduction is achieved in the presence of alkali metal. For $Re/A\ell_2O_3$ this was reported when the potassium salt was used for impregnation[13] while Larkin et al.[20] have reduced 3.5 wt percent Pt ions exchanged onto La-Y zeolite and Na-Y zeolite. The extent of reduction into three platinum states at 350°C was

	Pt^o	Pt^+	Pt^{2+}
Pt/La Y (percent)	47	32	21
Pt/Na Y (percent)	69	23	8

6.2 Activation by reduction to lower oxides

Two closely related catalyst systems are incompletely reduced in hydrogen, $CoMo/A\ell_2O_3$ and $NiW/A\ell_2O_3$[28,15]. Both are systems in which there is strong interaction between the alumina and the support. The results for molybdenum in $CoMo/A\ell_2O_3$ are shown in Fig. 1.

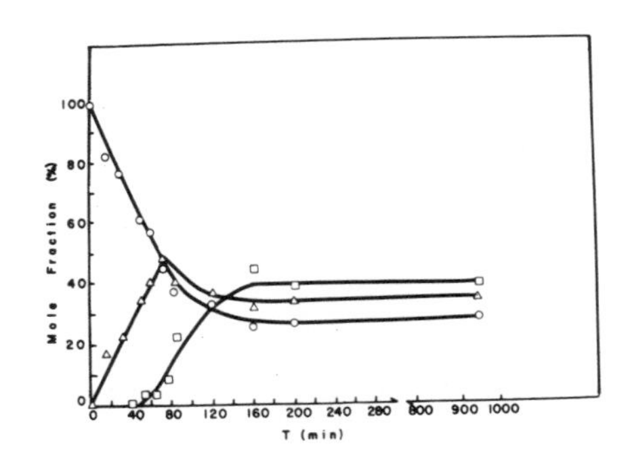

Fig. 1

Speciation of molybdenum on the catalyst surface as a function of reduction time at 500°C: (○) Mo(VI); (△) Mo(V); (□) Mo(IV).

Reproduced with permission from T.A. Patterson, J.C. Carver, D.E. Leyden and D.M. Hercules[28].

This shows that the Mo(VI) concentration decreases almost linearly with reduction time up to 90 minutes. At this point there is a marked change in slope and much slower depletion of Mo(VI) but reduction still continues thereafter. Two other states of molybdenum can be detected from the Mo(3d) signal;

Mo(V), whose concentration increases linearly until a maximum is reached at 70 minutes and then decays to a plateau just above 30 percent, and Mo(IV) which is not detected before 45 minutes reduction and then increases rapidly to reach a plateau of approximately 40 percent. Reduction at 300°C increases the Mo(V) maximum and plateau by ~ 20 percent. Similar results were observed by Cimino and De Angelis[11] for CoMo/$A\ell_2O_3$ and are compared with reduction of MoO_3, MoO_3/SiO_2 and $CoMoO_4$. For each of these systems reduction proceeds to the metal.

Surprisingly, there has been little comparable study of the state of cobalt following reduction.

Results for the reduction of a 6 wt percent Ni, 19 wt percent W/$\gamma A\ell_2O_3$ [Ni4303] catalyst are shown in Fig.2[15].

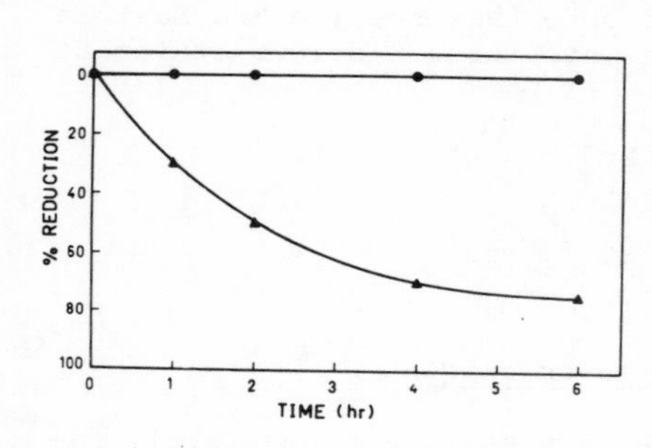

Fig. 2

Percentage reduction of tungsten (●) and nickel (▲) as a function of reduction time for catalyst Ni 4303. Reduction at 450°C with H_2 at a flow 80 cm^3/min.

Reproduced with permission from K.T. Ng and D.M. Hercules[15].

At 450°C, even after 6 hours there is no evidence for the reduction of tungsten to states below W(VI). Nickel species are readily reduced to the metal. However, reduction is incomplete and, even after prolonged hydrogen treatment, some 25 percent of the nickel remains unreduced. This is believed to be $NiA\ell_2O_4$ which is fairly stable towards reduction. This result for tungsten is in complete agreement with that of Bileon and Pott[10] who found that a 13 percent $WO_3/\gamma A\ell_2O_3$ catalyst is not reduced in hydrogen at 550°C.

Thus, even in apparently similar systems, reduction produces quite different chemical states.

6.3 Activation by sulphiding

CoMo/$\gamma A\ell_2O_3$ and NiW/$\gamma A\ell_2O_3$ systems are usually activated by sulphiding prior to use as hydrodesulphurisation and hydrogenation catalysts. Consequently, the sulphiding of these systems has been studied extensively. There is little controversy or divergence on the final state of either molybdenum or tungsten. It is MoS_2 and WS_2, respectively,[14,15,28-32]. However, sulphiding is never complete. This is illustrated in Fig. 3.

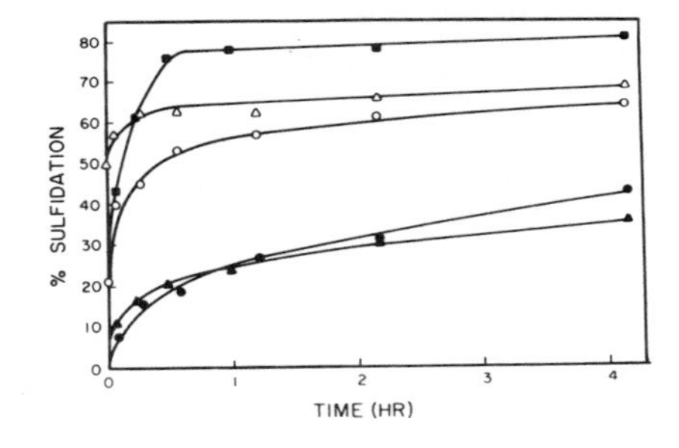

Fig. 3

Percentage sulfidation as a function of sulfiding time. Sulfiding was carried out at 350°C with 9.2 % H_2S/H_2 mixture at a flow rate of 100 cm^3/min. Solid points represent tungsten. Open points represent nickel.
(■) Aluminium tungstate;
(●) Ni 4303;
(▲) W 0801; (○) N 4303;
(△) Ni 0301.

Reproduced with permission from K.T. Ng and D.M. Hercules[15].

Catalyst Ni 0301 contained 11 wt percent Ni/γAl_2O_3 and W 0801 8 wt percent W/γAl_2O_3. Considering Ni 4303 alone the sulphiding of the nickel species is 65 percent complete after 4 hours at 350°C in 9 percent H_2S/H_2, while the W species is only 40 percent sulphided but has not reached a plateau.

Similar results are reported for CoMo/γAl_2O_3. Here the controversy lies in the exact state of cobalt following sulphiding. Prior to activation there are two cobalt species on the surface: a pseudo-aluminate favoured at low cobalt contents when the cobalt is well-dispersed on a high area support and when the calcination temperature is high[14], and an oxide phase, Co_3O_4[14], which predominates at higher loadings. Only the oxide phase can be sulphided. Cobalt ions in the pseudo-aluminate phase do not seem to form a sulphide. However, there are two schools of thought on the fate of the oxide phase following sulphiding. The school of Delman favours a sulphide; other schools favour the existence of metallic cobalt. Declerck-Grimée et al.[30], with samples sulphided at 400°C in a 15 percent H_2S-H_2 mixture, obtained the spectrum shown below, Fig. 4. The binding energy corresponding to the Co($2p_{3/2}$)-F species, 778.3 eV, is different from the value found for Co^o, 776.4 eV and is assigned to Co_9S_8. It is of particular importance that this line does not exhibit a spin satellite peak which would be expected for CoS. The other peak Co($2p_{3/2}$)-1 with a satellite and whose binding energy is 781.2 eV is unsulphided and assigned to the pseudo-aluminate. The interpretation for the Co($2p_{3/2}$)-F species is supported by studies on Co/Al_2O_3[33] where, after reduction and sulphiding, a line at 776.4 eV is detected. This is assigned to metallic cobalt on the basis of the magnitude of its bind-

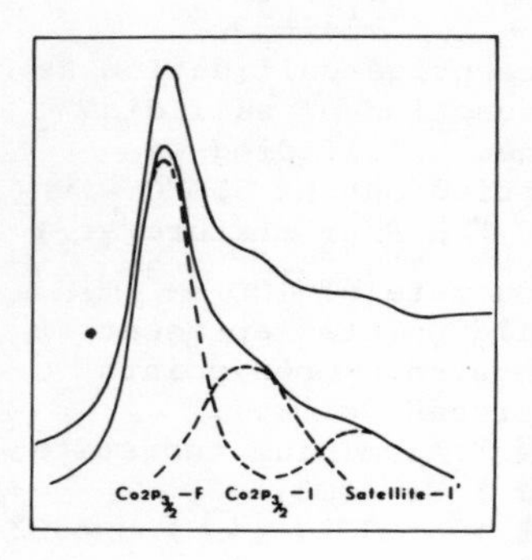

Fig. 4

XPS profile of the Co $2p_{3/2}$ level for C_3 catalyst after sulfidation. Reproduced with permission from R.I. Declerck-Grimée, P. Canesson, R.M. Friedman and J.J. Fripiat[30].

ing energy; also, even after sulphiding, the S(2p) line has very low intensity. Finally, in the reduction and sulphiding of unsupported cobalt compounds in the absence of molybdenum, treatment by H_2S-H_2 leads to greater quantities of metal cobalt species than a simple reduction treatment by H_2. The other interpretation of the sulphidable cobalt signal has been discussed by Brinen et al.[31] and Okamoto et al.[32]. They argue that the chemical state of the cobalt is as metal and make this assignment on the basis of the spin orbit splitting between Co($2p_{3/2}$) and Co($2p_{1/2}$). A value of 15.1 eV is measured following sulphiding. Paramagnetic Co^{2+} species are characterised by a spin orbit splitting of 16 eV for the Co(2p) lines and are generally accompanied by pronounced satellite structure. A spin orbit separation of 15 eV is generally seen for diamagnetic Co^{3+} compounds and metallic cobalt with little or no satellite structure. The most likely assignment is to metallic cobalt since the reaction is performed in a reducing environment. In support of this interpretation Brinen et al.[31] analysed the S/Mo intensity ratio as a function of cobalt concentration. The ratio is independent of the total cobalt concentration for these catalysts, while a linear relationship was found between the Co/Aℓ intensity ratio and total cobalt up to 5 wt percent. These results verify that the cobalt in the activated state is not sulphided. A substantial increase in S/Mo would be expected if sulphiding had occurred. The differences between these two schools still awaits resolution. A little further evidence can be provided in favour of each hypothesis. For the sulphide state Massoth[35] has pointed out, on the basis of thermodynamic arguments, that the mole ratio of $H_2S : H_2$ must be less than 10^{-4} for cobalt metal to be the stable state, whereas sulphiding is normally performing with ~ 10 volume % H_2S in H_2, i.e. molar ratio of $\sim 10^{-1}$. For the metal state it must be noted that in at least one study[31] the ease of reoxidation of cobalt to the oxide is reported. If a sulphide had formed, re-oxidation is usually via the sulphate[34].

7. IDENTIFICATION OF THE ACTIVE CATALYST SITE VIA ESCA

The ultimate aim of any industrial study of a catalyst system using any technique is to identify a chemical state or phase which can be linked more or less directly with the activity or selectivity of the catalyst operating under real conditions in an industrial reactor so that the presence of this state can be maximised during catalyst preparation and activation. ESCA has had appreciable success in meeting this aim but, perhaps not surprisingly in such a commercially sensitive area, published detailed studies clearly relating increasing activity/selectivity to increasing concentrations of the identified component are rare. Table 6 lists the identified active state in a range of catalyst systems and the reaction for which the active state has been identified. The most complete studies have been on three systems, oxidative dehydrogenation of 4-vinyl cyclohexene-1 [18, 19], dehydrogenation and dehydration of butan-2-ol[38,39], and the hydrogenolysis of triophen and CS_2 [29], and we shall discuss only these. The results for the oxidative dehydrogenation of 4-vinyl cyclohexene-1 are interesting because both the active states have binding energies which cannot be linked directly with the binding energy of a known palladium compound - i.e. both states arise from a special interaction with the catalyst support. The higher BE state, 339.2 eV, is the state selective for ethyl benzene, while the lower BE state, maximised by reduction with hydrazine, is selective for the desired product styrene. In the dehydrogenation and dehydration of butan-2-ol the selectivity of the reaction is controlled by the extent of dehydration of $Mg(OH)_2$ itself, i.e. the basicity or electron pair donor/acceptor properties of the system[39]. This concept has been extended to other magnesium compounds to correlate their binding energies to selectivity[38]. Finally, in the much-studied $CoMo/A\ell_2O_3$ system the first-order rate constant for the hydrogenolysis of thiophen and the hydrogenation of CS_2 (measured in an atmospheric microreactor) has been related to the concentration of MoS_2, with a small contribution from another currently unidentified state (Mo^{5+} or Mo^{4+} oxysulphide)[29] after sulphiding. Here, although there is again interaction between the support and the catalyst as prepared before sulphiding, the binding energy of $Mo(3d_{5/2})$ in MoS_2 is very similar to the value in the bulk sulphide. The rôle of cobalt in this system has not been defined[30,32]. Overall Table 6 indicates the diversity of active states which have been identified, and stresses the fact that without ESCA they could hardly have been predicted and identified. Their

Table 6

Identification of the 'Active State' for Various Industrially Important Catalytic Reactions

Catalyst	Reaction	Active State	Ref.
$-V_2O_5-MoO_3-MO_x$	furan → maleic anhydride	V^{4+}	36
$NiO/Al_2O_3/SiO_2$	Olefin dimerisation	layered silicate	8
Rh/charcoal	Selective Hydrogenation	Rh_2O_3	37
Pd/Al_2O_3 $/TiO_2$ $/SiO_2$	Ethylene oxidation	Pd^o (Pd complexes reduce activity)	3
Pd/Al_2O_3	Oxidative Dehydrogenation 4-vinyl cyclohexene-1 → styrene → ethyl benzene	Pd state with $Pd(3d_{5/2}) = 337.3$ eV Pd state with $Pd(3d_{5/2}) = 339.2$ eV	18 19
Ni Y zeolite	Alkylation of benzene with ethylene → s-butyl benzene → ethyl benzene	cationic nickel Ni^o	4
0.33 NiNa Y zeolite	CO oxidation NO oxidation	Ni in oxidised state outside zeolite cage	4

cont.

Table 6 continued

Catalyst	Reaction	Active State	Ref.
$Mg(OH)_2/MgO$	Butan-2-ol		38
	Dehydrogenation	MgO i.e. high basicity	39
	Dehydration	$Mg(OH)_2$ lower basicity	
$CoMo/Al_2O_3$	Thiophen/CS_2 hydrogenation/ hydrogenolysis	MoS_2 (Mo^{4+})	29

identification, together with their method of production, particularly the hydrazine reduction, is an initial step in taking art and empiricism out of catalysis.

8. CONCLUSIONS

In this review the preparation and activation of industrial catalysts have been discussed and chemical states identified by ESCA related to activity and selectivity. There are several major conclusions.

1. In preparation three states can be identified in supported catalysts; the formation of a bulk compound between the added component and the support; the formation of a physical mixture between the added component and the support and, uniquely, the formation of a state with a binding energy indicating electron donation from the added component to the support. This is usually called 'support interaction' or simply interaction with support.

2. Such interaction is detected most frequently when the support material is γ-alumina (or a zeolite). This could be a major clue as to why this support is held in such favour for industrial catalysts.

3. It is normally not possible to forecast which of the three states, listed under 1, will be formed during catalyst preparation, and their respective presence can only be separated via ESCA.

4. 'Support interaction' can strongly influence subsequent activation steps. This is particularly true during reduction with hydrogen. It may be less significant with sulphiding under reducing conditions.

5. At low loadings, noble metals supported on alumina and zeolites form states which may not be reducible to metal, are not identifiable with known compounds and seem to be characterisable only *via* ESCA.

6. The form of the state, noted under 5, can be varied with the chemical nature of the reducing agent.

7. In one catalytic system, reaction selectivity has been related to the presence of these states.

Although some trends are evolving from the data, many of the results seem unique to individual systems. One most hope that this is due to the limited amount of data and that some of the art and empiricism can be removed from catalysis. Nevertheless, it cannot be forgotten that we are frequently dealing with complicated many-component systems which have defeated man's academic ingenuity for approaching 200 years. It should not be surprising that, with increasingly relevant techniques in our grasp, understanding still requires time.

ACKNOWLEDGEMENT

Permission to publish has been given by the British Petroleum Company Limited.

REFERENCES

1. D.L. Perry, unpublished data.
2. J.K. Gimzewski, B.D. Padalia and S. Affrossman, J. Catal., 55, 250 (1978).
3. F. Bozon-Verduraz, A. Omar, J. Escard and B. Pontvianne, J. Catal., 53, 126 (1978).
4. K.M. Minachev, G.V. Antoshin, E.S. Shpiro and Yu A. Yusifov, Proceedings 6th Intern. Catalysis Congress, p.621 (1977).
5. D.P. Burke, Chemical Week (28/3/79) p.42.
6. D.P. Burke, Chemical Week (4/4/79) p.46.
7. R.B. Shalvoy and P.J. Reucroft, J. Electron Spect., 12, 351 (1977).
8. P. Lorenz, J. Finster, G. Wendt, J.V. Salyn, E.K. Zumadilov and V.I. Nefedov, J. Electron Spect., 16, 267 (1979).
9. A.W. Miller, W. Atkinson, M. Barber and P. Swift, J. Catal. 22, 140 (1970).
10. P. Bileon and G.T. Pott, J. Catal., 30, 169 (1974).
11. A. Cimino and B.A. De Angelis, 36, 11 (1975).
12. J.C. Vedrine, H. Praliand, P. Meriandeau, M. Che, Surface Science, 80, 101 (1979).
13. E.S. Shpiro, V.I. Avaev, G.V. Antoshin, M.A. Ryashentseva, K.M. Minachev, J. Catal., 55, 402 (1978).
14. P. Gajardo, R.I. Declerck-Grimée, G. Delvaux, P. Olodo, J.M. Zabala, P. Canesson, P. Grange and B. Delman, J. Less Common Metals, 54, 311 (1977).
15. K.T. Ng and D.M. Hercules, J. Phys. Chem., 80, 2094 (1976).
16. R.M. Friedman, J.J. Freeman and F.W. Lytle, J. Catal., 55, 10 (1978).
17. J.C. Vedrine, M. Dufaux, C. Naccache and B. Imelik, J.Chem. Soc. Faraday Trans I, 74, 440 (1978).
18. G. Mattagno, G. Polsonetti and G.R. Tauszik, J. Electron Spect., 14, 237 (1978).
19. A. Castellan and G.R. Tauszik, J. Catal., 50, 172 (1977).
20. F.P. Larkins, M.E. Hughes, J.R. Anderson and K. Foger, J. Electron Spect., 15, 33 (1979).
21. R. Bouwman and P. Bileon, J. Catal., 48, 209 (1977).
22. V.A. Ushakov, E.M. Moroz, P.A. Zhdan, A.I. Boronin, N.R. Bursian, S.B. Kogan and E.A. Levitskii, Kin. i Katal., 19, 587 (1978).
23. J. Escard, B. Pontvianne, M. Chenebaux and J. Cosyns, Bull Soc. Chim. France, 349 (1976).

24. R.J. Bertolacini, Nature, 192, 1179 (1961).
25. P.C.H. Mitchell and T. Edmonds, to be published.
26. J. Escard, B. Pontvianne and J.P. Contour, J. Electron Spect., 6, 17 (1975).
27. J.S. Brinen, J.L. Schmitt, W.R. Doughman, P.J. Achorn, L.A. Siegel and W.N. Delgass, J. Catal., 40, 295 (1975).
28. T.A. Patterson, J.C. Carver, D.E. Leyden and D.M. Hercules, J. Phys. Chem., 80, 1700 (1976).
29. G.C. Stevens and T. Edmonds, J. Catal., 44, 488 (1976).
30. R.I. Declerck-Grimee, P. Canesson, R.M. Friedman and J.J. Fripiat, J. Phys. Chem., 82, 889 (1978).
31. J.S. Brinen and W.D. Armstrong, J. Catal., 54, 57 (1978).
32. Y. Okamoto, T. Imanaka and S. Teranishi, J. Catal., 54, 452 (1978).
33. R.I. Declerck-Grimee, P. Canesson, R.M. Friedman and J.J. Fripiat, J. Phys. Chem., 82, 885 (1978).
34. T. Edmonds, unpublished data.
35. F.E. Massoth, J. Catal., 54, 450 (1978).
36. Yu M. Shul'ga, L.N. Karklin, M.V. Shimanskaya, L. Ya Margolis and Yu G. Borod'ko, Zhur. Fiz. Khim., 51, 1234 (1977).
37. J.S. Brinen and A. Melera, J. Phys. Chem., 76, 2525 (1972).
38. H. Vinek, H. Noller, M. Ebel and K. Schwarz, J. Chem.Soc. Faraday Trans. I, 73, 734 (1977).
39. H. Vinek, J. Latzel, H. Noller and M. Ebel, J. Chem. Soc. Faraday Trans. I, 74, 2092 (1978).

CATALYST CHARACTERISATION: PRESENT INDUSTRIAL PRACTICE

By

G.J.K. Acres, A.J. Bird, J.W. Jenkins, F. King

1. INTRODUCTION

During the past ten years substantial progress has been made in the development of physical techniques for the characterisation of heterogeneous catalysts. Of particular relevance to industrial practice are the advances that have been made in physical techniques that can be applied to catalysts used in everyday processes, i.e. those that industrialists call real catalysts. These developments are of particular significance, since, if properly applied, they can largely eliminate the worst features of recipe-based industrial catalytic processes.

Of particular importance in industrial practice is the application of physical techniques in the three stages which make up an industrial catalytic process, namely: process design, catalyst design and production and process trouble-shooting.

Process Design

At this design stage the importance of relating appropriate process variables to the physical parameters of the catalyst as they influence activity and selectivity cannot be overstressed. Diffusion and mass transfer play an important part in the use of industrial catalysts and can be determined and optimised by physical characterisation.

Catalyst Design and Production

With a full understanding of the role of the catalyst in a particular process as can now largely be obtained with existing physical characterisation techniques, past problems of catalyst production and scale-up can largely be eliminated.

Process Trouble-Shooting

In the past, with relatively little known about the factors affecting the activity, selectivity and durability of industrial catalysts, it is not surprising that, in the event of a process malfunction, the catalyst would be considered as prime suspect. With advances in methods of physical characterisation this situation can be evaluated with the end result that operator or plant malfunctions are at least as frequently found to be the cause of the problem.

Physical Characterisation Techniques

Highlighted in Table 1 are the principal physical characterisation techniques used by industry.

Table 1
Principal Physical Techniques
Used in Industrial Catalyst Characterisation

Technique
Gas Adsorption
Surface Area
Pore Size Distribution
Metal Area
Mercury Porosimetry
Temperature Programmed Reduction
Differential Scanning Calorimetry
Infra Red Spectroscopy
X-ray Diffraction
X-ray Fluorescence
Electron Probe Analysis
ESCA
Transmission Electron Microscopy
Scanning Electron Microscopy
Chemical Characterisation of Functional Groups

Of particular significance in industrial practice is that these techniques are applicable to real catalyst systems. They are, in the main, adaptable to equipment which can be used for fast evaluations which are necessary when studying a wide range of catalyst and process variables as frequently encountered in industrial situations.

A major factor in the success of the use of physical techniques in industry is to combine the results obtained from a number of them so as to build up a full picture of the characteristics of a catalyst. This practice is illustrated by examples from current industrial practice in the sections which follow.

2. PROCESS DESIGN

The physical characterisation of catalysts is not an end in itself. It is sometimes forgotten that physical parameters allied with activity-selectivity measurement are the basic tools of catalyst design and process development. It is the intention of this section to examine the effect of catalyst physical parameters on process design. Catalyst design is obviously related to process development, and the former usually precedes the latter. The major feature of catalyst design is the utilisation of the chemical properties of metal and support so as to vary the dispersion, distribution and degree of reduction of the catalytic centres, while process development is concerned with making use of reaction chemistry, reactor operation and catalyst design to maximise the amount of product formed and, if possible, to suppress side reactions. In fat hardening, which is chosen to illustrate the interaction of process development and physical characterisation, the reaction chemistry is complex, involving removal of one species (linolenate) while retaining two others (oleate and linoleate) in a preferred geometrical isomeric form (cis configuration) and ensuring that an increase in the fully saturated species (stearate) does not occur. Fig. 1.

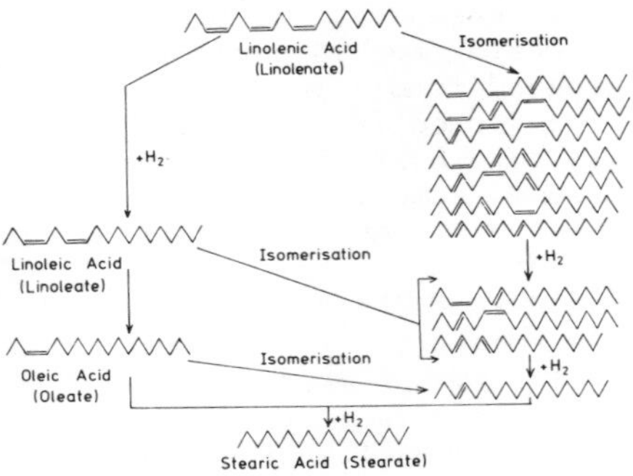

Fig. 1

Schematic of fat hydrogenation reactions

The principal aim of touch hydrogenating triglycerides is to tailor the physical and chemical properties of the fat to an end use. The major use of touch hydrogenated fats is in the frying industry, and for a fat to be suitable for this use it must have the following properties :

(i) be liquid down to -5 °C and show no turbidity arising from crystallisation of fully saturated triglycerides;

(ii) have a high oxidative stability at both room temperature (shelf life) and at frying temperatures (150-200 °C);

(iii) have a high stability to polymerisation;

(iv) have a high nutritious value.

The normal composition of a liquid vegetable oil such as soyabean is:

Palmitate 10 percent, Stearate 5 percent,
Oleate 25 percent, Linoleate 53 percent,
and Linolenate 7 percent.

Palmitate and stearate (fully saturated) are high melting point triglyceride esters and are responsible for turbidity and solidification of fats. Oleate, linoleate and linolenate are respectively the monoenoic, dienoic and trienoic triglyceride esters. They have low melting points and fats high in these esters are liquid to temperature below -5 °C. All three acids are normally present as the cis isomers, and in this form are easily assimilated by the body. The trans form has no nutritious value: it passes through the body without being assimilated. The oxidative and polymerisation stability of oleate and linoleate is much higher than that of linolenate which is responsible for rancidity and polymer darkening of fats. Hence the ideal frying oil would have the composition :

Palmitate 10 percent, Stearate 5 percent,
Oleate + Linoleate 83 percent,
and Linolenate 2 percent

where palmitate and stearate have not increased, the oleate/linoleate fraction has slightly increased from 78 percent to 83 percent and the linolenate has decreased from 7 percent to 2 percent, i.e. the linolenate has been converted to oleate/linoleate. The trans acid content would ideally remain at zero percent, but it is almost impossible to prevent some isomerisation during hydrogenation over heterogeneous catalysts, and a generally accepted low figure is 20 percent trans.

The hydrogenation operating parameters are set by the type of plant in use by the fat hardening industry[1-3]. The maximum hydrogen pressure is 0-30 psig with an upper temperature limit of 250°C and a lower temperature limit of 80°C (set by the pre-hydrogenation refining stage) and a maximum agitator speed of 2000 rpm. The feedstock is low in catalytic poisons such as sulphur or phosphorous compounds.

If the current nickel catalyst is to be replaced then the replacement must be more cost effective (Beasley et al.[4]). With the relatively higher cost of palladium catalyst this can only be achieved by the catalyst being more active at a lower temperature, thus saving on catalyst costs, and by recycling it a larger number of times, thereby reducing catalyst costs.

Nickel operating at 180°C has an activity of 6 mls H_2 min^{-1} mg^{-1} Ni. While at 120°C it is virtually inactive. At 180°C a nickel catalyst can be recycled about three times, giving a total of four hydrogenations before the catalyst is spent. Hence, for a palladium catalyst to be adopted, it must operate at 80-120°C with a hydrogenation rate of 6 mls H_2 min^{-1} Pd and recycle more than four times. These then will be the preliminary parameters which will be set for the activity of the palladium catalyst.

Nickel catalysts operating at 160-180°C have a reasonably high selectivity for the removal of linolenate, but produce rather a high level of oleate; they also isomerise the carbon-carbon double bonds to give high levels of trans acids. Typically an analysis for a nickel hydrogenated reaction is :

Palmitate 10.5 percent, Stearate 4.9 percent,
Oleate 60.8 percent, Linoleate 22.2 percent,
Linolenate 1.6 percent, Trans Acids 28 percent.

With the reaction chemistry set out above and the process operating conditions in mind, it is necessary to choose a catalyst to give the highest activity-selectivity relationships. It would be expected that maximum activity will be obtained by choosing the catalyst with the highest metal area. High metal areas may be obtained in two ways - either by increasing the total amount of active metal (i.e. increasing metal loading), or by modifying the method of preparation so that the metal is more completely dispersed. Fig. 2 shows the effect of increasing metal loading for two methods of preparation, designated type 1 and type 3. It will be noted that the metal area of type 3 catalyst for a given metal loading (up to 6 percent metal) is three times that for type 1 catalyst. With metal loadings above 6 per cent, type 3 catalyst shows only a very moderate increase in area, while type 1 catalyst continues to increase linearly. It is to be expected, therefore, that type 3 catalysts will show the highest activity, approximately three times that of the type 1 catalyst if conditions of kinetic surface control prevail.

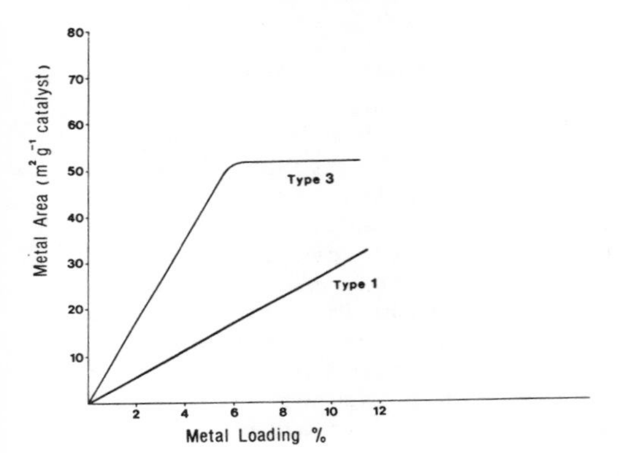

Fig. 2

Effect of metal loading on metal area for different methods of catalyst preparation.

However, when the two catalysts are tested against soyabean oil, it is found that an unexpected activity pattern is obtained (see Fig. 3). Type 1 catalyst is at least an order of magnitude higher in activity than type 3 for any particular metal loading; and where type 3 catalyst is expected to show an almost constant activity, a sharp increase in activity is experienced. Fig. 3 shows the activity-metal loading curves for the two types of catalyst. Although this would appear to be contrary to the original postulate that the activity of a catalyst should be proportional to its metal area, closer examination of Fig. 3 shows that the postulate is correct for, with each type of catalyst,

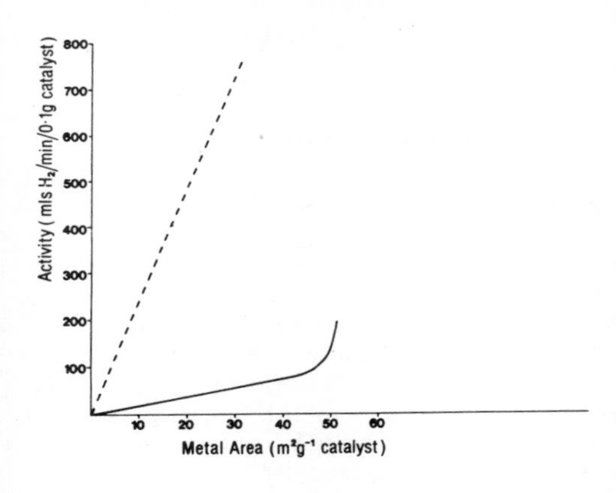

Fig. 3

Effect of metal area on the activity of type 1 and type 3 catalysts.

activity is proportional to metal area. What has to be explained is how type 1 catalyst has a higher activity per unit of metal area (24.5 mls H_2 $min^{-1}[m^2]^{-1}$), compared to type 3 catalyst (2.6 mls H_2 $min^{-1}[m^2]^{-1}$, and how type 3 catalyst increases in activity per unit area of metal when the metal area is almost constant. At the point of activity increase the overall rate of catalyst activity is 3.8 mls H_2 min^{-1} $[m^2]^{-1}$, but the slope of the tangent to the curve is 25.6 mls min^{-1} $[m^2]^{-1}$, very close to that for type 1 catalyst, i.e. the tangent to the curve for type 3 catalyst is parallel to the activity curve for type 1 catalyst. This strongly suggests that the mechanism responsible for the high activity of type 1 catalyst is also responsible for the sharp increase in activity of type 3 catalyst at high metal loadings. The answer to the mechanism is revealed by transmission electron microscopy and ESCA. Transmission electron micrographs of the two catalysts are shown in Figs 4a and 4b, from which it is seen that the metal is located differently in each. Type 1 catalyst has the metal located on the outer surface of the carbon particle, while type 3 has the metal located deep within the pore system. Hence there is a large in-pore

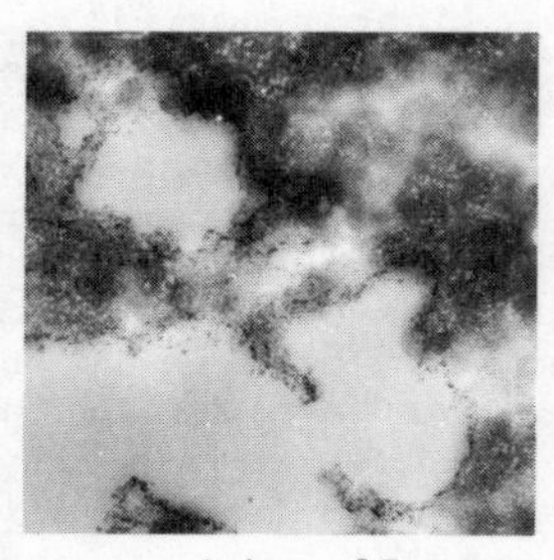

(a) x 150,000

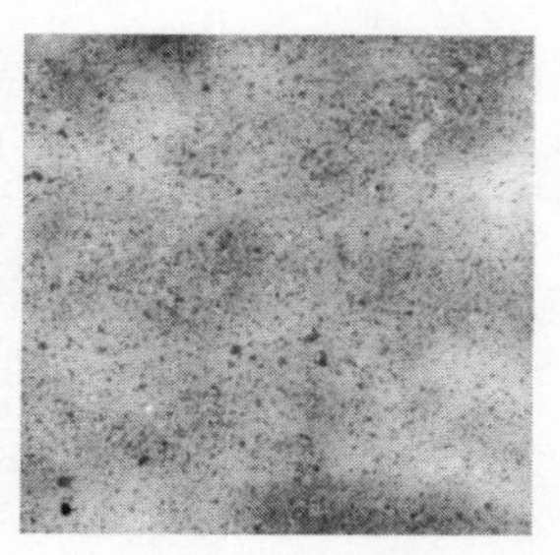

(b) x 200,000

Fig. 4

Transmission electron micrographs of type 1 and type 3 catalysts.

diffusional resistance to hydrogen and triglyceride molecules with type 3 catalyst, but a negligible resistance to diffusion for type 1 catalyst. If this interpretation is correct, then an increase in the amount of metal on the exterior of the type 3 catalyst should be seen where the sudden increase in activity occurs. But electron microscopy is not always reliable quantitatively, as it is subject to a number of sampling errors (see also the Chapter by Howie in this volume). What is required is a tool which measures the total amount of metal exposed on the exterior of a statistically significant number of carbon particles. For this purpose ESCA is ideally suited (see also the Chapter by Edmonds in this volume). By measuring the total palladium 3d electron flux for the catalysts it should be possible to obtain a relationship which will relate catalyst activity to the exposed palladium metal area and to see how this differs from the total palladium metal area. The Pd_{3d} electron counts versus the metal loading show that the type 1 catalyst has a much higher concentration of metal on the carbon surface than the type 3 catalyst. Further, there is a linear relationship between metal loading and the ESCA electron counts for the type 3 catalyst and a non-linear relationship for the type 1 catalyst (see Figs 5 and 6). This suggests that the amount of metal on the outside of the support particles is in direct proportion to the metal loading for type 3 catalyst, while the metal on type 1 catalyst penetrates the pore system to be hidden to ESCA as the metal loading increase. Fig. 5, where ESCA electron flux for

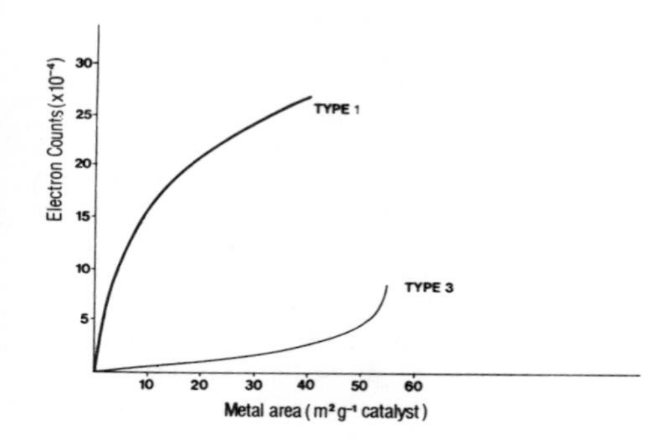

Fig. 5

The degree of surface placement of palladium metal given by two different methods of preparation illustrated by the effect of metal area on the Pd_{3d} electron count.

the Pd_{3d} line is plotted against metal area, still exhibits the non-linear curve for type 1 catalyst, but it will be seen that the curve for the type 3 catalyst shows a considerable increase in ESCA signal for a small metal area increase; hence it can be concluded that this metal is indeed located on the outside of the carbon particle. If superficial metal area alone is responsible for activity in this reaction, then a linear relationship should exist for a plot of catalyst activity versus ESCA Pd_{3d} electron counts. Fig. 6 shows such a relationship. The intercept on

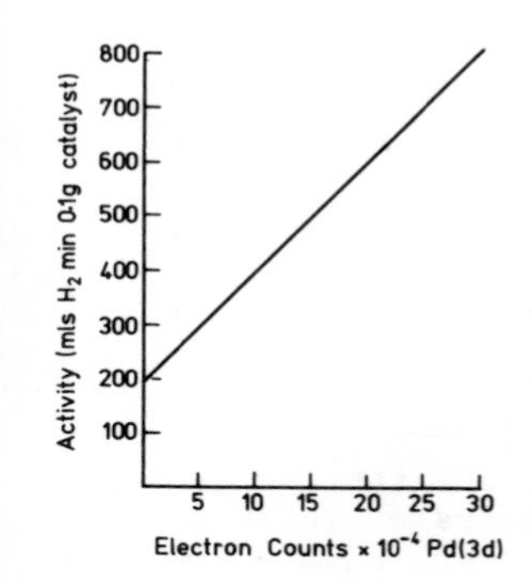

Fig. 6

Activity versus Pd_{3d} electron counts showing the effect of surface placement of palladium metal on catalyst activity.

the activity axis indicates that some of the metal is available to the reactants despite being hidden from ESCA examination. This metal is probably located within the surface convolutions (shown in the electron micrographs (Fig. 4)) and is easily accessible to the reactants.

The activity pattern shown by the two catalysts can now be seen as a function of metal placement primarily and secondarily as a function of metal dispersion. Both catalysts fulfil the requirement of having an activity greater than 6 mls H_2 min^{-1} mg^{-1} Pd, type 1 catalyst is 132 mls H_2 min^{-1} mg^{-1} Pd and type 3 catalyst is 18 mls H_2 min^{-1} mg^{-1} Pd both at a 5 percent palladium loading and 180°C, while for 120°C the figures are 108 mls H_2 min^{-1} mg^{-1} Pd for type 1 catalyst and 8 mls H_2 min^{-1} mg^{-1} Pd for type 3 catalyst.

The physical picture which has emerged for the two catalysts is equally important to the selectivity[4]. Where metal is held within the pore structure, large diffusional resistances to the reactants and products are to be expected. But hydrogen, being a smaller molecule, and being also consumed in the reaction, reaches, and reacts with, the metal quite readily. The residence time for a triglyceride molecule within the pore structure will therefore be high compared to that of the outside surface of the particle, and the types of catalyst should show differences in the distribution of fatty acids when used to hydrogenate a soyabean oil. With longer residence times an increased degree of saturation can be expected, together with high trans acid concentrations. Thus type 3 catalyst would be expected to have a higher stearate/oleate fraction than the type 1 catalyst and a higher trans acid concentration. The predominence of mass transport limitation should also blur the high selectivity of palladium for the hydrogenation of the trienoate (linolenate).

Table 2

Comparison of Hydrogenation Product Compositions Obtained with Palladium on Carbon Catalysts

Catalyst	Temp.	Product Composition *					IV
	(°C)	P	S	O	L_2	L_3	
Type 1	160	10.9	4.0	49.3	33.9	1.9	105
	100	10.6	4.9	47.3	35.2	2.0	106
	80	10.4	5.2	51.6	30.9	1.9	102
Type 3	160	10.7	4.4	48.1	33.9	3.0	104
	120	10.6	5.2	39.3	41.0	3.9	113
Untreated Soyabean Oil		10.5	3.9	23.2	53.4	9.0	135

* P = Palmitate S = Stearate O = Oleate
L_2 = Linoleate L_3 = Linolenate

Table 2 shows the product distribution for the two catalyst types (5 percent metal loading) at temperatures of 160°C and 180°C. All three results with type 1 catalysts have a low linolenate level (~ 2 percent), while the type 3 catalyst is higher at 3-4 percent. For comparable iodine values the type 1 catalyst also shows a lower stearate level and a higher linoleate level. From the overall physical picture of the type 3 catalyst and the product distribution shown in the table, it seems unlikely that further research work will improve it for this reaction: type 1 catalyst, however, with its superficially located metal, is amenable to fine tuning using the process operating conditions to modify its behaviour. Three process operating variables can be modified - the amount of catalyst used, the temperature of operation, and the agitation. Table 3 shows the results of such experimentation. Increasing mass transfer by decreasing agitator speed (at constant temperature) decreases selectivity; note that the linolenate content is higher for the low speed run and the linoleate content is lower. In both runs the trans acid content is high at 24-26 percent. However, by lowering temperature by 20 percent, metal loading by 20 percent and agitator speed by 15 percent, conditions of near-perfect mass transfer are achieved with a low linolenate (2.0 percent) and stearate (4.5 percent) content, a high linoleate (35.2 percent) and a high unsaturates level (82.5 percent), IV (106). The trans acid level is remarkably low at 14 percent, half that of the nickel catalyst. This result shows what can be obtained in the area of process development by a proper understanding of the physical nature of the catalyst allied with an appreciation of the operating parameters and how they affect mass transfer.

One other development which follows as a consequence of the physical properties of the two catalysts is the attainment of a catalyst having the activity of type 1 catalyst and the poison resistance of type 3 catalyst. That is, how can a catalyst with the metal within the pore system be endowed with the high activity of catalyst with the metal located superficially ? Basically this is a question of increasing the effectiveness factor of the catalyst. The level of mass transfer resistance offered to a reactant by a pore system is the product of the pore diameter and the degree of tortuosity. If either the pore diameter can be enlarged or the degree of tortuosity lessened, then higher activities can be obtained[4]. As the molecules which would be hydrogenated with such catalysts would be large (C_{12}-C_{50}) and the poison molecules associated with them even larger, poison resistance would be retained. The pore structure of a carbon with specially enlarged pores is shown in Fig. 7 in comparison with a type 3 charcoal. When catalysts are prepared on these two charcoals by the same preparation method and used to hydrogenate an unsaturated C_{36} acid, the catalyst on the wide pore carbon is some six times more active despite having only half the metal area of the narrow pore catalyst. Fig. 8 compares the two catalysts in terms of double bond saturation with time. The activity comparison given above is based on overall time to IV 15.

Table 3

Selectivity Effects Upon Varying the Mass Transfer Resistances to the Catalyst

Catalyst	Temp.	Metal Loading On Oil (%)	Soyabean Oil Activity $cmH_2\ min^{-1}\ 0.1\ g^{-1}$	Agitator Speed (rpm)	Fatty Acid Distribution (%)					Trans-isomer (%)	IV
					C16	C18	C18:1	C18:2	C18:3		
Type 1	120	0.005	346	2000	10.0	5.3	50.3	33.1	1.3	24	104
	120	0.005	288	1650	9.9	5.0	58.3	25.1	1.7	26	108
	100	0.004	202	1400	10.6	4.5	47.3	35.2	2.0	14	106
Untreated Soyabean Oil					10.1	4.7	25.2	54.2	7.5	0 - 4	139

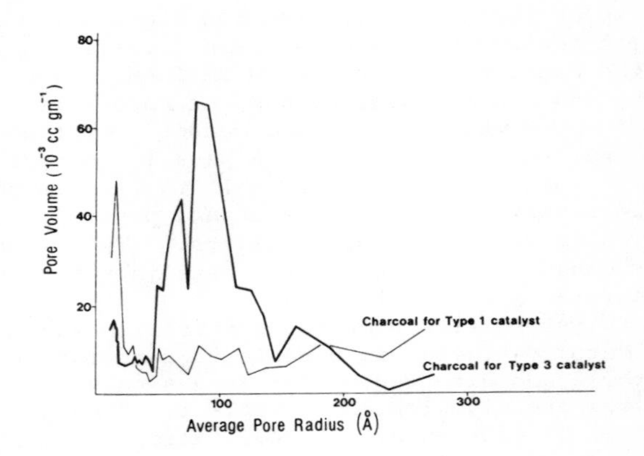

Fig. 7

Pore size distribution for two dimer acid hydrogenation catalysts.

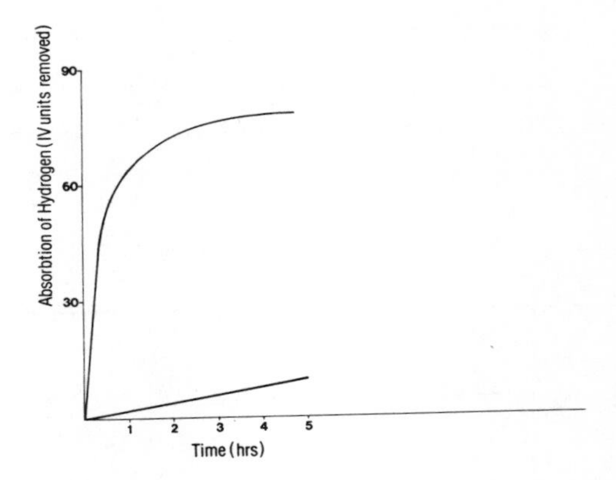

Fig. 8

Hydrogenation of C_{36} dimer acid over two catalysts prepared on narrow pore and wide pore carbons by the same method of preparation.

3. CATALYST DESIGN

An industrial catalyst has to be optimised to obtain the best activity, selectivity and lifetime for a given process. Suitable physical techniques are chosen specifically to follow the change in the catalyst performance with formulation. Furthermore, the industrial catalyst has, of necessity, to be studied in the form in which it is used, under conditions relevant to the process, so that measurements have to be made in a relative short period of time; and, to this end, it is often expedient to alter substantially conventional physical techniques for the characterisation of catalysts. Two examples relating to car exhaust emission control catalysts have been chosen to illustrate these points.

Current catalyst systems used for controlling exhaust emissions from motor vehicles are capable of removing carbon monoxide, hydrocarbons and nitrogen oxides in a single unit. The reactions involved are highlighted in Table 4. The performance of the catalyst is a function of air/fuel ratio as illustrated in Fig. 9.

Table 4
The Three Way Catalyst

Polluting Gases				
Unburnt Fuel				Clean Exhaust
Carbon Monoxide		$CO + O_2 \rightarrow CO_2$		Carbon Dioxide
Nitrogen Oxides	$\rightarrow$	$CO + NO \rightarrow CO_2 + N_2$	$\rightarrow$	Nitrogen
+		$HC + NO \rightarrow CO_2 + H_2O + N_2$		Oxygen
Carbon Dioxide		$HC + O_2 \rightarrow CO_2 + H_2O$		Water
Nitrogen				
Oxygen				

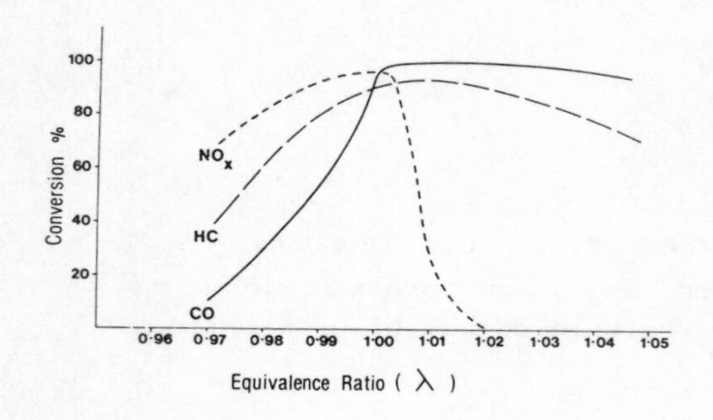

Fig. 9
Three way catalyst selectivity test.

Physical characterisation has played a major part in the development of this technology, as illustrated by the following two examples.

Metal Area Determination

Efficient dispersion of a supported precious metal catalyst, which maximises the available active surface for incoming reactants, is a very important requirement. Universally applicable standard methods of catalytic surface area determination do not, unfortunately, exist, but certain procedures are commonly used, particularly for supported metals. Three types of apparatus are usually employed :

(i) Static vacuum system[5,6].

(ii) Pulse flow method[7].

(iii) Continuous flow apparatus[7,8].

The static vacuum method is not often used industrially because it is a time-consuming technique, but the other two are.

Whereas the pulse flow method, involving injection of slugs of carbon monoxide is routinely used for the determinations of the metal area of supported precious metal catalysts, in this instance it is not applicable. Rhodium is an essential component and one of its functions is to promote the reduction of NO_x. The metal is effective at low concentrations, but its precise efficiency is a function of its dispersion. In order to measure the adsorption of NO at low Rh dispersions, a modified gas adsorption system was developed (see Fig. 10). The apparatus is a modified version of the continuous flow system of metal area measurement but the detector employed is a quadrapole mass spectrometer. The sample is placed in the furnace at an appropriate temperature, purged with argon and then contacted with a stream of gas composed of an adsorbate and inert carrier. The amount of gas adsorbed as a function of time is determined and a comparative metal area calculated. The sensitivity of the equipment and its ability to handle corrosive gases are important considerations.

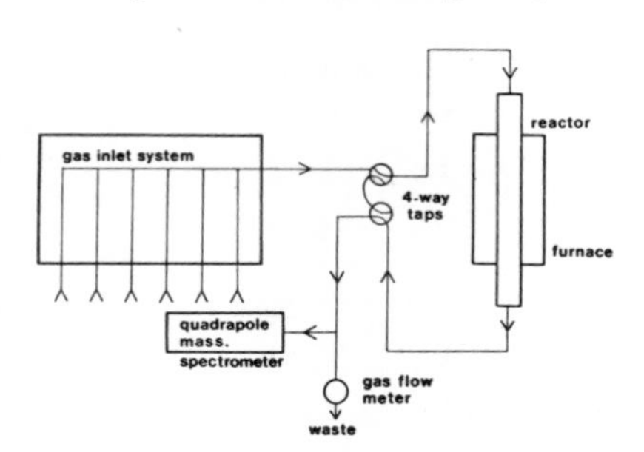

Fig. 10

Mass Spectrometer Apparatus.

The extent of chemisorption of nitric oxide on rhodium catalysts is shown in Fig. 11, where preparations are illustrated having significantly different metal dispersions. The effect of metal loading on Rh dispersion is to be noted. The improvement in activity resulting from the better dispersion of rhodium via preparation (1) is borne out by the results of 'steady state' engine tests conducted on the 'complete' catalyst (see Table 5).

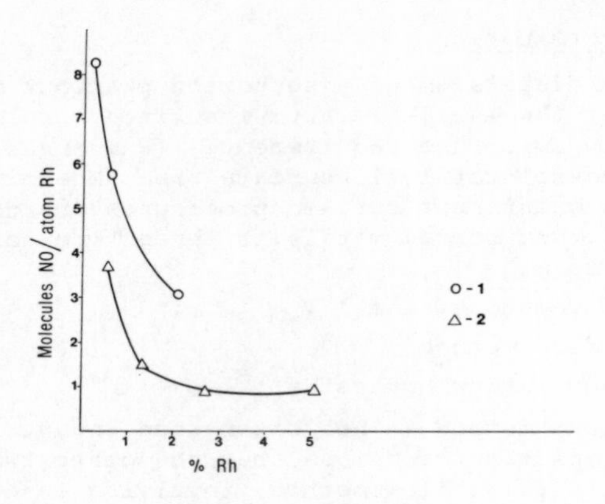

Fig. 11

The effect of catalyst preparation on rhodium dispersion.

Table 5
Steady State Engine Tests on Different Rhodium Treatments

	0.986			0.995		
	% Conversion			% Conversion		
Concept	HC	CO	NO_x	HC	CO	NO_x
1	60	55	96	74	74	97
2	46	32	73	76	59	97

Dynamic Gas Absorption

In some practical catalyst systems the rate at which gas is absorbed/desorbed is important. TWC systems are an example, and have been studied using the mass spectrometer system outlined above. The catalysts should, ideally, have a selectivity on the air/fuel ratio scale which is as wide as possible and at the same time have high activity for the removal of the appropriate gases. One method of achieving this is the inclusion in the catalyst design of an oxygen storage component (OSC), which stores, under transient oxidising conditions and subsequently releases under 'rich' conditions. This enables the oxidising and reducing capabilities of the catalyst to be extended. In order to design three-way catalyst systems for various fuel management systems used on cars, three oxygen storage components were developed. These three, labelled A, B and C, were optimised by using the mass spectrometer flow system to measure the uptake of oxygen. The system enabled the total uptake and the rate of uptake of oxygen to be monitored. As shown in Table 6, system A has a slow response but large capacity, whereas C has a low capacity but rapid response.

Table 6

Oxygen Storage Capacities
(ml STP g^{-1})

Treatment	A		B		C	
	Total Uptake	1 Min. Uptake	Total Uptake	1 Min. Uptake	Total Uptake	1 Min. Uptake
Fresh	3.36	1.29	2.55	1.41	0.13	0.12
Air Aged at 700°C	3.89	1.52	1.12	0.81	0.26	0.22

System B lies between A and C. The effect of the different oxygen storage components on the selectivity of a complete catalyst is amply illustrated by the results obtained on an actual car (see Table 7). Component B is better than A and considerably better than C for this particular vehicle and fuel management system, but for other vehicles, having different fuel management systems, the order may be reversed.

Table 7

Tests on Alternative Oxygen Storage Components. (See text.)

	% Conversion		
	HC	NO_x	CO
A	78	86	68
B	82	88	75
C	69	88	58

4. CATALYST PREPARATION AND SCALE-UP

The industrial preparation of catalysts can be discussed conveniently in terms of the objective, the problem and the solution. The objective is to produce catalysts on a large scale of a consistently high quality in an economically viable manner from readily available raw materials. The problem is that any catalyst preparation involves several successive unit operations such as impregnation, drying and activation, each of which yields a catalyst intermediate or precursor of somewhat undefined nature. It is necessary to be able to characterise these precursors if the overall catalyst manufacture and final product is to be successful and meet specifications. For the purpose of quality control such characterisation must be done simply and rapidly, even whilst the catalyst is being manufactured. The differing nature of these catalyst precursors arises from the chemical and physical changes that occur during preparation and the interactions between the active material and the catalyst support. In the following examples we describe one solution to this manufacturing problem, namely the use of temperature programmed reduction (TPR), which is described fully in this monograph by McNicol (see also ref. 9).

Although full details of TPR are given elsewhere, we recall here that, by following the change in hydrogen concentration in a nitrogen gas carrier as it passes over the catalyst sample, the latter's rate of reduction can be continuously recorded. If the sample temperature is raised at a uniform rate, the different components on the catalyst surface reduce at different temperatures, and the record of the rate of reduction typically will contain several reduction peaks similar to a conventional chromatographic analysis. From the temperatures at which these peaks occur, the composition can be inferred, and from the sizes of the peaks the relative amounts may also be deduced. Since the apparatus is very sensitive to small uptakes of hydrogen, the hydrogen chemisorbed on a metal surface or adsorbed into the bulk of a metal such as palladium can also be measured quantitatively. We thus obtain not only a measure of the changes in surface composition but also a measure of metal dispersion of the final reduced catalyst. Three examples will suffice to illustrate the type and usefulness of the information obtained.

i) A calcination study of a series of impregnated platinum-silica catalysts.

From small scale trial preparations it was found that the quality of the catalyst was very dependent upon calcination time and conditions. In Fig. 12 are a series of TPR fingerprints of such a catalyst taken after calcination for different lengths of time. The metal is mobile under such conditions and it can be seen that the final metal dispersion first decreases and then increases. These changes are paralleled by rather marked changes in the TPR fingerprint. By following the systematic variations in the TPR fingerprint with time the catalyst manufacturer can control and optimise his final product and substantially reduce the chances of producing an inferior batch.

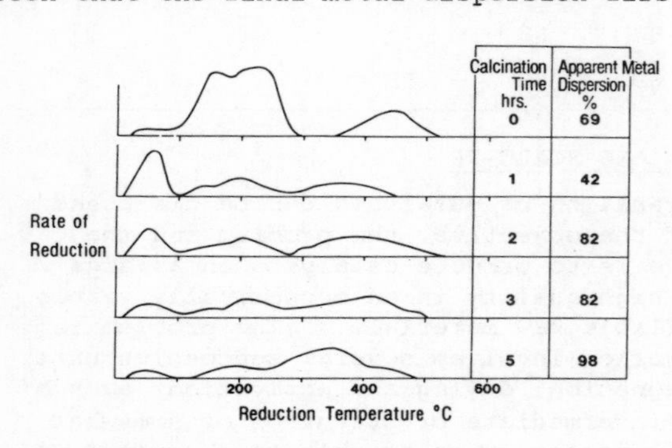

Fig. 12
Temperature programmed reduction of Pt/SiO_2 catalysts.

ii) The second example is also concerned with quality control, but from a somewhat different aspect. The careful experimentalist can afford the luxury of using nothing but the highest quality reagent grade materials for his catalyst preparations. Catalyst manufacture on a large scale, however, may necessitate the use of technical grade raw materials of uncertain and variable quality with substantial variations from one batch of raw materials to the next. Such undesirable impurities frequently give bands in the TPR fingerprint. Fig.13 shows TPR fingerprint of a good and bad catalyst as indicated by the hydrogen chemisorption values. Quite clearly the bad sample has an extra component. In this case we know it is sulphate - a common impurity in some alumina supports.

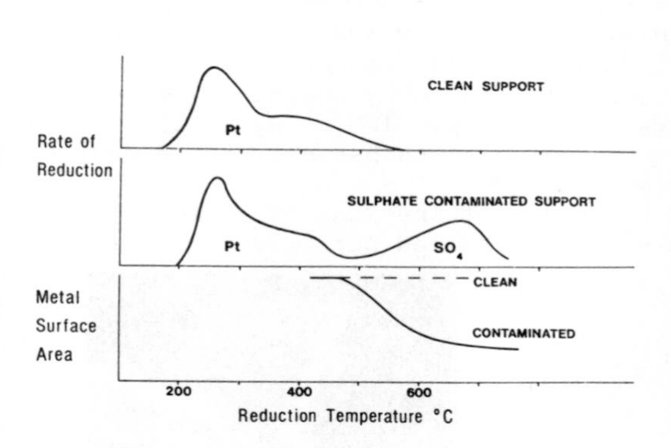

Fig. 13

Temperature programmed reduction of $Pt/A\ell_2O_3$ catalysts.

iii) The third TPR example is a case of a catalyst 'post mortem'. An unusual example of such an event is shown in Fig. 14 which shows the different TPR fingerprints of specimens of bimetallic platinum/tin on alumina catalyst. The catalyst as made had the platinum and tin combined as shown by a single reduction peak. However, after use the bimetallic combination has been destroyed and two separated platinum and tin peaks show up. This could have arisen during air regeneration, particularly if wet air had been used.

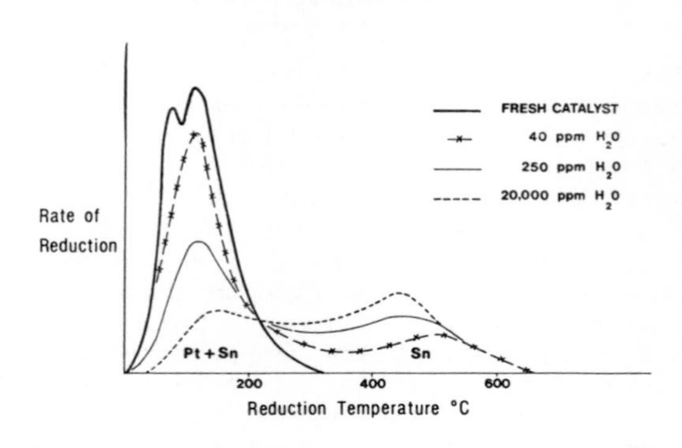

Fig. 14

Temperature programmed reduction of $Pt\text{-}Sn/A\ell_2O_3$ catalysts.

5. PROCESS 'TROUBLE-SHOOTING'

As an example of the combined use of physical techniques in catalyst process 'trouble-shooting', a description of the use of XRD, EPA, SEM and TEM in characterising Rh/Pt gauzes, used as catalysts in the oxidation of amonia, will be given. On occasions Rh/Pt gauze packs, after a short period of plant operation, have been reported to exhibit anomalously low conversion efficiencies[10,11]. This effect is normally associated with medium and high pressure plants and until recently could not be accounted for either as a catalyst or plant problem. By application of EPA, SEM and XRD, however, an explanation has now been found as illustrated in the Figures 15 to 19.

Figs 15 and 16 show the EPA distributions of Rh in a gauze with normal activity and one with reduced activity. These

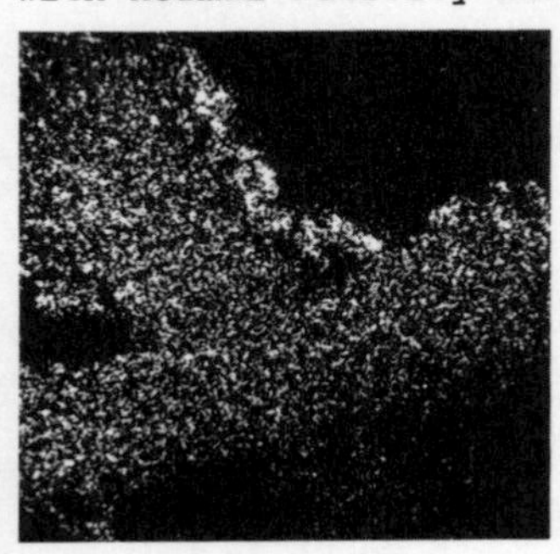

Fig. 15

Electron probe analysis showing the rhodium distribution of an active gauze.

Fig. 16

Electron probe analysis showing the rhodium distribution of a poisoned gauze.

results immediately suggest a Rh enrichment factor. SEM photographs of three gauzes having normal, intermediate and low activity, Figs 17, 18 and 19, indicate two forms of surface structure namely platelets and needles. Additional XRD data shows the platelets to be essentially 10% Rh/Pt, whilst the needles are rhodium oxide (Rh_2O_3). An explanation of the formation of rhodium oxide has been given by Minzi and Schmahl[12], who studied the relationship between the decomposition of rhodium oxide and oxygen pressure and temperature, for Rh/Pt alloys. This work showed that rhodium oxide is only stable below temperatures in the 800-900°C range depending upon the partial pressure of oxygen. Thus the problem becomes one of plant operating conditions of temperature and ammonia to air ratio which must be kept in the range where rhodium oxide is not formed.

Fig. 17

Scanning electron micrographs - normal gauze.

Fig. 18

Scanning electron micrographs - intermediate activity.

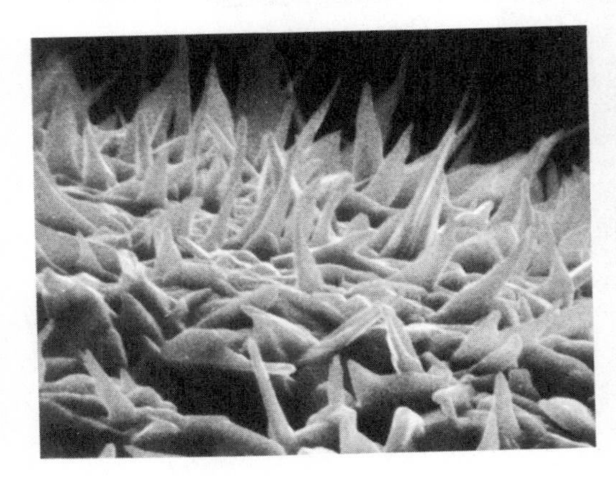

Fig. 19

Scanning electron micrographs - low activity.

Time lapse (or real time) microcinematography can be of considerable value for the TEM study of the surface re-arrangement of, say, platinum and platinum/rhodium during their use as catalysts for the oxidation of ammonia. Much may be gained about the broad aspects of how surface re-arrangement takes place and in particular the mechanism of whisker formation. The main features of such a film have been described[13], but, by way of illustration, two micrographs are shown here (Figs 20 and 21) which strikingly reveal the nature of the distribution of rhodium in a Rh/Pt gauze catalyst.

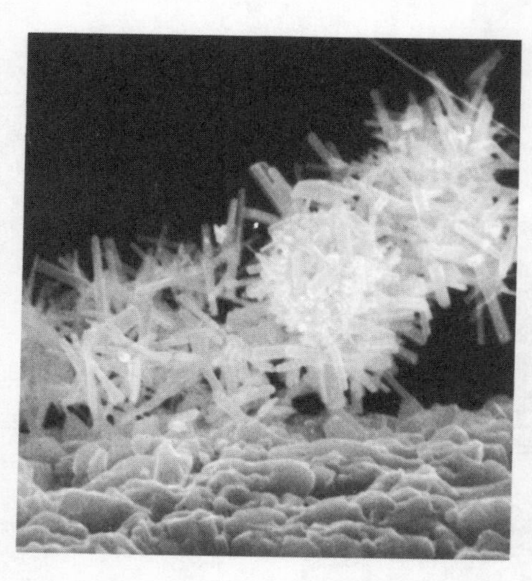

Fig. 20

Scanning electron micrograph of whisker growth on a Rh/Pt gauze.

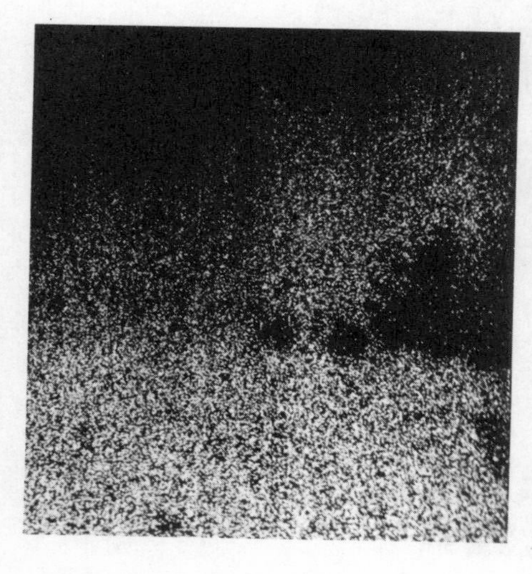

Fig. 21

Electron probe analysis showing the distribution of rhodium in the whisker shown in Fig. 20.

REFERENCES

1. R.R. Allen, J. Amer. Oil. Chem. Soc., 39, 457-458 (1962).
2. J.W.E. Coenen, J. Oil Technol. Assoc. India, 16-27 (1969).
3. J.W.E. Coenen, J. Amer. Oil. Chem. Soc., 53, 382-389 (1976).
4. P.F. Beasley, A.J. Bird and M.C. Sweeney, J. Italian Fats and Oils Soc., to be published.
5. D.A. Hayward and B.M.W. Trapnell, Chemisorption, Butterworths, London (1964).
6. R. Anderson, Experimental Methods in Catalyst Research, Academic Press, New York (1968).

7.a P.E. Eberly and E.H. Spencer, Trans. Farad. Soc., 57, 287 (1961).

7.b P.E. Eberly, J. Phys. Chem., 65, 68, 1261 (1961).

8. F. Nelson and F. Eggerton, Anal. Chem., 30, 1387 (1958).
9. J.W. Jenkins, B.D. McNicol and S.D. Robertson, Chemtech. 7, 316 (1977).
10. N.H. Harbord, Platinum Metals Rev., 18, 97-102 (1974).
11. F. Sperner, Platinum Metals Rev., 20, 12-20 (1976).
12. E. Minzi and N.G. Schmahl, Z. Phys. Chem., 41, 78 (1964).
13. R.T. Baker, R.B. Thomas and J.H.F. Notton, Platinum Metals Rev., 18, 130-136 (1974).

V

THE APPLICATION OF CATALYST CHARACTERISATION TO PROBLEMS ARISING IN THE DEVELOPMENT AND USE OF CATALYSTS

By

R.J. Sampson

1. INTRODUCTION

Society, by the use of economic forces, is placing more and more intense pressure on the chemical industry to use raw materials more efficiently (i.e. selectively) and to reduce its requirement, per unit output, of both energy and capital. At the same time the industry is seeking to make its processes safer and environmentally more acceptable. Since the technology of chemical conversion is in large measure centred on catalysis, an increasing proportion of industrial chemical research and development is being devoted to the catalysis area to provide answers to the needs outlined above.

In describing the characteristic feelings of the typical practising technologist when he first encounters the world of the heterogeneous catalyst, Satterfield[1] writes:

"..... his first impression..... is apt to be that of a vast and confusing field, replete with an enormous quantity of perhaps significant but empirical facts, interspersed with perhaps useful theories."

This situation is not altogether surprising, for the majority of heterogeneous catalysts are highly complex composites, usually poorly characterised and not readily characterisable. A mixture of inspiration, science, technology and empiricism is used in their design and development. Their purpose is to bring about, as rapidly as possible, a desired conversion of the reactants, while minimising the rates at which these substances undergo other (sometimes thermodynamically more favoured) reactions, yet in most cases the mechanisms of the desired and unwanted reactions are understood in broad outline only. They must of course have physical properties (e.g. mechanical strength) which are adequate for the duty to which they are to be put, and their chemical performance must be sufficiently unchanging to meet technical and economic viability. A successful catalyst is thus required to meet enormous and varied demands.

Characterisation techniques, of the type discussed elsewhere in this monograph, are increasingly aiding the solution of the problems posed in the opening paragraphs. The value and scope of these techniques are being continually increased by refinement, while new methods are being devised and, where appropriate, included in the armoury. Slowly, but surely, the feelings expressed by Satterfield are being dispelled, thanks in part to these characterisation activities. But much remains to be done, there being stimulating opportunities for the physicist and the chemical physicist. Their efforts aid the catalysis scientist in three ways:

(1) the provision of means of monitoring the development of catalyst preparation methods, presupposing the catalyst developer knows (or can establish by correlation with performance) what he wants and what he must avoid;

(2) the identification of changes occurring during the use of a catalyst. This information may assist the answering of questions such as "Why does the catalyst's performance change during normal use?" and "Why has performance been lost prematurely?". Answers to questions such as these permit a scientific statement of the problem, a prerequisit for a scientific approach to its solution;

(3) the acquisition of information on the mechanism of a reaction in relation to the surface structure. Collection of information of this kind often presents special difficulties, since it ideally requires study of the working catalyst (though study of the catalyst/reactant system under conditions far removed from "real life" is often helpful). However, it leads to fundamental understanding, and thence to rigorously-based new ideas for catalyst design.

This article outlines in turn each of the major classes of information of value to the catalysis scientist, and seeks to show why he requires it. It will not deal in depth with methods that are described elsewhere in this volume, apart from the occasional example.

Somewhat arbitrarily, the subject matter is organised into six categories, viz:

Surfaces and sorbates.
Catalytic phases.
The dispersion of active phases on support surfaces.
Intrinsic properties of support materials.
Support aggregate characteristics.
"Inadvertent" surface species.

These categories are not independent. To achieve a successful catalyst the developer is almost always compelled to compromise: he trades-off one aspect of performance to bring another up to at least the minimum acceptable level.

2. SURFACES AND SORBATES

In catalysis, the reactant molecules transfer through the continuous phase to the surface of the catalyst where adsorption occurs; during or after adsorption on an "active" site, their bonds may rearrange. Finally, new species desorb, re-entering the continuous phase as products. Clearly, a key step is the nature of the bond rearrangement, which involves both surface and sorbate. The detailed nature of the surface determines what rearrangement or rearrangements occur, and also the rates of these changes. This detail thus determines the intrinsic activity and selectivity levels of the catalyst surface. Table 1 outlines the uses to which information about surfaces and sorbates may be put.

Table I

Surfaces and Sorbates

Classes of Information	Potential Uses for the Information
The general natures, detailed structures, concentrations and distributions of (a) Potential Sorption Sites (b) Sorbed Species	A. The generation of hypotheses which may be used as a basis for (i) ideas for improving the surface (ii) postulation of surfaces capable of catalysing other desired reactions
	B. Helps to answer the question "Is the actual surface being investigated the surface desired? If so, does it behave as predicted?"

The reaction of ethylene and oxygen to form ethylene oxide over a supported silver catalyst provides an example. An infra-red (IR) study of supported silver, to which oxygen had been admitted, followed by ethylene, at a temperature below that required for catalysis, revealed the formation of a species the spectrum of which was compatible with the presence of a peroxy structure.[2] On raising the temperature, the peroxy band disappeared and a new spectrum, corresponding to that of ethylene oxide adsorbed on silver, appeared. It was inferred that the mechanism of ethylene oxide formation involves an intermediate - $AgOO(C_2H_4)$ was postulated. In turn this suggests that an adsorbed dioxygen species reacts with ethylene (it was known that little, if any, ethylene adsorbs on silver, though oxygen is readily adsorbed). The hypothesis was strengthened by a series of experiments[3], (a)-(d), in which ethylene was added to $^{16}O_2$, (a), $^{18}O_2$, (b), a mixture of $^{16}O_2$ and $^{18}O_2$, (c), and a mixture of $^{16}O_2$, $^{18}O_2$ and $^{16}O^{18}O$, (d) in each case preadsorbed on silver at 95°. One, (a) & (b) two (c) or three (d) IR bands were observed in the peroxy region noted by Gerei et al.[2] and attributed to the formation of one, two or three isotopic species containing two oxygen atoms. The results were in accord with expectation if no isotopic scrambling had occurred before or during the formation of the species from dioxygen. On warming to 110°, one, (a) & (b), or two, (c) & (d), bands only resulted, at one or other of two wave numbers. This result was compatible with the formation of adsorbed $C_2H_4{}^{16}O$ and/or $C_2H_4{}^{18}O$, according to the original dioxygen isotopic composition. A further experiment showed that $^{16}O_2$ and $^{18}O_2$, co-adsorbed at 160°, and cooled prior to addition of ethylene, resulted in three IR bands, attributed to the three isotopic dioxygen species, implying that at higher temperatures, nearer to those required for catalyses, scrambling occurs by dissociation or some other mechanism. The relative intensities of the bands in all the experiments were also compatible with these interpretations.

For many years it has been known that addition of controlled levels of a few ppm of an organochloride (e.g. ethylene dichloride) to the reacting gas enhances the selectivity of ethylene oxide formation from, under favourable conditions, around 60 per cent to around 70 per cent:

$$C_2H_4 + O_2 \rightarrow \underset{O}{CH_2\text{-}CH_2} \quad \text{(ethylene oxide)}$$

$$C_2H_4 + O_2 \rightarrow CO_2 + H_2O$$

The presence of the chloride also results in a reduction in the rate of ethylene oxidation. Kilty, Rol and Sachtler[3] also reported adsorption studies of oxygen on a supported silver catalyst. This revealed three types of sorption, one being fast and non-activated, the second having an activation energy of 33 kJ mol^{-1} and the third, 58 kJ mol^{-1}. Preadsorption of chloride at increasing coverages was found to reduce progressively the first type of sorption, each Cl replacing one O, until ultimately it was wholly precluded. Further increases in the pre-adsorbed Cl dose resulted in reduction of the 33 kJ mol^{-1} chemisorption, though now each chloride replaced two oxygens. These findings suggest that the non-activated process is a dissociative chemisorption and that its prevention during catalysis (by the maintenance of an appropriate adsorbed chloride level) increases the selectivity of the oxidation, i.e. reduces the extent of reaction (5) below.

Together, the IR, oxygen isotope and oxygen/chloride adsorption experiments suggest the scheme (1) to (5) :

$$O_2 \rightleftarrows O_{2ads} \qquad (1)$$

$$O_{2ads} \rightarrow 2O_{ads} \text{ (blocked by Cl)} \qquad (2)$$

$$O_{2ads} + C_2H_4 \rightarrow (OOC_2H_4)_{ads} \qquad (3)$$

$$(OOC_2H_4)_{ads} \rightarrow \underset{O_{ads}}{CH_2\text{-}CH_2} + O_{ads} \qquad (4)$$

$$O + C_2H_4 \text{ (or } C_2H_4O) \rightarrow CO_2 + H_2O \qquad (5)$$

Reaction (4) implies that the formation of ethylene oxide and of the non-selective oxidant, O_{ads}, are coupled. While selectivity may be enhanced by the use of a sufficient coverage of chloride for retardation of (2) but not of (1), a further improvement will require an opportunity for O_{ads} to react in a manner alternative to (5), e.g. by modifying the surface to permit the reverse of (2) to occur at a sufficiently high rate to compete with (5) as the fate of O_{ads}.

This discussion of ethylene oxidation includes a few of the many types of questions which arise concerning surfaces and sorbates, and illustrates how answers to these may suggest catalyst improvement opportunities. A more general, but still barely representative, list of such questions follows: "Which are the normally exposed crystal faces?". "Which, if any, of the specific faces bring about especially active or selective catalysis?". "Does the inclusion of a second species, e.g. as in an alloy, influence the predominantly exposed faces, or the concentration of surface defects?". "Is there surface enrichment?". "Are there 'ensembles' (i.e. specific groupings of atoms on this surface)?". "Are useful chemisorption sites to be found at line or point surface defects, such as steps, step adatoms or kink sites?". "Do the reaction conditions or reactants bring about facetting (i.e. the preferential development of specific faces), or do they lead to annealing?". "Does a reactant adsorb in an associative or in a dissociative manner?". "Does a given reactant adsorb in more than one mode, and, if so, what are the associated energetics and which has relevance to the desired, and which to the unwanted, catalysis?". "Can sites of an unwanted type be blocked by incorporating another substance in the reactants or in the surface?". "Where acid sites are present, what is their nature, acid strength (and acid strength distribution) and concentration?". "Does an unusual but desirable coordination occur around a cation in the surface of a specified binary system?". "To what extent is there exchange between lattice oxygen and sorbed oxygen?". The characterisation methods used in the ethylene oxidation example are of value in providing an understanding of that system which is sufficient to act as a stimulus for ideas on improving the system. However, they lead to only superficial understanding of the detailed nature of the system - they provide only ideas on the stoichiometry and competition between the perhaps more important sorbate species. Even for this depth of understanding the IR technique necessitated the use of model solids, rather than actual catalysts. Many of the more detailed questions, such as some of those listed above, clearly require the use of model surfaces, such as specific crystal faces or polycrystalline films, either for ease of interpretation or because of the limitations of the techniques available. Ultra-high vacuum (UHV) conditions and clean surfaces are often required - both far removed from real catalysis. Nevertheless, progress in the investigation of surfaces and sorbates is rapid, and will undoubtedly have a steadily greater influence on the understanding, and hence the advancement, of real catalysts.

3. CATALYTIC PHASES

The major uses to which phase information may be put are summarised in Table 2.

Table 2

Uses for Phase Information

Phase Information	Major Uses
The nature and proportions of each phase present	The identification (with or without structural determination and stoichiometry) of phases which possess desirable sites and also of those which are harmful (e.g. which result in selectivity loss by catalysing sequential or parallel chemistry).
	The correlation of phases present with preparative methods.
	The correlation of time-dependent performance changes with changes in the phases present.
	The structural determination of phases with open crystal structures into which reactants may diffuse (especially zeolites).

The oxidation of n-butane to maleic anhydride provides an example. Chevron Research (1972)[4] established that unsupported vanadium/phosphorus oxide composites possess selectivities for this reaction directly related to the content of a specific phase (designated 'B'), this phase being characterised both qualitatively and quantitatively by x-ray diffraction. In addition, it was found that catalyst activity was related to the content of this phase. Methods of composite preparation which resulted in high proportions of 'B' were developed, and these composites possessed highly advantageous performances. ICI (1978)[5] showed that deleterious effects arise from the presence of a phase 'E', which had previously been shown to possess composition $VO(H_2PO_4)_2$. A method was established for the removal of this phase from the composite precursor prior to calcination, and also for removal of amorphous mixed oxides. The resultant catalysts had excellent performances, and contained a phase 'X': the best performances were associated with composites containing high levels of 'X' and in which phases other than 'B' (and/or a related phase, 'B') were absent.

One of the reasons for time-dependent changes in the performance of some catalyst types has been found to be the conversion, under reaction conditions, of a catalytically desirable phase to an inert or less desirable phase (or the converse). Often, a new phase of different oxidation state is formed during selective hydrocarbon oxidation using oxide catalysts. For example, Simard, Serrese, Clark and Berets[7] and Ioffe, Klimova and Makeev[8] have reported that deteriorating performance of vanadia catalysts, used for aromatic hydrocarbon oxidation, is due to a phase change. In some of these cases it is possible to avoid or retard unwanted phase changes by altering the reactant redox balance (i.e. the oxygen to hydrocarbon ratio). Phase structure determination is a valuable tool in the development of zeolites for catalytic purposes. It is the most valuable guide for zeolite preparation development. Also the size/shape of molecules which may be admitted into, or formed within a channel is of course dependent upon the dimensions of the channel. The presence of intersecting channels, as well as the channel size, influences diffusivity within the zeolite crystallite and may also influence the extent to which the structure is prone to blocking by "coke"-formation. The zeolites which are widely used for catalytic cracking have a highly interpenetrating network of channels which no doubt contributes to their enormous success.

Occasionally, sub-surface defects or inhomogeneities may be important - for example, in some oxidations the participating oxygen is lattice oxygen, which is continually replenished from the gas phase: since anion vacancies and crystallographic shear planes may facilitate oxygen mobility (and so perhaps control rate), ability to characterise them is important.

4. DISPERSION

Dispersal of the active phase over the interior surface of a high area, porous support is often necessary to achieve adequate surface area (and so activity) per unit volume of reactor space. The support must, of course, be able to maintain the dispersion under reaction conditions for a time which is long enough to allow an adequate catalyst life: the rate of growth of larger crystallites of the active phase at the expense of smaller ones (i.e. sintering) is one criterion for an acceptable supported catalyst. Additionally, supports often play more subtle roles, as in the case of dual-function catalysts. Table 3 shows the major reasons why dispersion characterisation is of great value.

Table 3
Dispersion

Dispersion Characteristics	Applications of Information on Dispersion
Crystallite size. Crystallite size distribution. Crystallite shape. Macro-uniformity of the dispersion. Support-phase interaction.	To correlate with method and details of incorporation into the support.
	To correlate with catalytic performance.
	To correlate with the nature of the support.
	To correlate dispersion loss with time-dependent performance changes.
	For use in the development of regeneration by redispersion.

Occasionally, support-free dispersions may be successfully achieved by comminution. For example, ICI (1979)[9] describes application of this method to vanadium/phosphorus oxide composites in which x-ray diffraction was used to characterise crystallite sizes. The characteristics of a dispersion are influenced by the nature of the support surface and by the dispersal method used to obtain it. Various impregnation and co-precipitation methods are frequently used often including carefully-developed steps of "ripening", drying and pre-treatment (e.g. calcination and reduction). Guidance by direct measurement of dispersion is not only desirable in the development of these procedures, but is also invaluable as a means of quality-control during catalyst manufacture.

The nature of the support may influence not only the degree of dispersion but also the dispersed phase crystallographic planes which are predominantly exposed, i.e. it may bring about epitaxial growth. The characteristics of the dispersed crystallites is sometimes influenced by the presence of a third component. A comparison of the state of dispersion of the active phase in a used catalyst, relative to the fresh material, shows whether sintering may have contributed to the loss of activity. If so, measurements of this kind are needed to help establish the sintering mechanism. Is it purely physical, or is 'mineralisation' more important? (Mineralisation in the present context is the term used to describe the reversible formation of a more mobile compound, the equilibrium concentration of which is high enough to provide a pathway for sintering, even though the active phase may not undergo a bulk phase change.) Having established the mechanism, it may be possible to devise means of retarding, or even preventing it. Sintered catalysts may sometimes be re-dispersed, for example by temporarily converting the active phase to a substance which has a very high affinity for the support. Characterisation of the state of dispersion is clearly likely to be helpful in devising optimum redispersion conditions.

5. SUPPORT MATERIAL: INTRINSIC PROPERTIES

A suitable support has porosity characteristics which permit the formation of a stable dispersion of the active phase. This, however, may not be adequate: the pore structure should also allow the reactants and products to diffuse within the pores sufficiently rapidly to ensure that no significant concentration gradients are established within the aggregate due to replacement of reactant by product. If such gradients are present, the active sites of the interior will be starved of reactants, so that their full chemical potential will not be utilised. Worse, if a sequence of chemical reactions (which leads to selectivity loss) is possible, the pore diffusion limitation may lead to a selectivity much lower than that which would be expected from the intrinsic chemical rate constants. Thus, in the development of a support material, one aim is to achieve the maximum useful support surface area in relation to its use in the reaction concerned. A bimodal pore distribution is sometimes beneficial in this respect.

In the isomerisation (6) of mixed xylenes, which, in

$$oX \rightleftarrows mX \rightleftarrows pX \qquad (6)$$

combination with separation and recycle, is used to produce the specific isomers o-xylene and p-xylene, amorphous silica alumina catalysts may be used. The mechanism of the acid-catalysed reaction may be represented by equations such as (7) :

$$\text{Me-C}_6\text{H}_4\text{-Me (meta)} \underset{}{\overset{+H^+}{\rightleftharpoons}} [\text{arenium ion}]^+ \rightleftharpoons [\text{arenium ion}]^+ \underset{}{\overset{-H^+}{\rightleftharpoons}} \text{Me-C}_6\text{H}_4\text{-Me (para)} \qquad (7)$$

However, a parallel set of reactions (8) also occurs, leading to disproportionation, and so to selectivity loss :

(8)

TOLUENE + TRIMETHYLBENZENE

Reactions of type (8) are slower than the isomerisations. However, the use of amorphous silica aluminas, as normally prepared, leads to a ratio of isomerisation to disproportionation which is much lower than could be achieved theoretically on the basis of the relative chemical rate constants. The reason is that the hydrocarbons do not transfer sufficiently rapidly between the interior pore structure and the continuous phase. This results in incomplete exploitation of the chemical potential of the surface, both rate and selectivity being restricted by pore diffusion. Steam treatment may be applied to these silica aluminas to redistribute the pore-volume into a range of larger pore sizes, resulting in more acceptable performance, even though the total surface area is reduced. More recently, superior treatments, such as that reported by ICI (1973)[10] result in redistribution of the pore volume into pores with a tight pore-size distribution near to the optimum. The performance is improved significantly by minimising both pore-volume contained in pores of dimensions below the critical level at which selectivity loss become significant and, simultaneously, the pore-volume due to pores which are so large that they contribute little to the total surface area. Porosity characterisation is virtually indispensable in developments such as this.

For some of the highly exothermic or endothermic reactions it is possible that diffusion from the continuous phase through the boundary layer into the pore structure needs to be faster than required for the catalytic chemistry in order to maintain the interior of the catalyst aggregate sufficiently nearly isothermal, i.e. the heat conduction through the catalyst solid itself needs to be augmented by rapid transport of reactant and product molecules between the continuous phase and the catalyst interior. Table 4 summarises the areas of application of porosity characterisation, and of other properties of support material.

Table 4
Characterisation of Support Materials

Properties	Areas of Application of Characterisation Measurements
Porosity characteristics: pore volume mean pore diameter pore size distribution pore shape	The development of support preparation methods to produce material which meets assumed diffusion needs, strength, etc.
Mechanical characteristics	To correlate support properties with actual catalyst performance (e.g. to establish experimentally whether diffusion is limiting).

Mechanical strength is an important factor. The aggregate must be able to withstand the stresses which it will receive in the ultimate catalytic use and also, where applicable, during preparation steps, such as impregnation. For fixed-bed use a minimum level of crushing strength is necessary. In fluidised beds attrition resistance must be adequate. There may be very special requirements, as in the case of some polymerisation catalyst supports: here the aggregate should be weak enough for growing polymer to cause it to break down to the ultimate crystallites, so that monomer molecules are able to diffuse sufficiently rapidly to the active centres, thereby ensuring that the ultimate polymer material contains only a very low level of catalyst residue.

Supports are often composed of small crystallites, or less usually, small amorphous particles. The porosity is at least partly due to the interstices between them (and so depends upon aggregation method). The intrinsic mechanical properties are due to the adhesive forces between them; and, clearly, an understanding of these forces would be valuable. The use of cements is sometimes necessary. Many supports are prepared by the precipitation, aging and dehydration of gels. Techniques for investigating these stages are in principal valuable, since the characteristics of the crystallites which will form the support are controlled by changes which occur during these operations.

6. SUPPORT AGGREGATE CHARACTERISTICS

The mode of aggregation was previously mentioned as a factor which contributes to porosity and intrinsic mechanical properties. The shape and size of the aggregate are also of prime importance. Cylinders, rings, spheres and extrudates are most commonly used in fixed-bed applications. Their characterisation, however, is straightforward and often the aggregate can be produced in any desired shape and size.

Pressure drop across the bed is one factor which influences the choice of shape and size. Another is the ratio of aggregate superficial area to aggregate volume. If this is too small, there may be insufficient exchange between molecules in the boundary layer with those in the pore structure to satisfy the chemical potential. The relationship between aggregate size and reactor size should be such that there is an acceptably low level of bypassing by the reaction fluid. Also, aggregate crushing strength is dependent upon aggregate shape and size.

7. "INADVERTENT" SURFACE SPECIES

Species often appear on catalyst surfaces during preparation or use without deliberate action. They are often poisons and sometimes promoters. Sometimes they originate as an impurity in the catalyst phase itself, or in the support. In other instances impurities in the reactants are responsible. They may block active sites (e.g. bases adsorbed on acid sites). They may accelerate (or retard) sintering, or catalyse structural changes in the surface. Self-poisoning is a common phenomenon, arising when a reactant or a product is converted to a strongly sorbed species which blocks active sites. Polymeric hydrocarbon residues (known as "coke") are often laid down, sometimes in a form sufficiently massive to cause pore blockage. Phenomena of this nature are quite common. Characterisation is necessary to establish their chemical nature and spatial distribution within the catalyst aggregate. Information of this kind may give clues to their origin, so that appropriate action can be taken.

8. CONCLUDING REMARKS

Characterisation of catalysts clearly plays a key role in catalyst development and use. The diagram shows a typical sequence of activity from the identification of the need for a specific industrial process to the realisation of the need. From the earlier sections of this article it is clear that characterisation provides essential input at each of the four main steps.

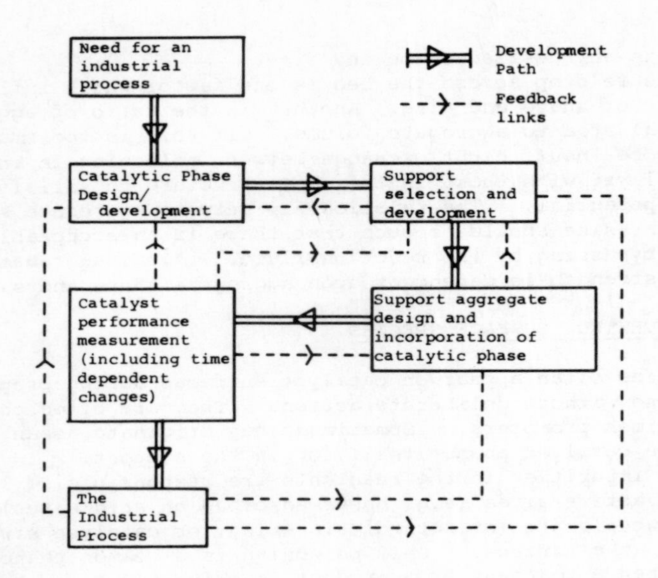

REFERENCES

1. C.N. Satterfield, "Heterogeneous Catalysis in Practice" (McGraw-Hill) (1980).
2. S.V. Gerei, K.M. Kholyavenko and M.Y. Rubanik, Ukv Khim Zh, 31, 449 (1965).
3. P.A. Kilty, N.C. Rol, W.M.H. Sachtler, Fifth Int. Cong. Catalysis (1972).
4. Chevron Research, U.S. Patent 3, 864, 280 (1972).
5. ICI Germ Off 2822322 (1978).
6. G. Ladwig, Z. Chem., 8, 307-308 (1968).
7. G.L. Simard, G.C. Serresc, H. Clark and D.J. Berets, Ind. Eng. Chem., 47, 1424 (1960).
8. I.I. Ioffe, N.V. Klimova and A.G. Makeev, Kinetics and Catalysis, 3, 165 (1962).
9. ICI European Patent Application 0003431 (1979).
10. ICI 1332303 (1973).

THE STUDY OF SUPPORTED CATALYSTS BY TRANSMISSION ELECTRON MICROSCOPY

By

A. Howie

1. INTRODUCTION

Electrons interact much more strongly with matter than do X-rays or neutrons and can thus be scattered appreciably by quite small atomic clusters. They can also be conveniently deflected and focused by electric or magnetic fields so that magnified real-space images can be formed in addition to simple diffraction patterns. The application of electron microscopy to the study of supported catalysts depends crucially on these factors and is exploited in a number of different instruments and imaging techniques. Some of these are well established from earlier work in materials science; others have been developed or adapted for application to catalysts. In the space available here we can do no more than indicate the present range of instrumentation and imaging techniques and provide an outline of the scattering mechanisms involved in the various cases. The reader may thus be able to form some idea of what is and what is not possible in the present state of the art. He may also be able to guard against some at least of the errors of image interpretation which threaten the course of most electron microscope investigations.

2. INSTRUMENTATION

2.1 The scanning electron microscope

Electron microscopes fall into two categories. In the scanning electron microscope or SEM (see Fig. 1a) the electron optics act before the specimen is reached to convert the beam into a fine probe which can be as small as 100 Å in diameter at the specimen surface. As the probe is scanned over the specimen surface by deflection coils, an image may be formed by collecting in a detector any suitable signal (backscattered electrons, emitted X-rays or optical photons) and displaying this signal in a raster synchronous with that of the probe. For accounts of the principles of imaging in the SEM the reader is referred to the books by Holt et al.[1] and Goldstein and Yakowitz[2]. Generally these machines employ conventional thermionic guns operating between 5 and 50 keV corresponding to electron wavelengths λ in the range 0.17 to 0.05 Å. The resolution d actually achieved can be considerably greater than the incident probe diameter because of beam spreading effects in the specimen. The X-rays, for instance, can be generated from a region as great as 1 μm in diameter in a bulk specimen. A great advantage of

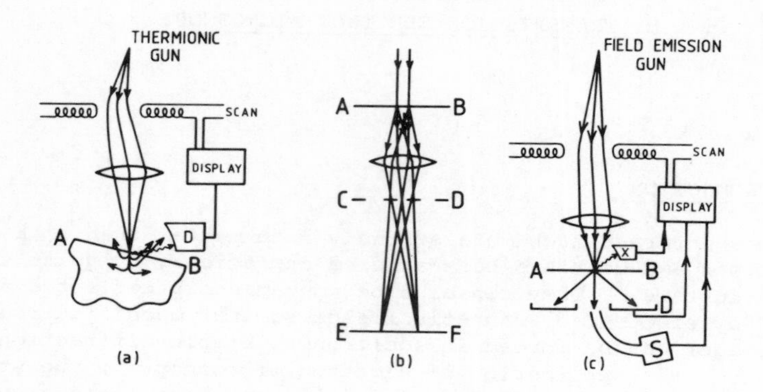

Fig. 1

Schematic diagrams showing the operation of (a) the scanning electron microscope (SEM), (b) the conventional transmission microscope (CTEM) and (c) the scanning transmission microscope (STEM) using a spectrometer S to collect the electrons in a small angle axial detector as well as an annular detector D and an X-ray or possibly backscattered electron detector X to form a variety of images.

these machines is the enormous depth of field $D \simeq d^2/\lambda \simeq 20\ \mu m$. This enables very impressive, in-focus images to be obtained from the highly irregular structures typical of catalyst specimens. Fig. 2 is an example from the work of Fisher and Szirmae[3] showing Fe catalysed oxidation pits in a single crystal of pyrolitic graphite. Further illustrations of the power of

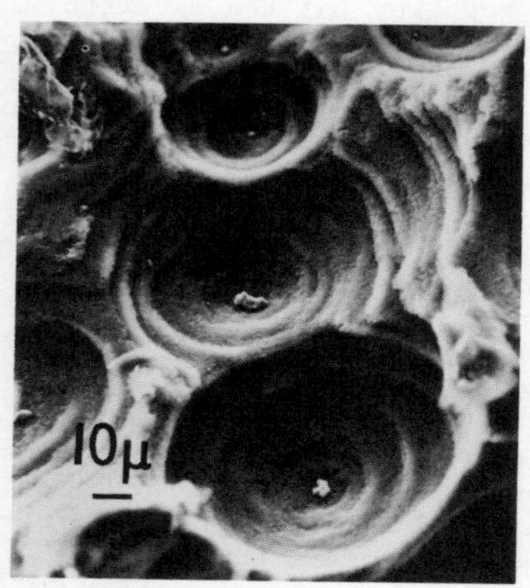

Fig. 2

SEM images of pits in carbon formed by Fe catalyst particles in wet hydrogen at 100°C.
(Fisher and Szirmae[3], with permission.)

the SEM, particularly when combined with microanalysis data, are given in the chapter by Acres. Scanning instrumentation has been developed, e.g. by Lee and Fisher[4], and Bishop and Poole[5] for completely automated location, identification and sizing of large numbers of particles in suitable samples.

Useful though such SEM methods are in catalyst studies, they fail to supply the high resolution detail increasingly demanded but which can only be obtained with two major modifications of imaging technique. In the first place, the beam spreading effect is eliminated by the use of thin samples ($t \leq 2000$ Å) and higher energy electrons ($E \simeq 100$ keV) which means that the strongest image signals are formed by transmitted electrons typically scattered at angles less than 10°. The resolution is now determined by the diameter of the incident electron probe which, in the case of thermionic sources, cannot be reduced much below 50 Å without leading to severe loss in probe current and consequently intolerably noisy images.

2.2 The conventional transmission electron microscope

The problem of probe current is overcome in the second category of electron microscope - the conventional transmission electron microscope CTEM shown in Fig. 1(b) where the scanning principle is abandoned in favour of an imaging mode closer to that used in optical microscopy. A large stationary illuminating spot is employed and magnified images making use of the tremendous efficiency of photographic recording are formed by lenses placed after the specimen. Diffraction patterns from selected areas of the object (picked out with an aperture in the image plane EF conjugate to the specimen plane AB in Fig. 1(b)) can be recorded by adjusting the final lenses to focus on the diffraction plane CD. From parts of the diffraction pattern selected by an objective aperture placed in the plane CD, images of various kinds can be formed by focusing the final lenses on the image plane EF. For a full account of these various imaging techniques and their application to thin crystalline samples the reader is referred to the book by Hirsch et al.[6]

The CTEM has now been developed to yield resolution in the 2 - 3 Å region, provided extremely thin ($t \leq 200$ Å) specimens are available. It has become one of the most successful tools for investigating the local structure of materials on the finest scale. Unfortunately, the sensitivity of the instrument to local composition is usually very small since the lack of any scanning mode means that, for instance, characteristic X-ray images are not normally available. A good deal of success has, however, been achieved in the last few years by fitting the high resolution CTEM with scanning attachments, X-ray detectors and electron energy loss spectrometers (see, for example, Hutchison and Lucas[7], Joy and Maher[8], and Colliex and Trebbia[9]). Information about local composition on a scale of perhaps 50 Å determined by the probe size can thus be obtained to supplement the high resolution transmission images.

2.3 The scanning transmission electron microscope

A much more radical and satisfactory solution to the whole imaging problem has become available recently with the development of the scanning transmission electron microscope (STEM) using a high brightness dark field emission gun to produce a probe of ~ 3 Å diameter (Crewe et al.[10]). This machine, shown schematically in Fig. 1(c), can produce all the high resolution images of the CTEM but has in addition the great flexibility of the SEM and can form images from energy loss electrons, emitted X-rays or light, etc. which provide information about local composition on a scale of 50 Å and perhaps much less. The STEM thus represents a synthesis of the two main lines of development in electron microscope technology and the new possibilities which it offers seem particularly relevant in the context of catalyst characterisation.

2.4 Other instrumental attachments

Several additional instrumental attachments and devices have been developed which extend the application of electron microscopy to a number of problems highly relevant to the field of catalysis. The rather dirty vacuum conditions of ~ 10^{-5} torr typically achieved by oil diffusion pumps in electron microscopes are equally remote from either the high pressure conditions prevailing in many catalytic reactions or the ultra high vacuum conditions desirable for clean surface studies.

Environmental cells have been developed[11-15] enabling reactions of various kinds to be followed in the CTEM in some cases at temperatures up to 900 C and at ambient gas pressures which can be as low as 10^{-9} torr or as high as ~ 0.5 atmosphere in high voltage machines. Particularly striking in situ observations have been made of the catalytic oxidation of graphite by Pt and Pd (Baker et al.[16]) and of the formation of filamentous carbon as a result of the decomposition of acetylene on Pt-Fe particles (Baker and Waite[17]). Sintering processes and surface structure changes have also been followed under well controlled conditions (Moodie and Warble[18]).

An alternative approach to the study of surface structure and reactions which avoids compromising electron microscope performance (but does not allow continuous in situ observation) is to employ some kind of transfer device (Ambrose[19]) to convey the specimen to the microscope from a UHV or other controlled reaction chamber where LEED/AUGER, RHEED or other facilities may be available to monitor the course of the reaction. This approach has been used by Ambrose et al.[20], and Levitt[21] to follow the oxidation of Cu, and is a useful supplement to in situ studies of the same reaction (see Heinemann[15]; Goulden[22]). With the full goniometer and high resolution facilities of the microscope still available it is possible to identify[21] the role of crystal defects in the oxidation reaction.

3. IMAGING IN THE TRANSMISSION ELECTRON MICROSCOPE

3.1 Basic techniques

Adequately representative samples of supported catalysts suitable for transmission electron microscopy are not always easy to prepare. Sufficiently thin platelets can sometimes be produced by grinding the material in a mortar and transferring a small quantity in suspension in an inert liquid to a thin carbon film. Microtomy methods can also be employed, not only on the original material, but also on powder produced by grinding or crushing which has then been embedded in an epoxy resin and allowed to set. Quite thin sections $t \leq 300$ Å can be cut with a diamond knife in the case of C, SiO_2 or γ-Al_2O_3 supports; α-Al_2O_3 supports are too hard to be treated in this way however.

As indicated above, the kind of image formed in the CTEM depends on the position of the objective aperture. If this includes the diffraction spot corresponding to the incident beam direction, a bright field image will be formed. Dark field images are obtained when the aperture is positioned to accept only scattered electrons. For high resolution imaging, when rather larger objective apertures must be used and instrumental effects such as spherical and chromatic aberration become important, it is usually best to leave the aperture concentric with the optic axis of the instrument and tilt the illumination in order to bring the scattered electrons of interest into the aperture.

Very useful results often follow from simple bright field and dark field images without detailed interpretation of image contrast effects. A rough idea of particle sizes and of the disposition of particles on the support can readily be obtained. The concentration of particles, particularly larger particles, near the surface of the grains of the support material has, for instance, been observed in several cases (Pope et al.[23]; Freeman et al.[24]; and Acres in this volume). Particle size distributions can also be determined in favourable cases when the particles are not too small, but for statistically reliable results this involves counting of a rather large number of particles in several samples. When the particles are small $D \leq 15$ Å it becomes increasingly difficult to be sure of detecting them above the substrate noise or to determine their size reliably without making a more detailed study of the image contrast. Some attention to scattering theory and image contrast is also essential if detailed information about the shape or structure of small particles is required. An account of some of the contrast effects observed at small catalyst particles which is rather more sophisticated than can be provided here is given by Flynn et al.[25].

3.2 Scattering Processes

The predominant source of image contrast and eventually structural information in transmission microscopy is elastic or quasi-elastic scattering when the fast electron scatters in the specimen without exciting any electronic excitation. At small scattering angles (where $S = 2\sin\frac{\theta}{2}/\lambda \simeq \theta/\lambda \leq 0.1$) corresponding to large impact parameters, the electron interacts coherently with many atoms at once and, in the case of a crystal, the scattering process is Bragg reflection. In a crystal of thickness t where only one Bragg reflection from planes of spacing d occurs, the scattering angle is $\theta \simeq 2\theta_B \simeq \lambda/d$ and the scattered intensity is (Hirsch et al.[6])

$$I_{Bragg} = \sin^2(\pi t\sqrt{1+W^2}/\xi_g)/(1+W^2) \quad (1)$$

where $W = s\xi_g = \Delta\theta\xi_g/d$ is a dimensionless parameter denoting the deviation from the Bragg condition when the crystal is turned by an angle from the exact Bragg angle, and ξ_g is the so-called extinction distance. Typical values of ξ_g for low order reflections in heavy metals are about 200 Å in the case of 100 keV electrons (Hirsch et al.[6]). In the case of light atom crystals such as Al, Al_2O_3 the corresponding figures for the strongest Bragg reflections would be about 600 Å. It can be seen from eqn. (1) that the Bragg reflected intensity oscillates with crystal thickness t reaching a maximum value of unity at the Bragg position W = 0 when $t = \frac{1}{2}\xi_g, \frac{3}{2}\xi_g$ ... and being proportional to t^2 in very thin crystals.

At larger scattering angles, corresponding to a small impact parameter with an individual atom, the main scattering process is thermal diffuse scattering with an intensity I_D

$$I_D = 1 - \exp(-2\pi t/\xi_o') \quad (2)$$

$$\simeq \frac{2\pi t}{\xi_o'} \quad \text{for } t << \xi_o' \quad (3)$$

The quantity $\xi_o'/2\pi \simeq 380$ Å for 100 keV electrons in Pd is the mean free path for diffuse scattering obtained by integrating the diffuse scattering distributions over all scattering angles (Hall and Hirsch[26])

$$\frac{1}{\xi_o'} = \frac{4\pi^2\hbar^2}{\Omega m^2 v^2}\int |f(S)|^2(1 - \exp(-2M_s))SdS \quad (4)$$

where v is the fast electron velocity, M_s the Debye Waller constant and Ω the atomic volume.

Inelastic scattering processes corresponding to the generation of electronic excitations in the specimen can also occur but usually have rather small scattering angles and mean free paths of at least 1000 Å so that they are usually of minor significance in image contrast of very thin samples. The inelastic component can however provide useful additional information about the specimen when an energy spectrometer of some kind is available to reveal the details of the loss spectrum (see below).

3.3 Bright field diffraction contrast

The simplest mechanism of bright field image contrast is amplitude contrast which arises because electrons are scattered outside the objective aperture and are lost to the usual bright field image which consequently appears dark in the strongly scattering region of the specimen. In the case of thin crystals where most of the amplitude contrast arises from Bragg reflection (particularly when apertures small enough to exclude the low order Bragg beams are employed), the effect is known as diffraction contrast and depends sensitively on crystal orientation (Hirsch et al.[6]). An example of diffraction contrast effects is shown in Fig. 3 of Pd catalyst crystals on a carbon support.

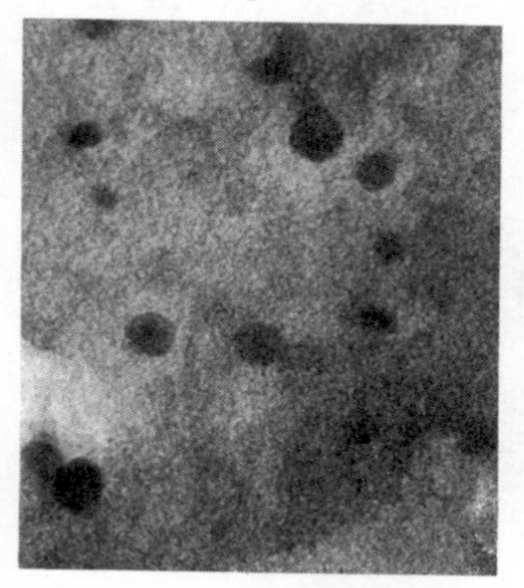

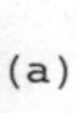

(a)

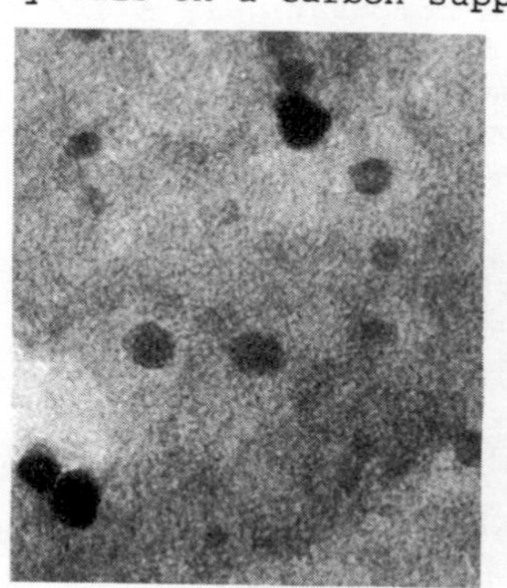

(b)

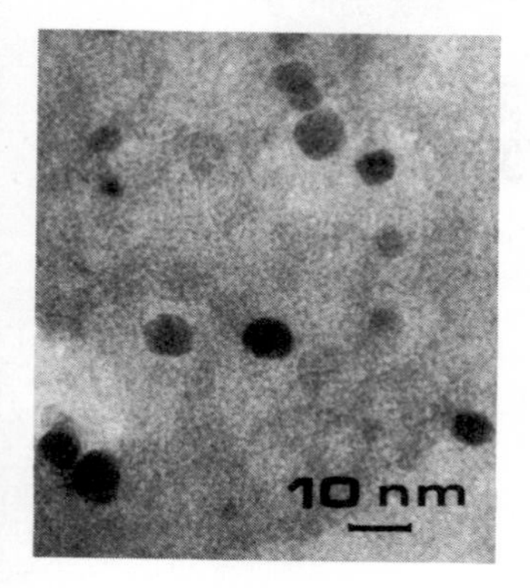

(c)

Fig. 3

Bright field diffraction contrast images of Pd single crystal particles on a charcoal support showing the influence of small changes of orientation in (a), (b) and (c) on the image contrast.
(Courtesy M.M.J. Treacy.)

The strongly diffracting particles appear almost black in the bright field image where the intensity $I \simeq 1 - I_{Bragg}$. Indeed, in some cases[27], thickness fringes can be observed which may be analysed with the aid of eqn. (1) and indicate that the particles have an almost spherical shape. Other particles which appear practically transparent in Fig. 3(a) are not in the Bragg diffracting orientation and are scattering by the much weaker diffuse scattering mechanism of eqn. (2). After a small tilt of the specimen the contrast is completely changed as shown successively in Figs 3(b) and (c), where the previously dark, Bragg reflecting particles are now quite transparent and vice versa. This simple diffraction effect, though well-known in materials science and demonstrated for catalyst particles by Flynn et al.[25] and Treacy and Howie[27], is still sometimes ignored. Images like those of Fig. 3(a) should not, for instance, be interpreted as indicating the simultaneous presence of quasi-spherical particles and thin disc or raft-like particles unless the effect of specimen tilting has been carefully examined.

3.4 Dark field images

At the comparatively small scattering angles available in CTEM, the dark field images are usually also dominated by diffraction contrast effects so that particles diffracting into the objective aperture appear bright against a low intensity background. Figs 4(a) and (b), including some multiply-twinned Au particles

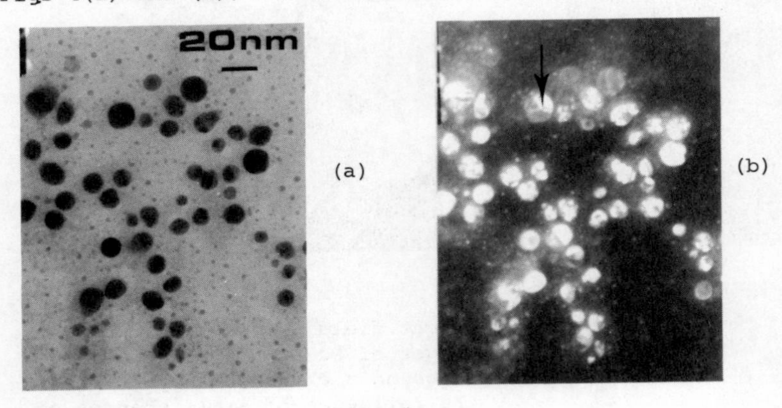

Fig. 4

Images of small Au particles in (a) bright field and (b) dark field. Most of the particles are not single crystal but contain defects like lamellar twins or have multiply twinned structures. These details, particularly the multiply twinned structure (arrowed) are much more clearly identifiable in dark field. (Courtesy L.D. Marks.)

similar to those recently found in some Ag catalysts (Marks and Howie[28]), show how the presence of the differently-oriented single crystal segments in the particles is much more readily apparent in dark field (b) than in the bright field image (a). The improved visibility of diffracting particles in dark field, particularly when the substrate is rather thick, makes the dark field method useful for detecting and counting catalyst particles. Corrections may be necessary to allow for the fraction of particles which are not in the diffracting condition however. Studies by Freeman et al.[24] using an annular illuminating aperture to produce dark field hollow cone illumination indicated that in the case of Pd catalysts practically all of the particles would be detected if an objective aperture capable of picking up both the 111 and 200 Bragg reflections were employed.

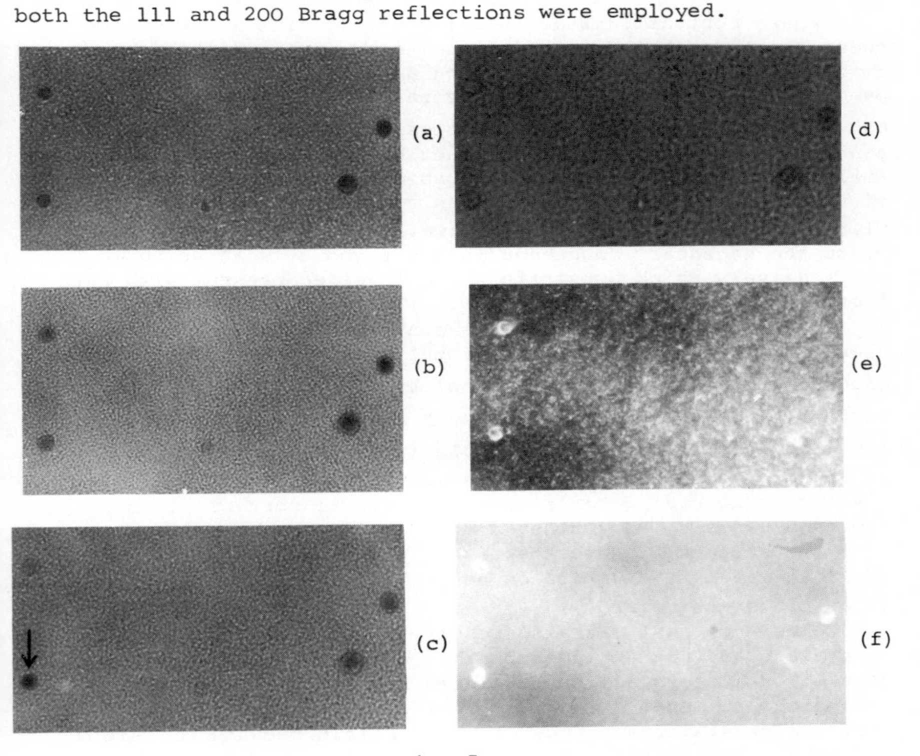

Fig. 5

Images of Pd particles on a charcoal support showing phase contrast effects at various defocus conditions in (a),(b),(c) and (d). Two faint diffraction images are visible as bright ghosts near the arrowed particle in 5(c). Figs 5(d) and (e) are coherent dark field and hollow cone dark field images of the same area. (Courtesy M.M.J. Treacy.)

Very small catalyst particles D ≤ 15 Å become increasingly difficult to distinguish reliably against the speckle effect from the disordered support material which is particularly evident in the standard CTEM dark field image taken with tilted illumination and a circular aperture. Computations of Krivanek and Howie[29] showed, for instance, that a 15 Å diameter diffracting crystallite is almost invisible when placed in or on a 60 Å thick amorphous support. The speckle from the support is somewhat suppressed by using hollow cone dark field techniques (see Fig. 5(e) and (f)), but the problem of reliably detecting very small catalyst particles is still a severe one.

3.5 Phase contrast images

Phase contrast arises in bright field from interference between the undeflected wave and scattered waves which are accepted by the objective aperture. Being an interference effect it depends on coherent wave amplitudes rather than intensities as in eqn. (1). The contrast thus varies like t rather than t^2 and tends to dominate over the diffraction contrast effect for very small particles. Detailed discussions of phase contrast imaging of small particles have been given by Hall and Hines[30], Flynn et al.[25], but the effects are not always observed in practice because the necessary coherence is destroyed by scattering in the comparatively thick substrate which is often present with catalyst specimens.

The simplest cause of phase contrast is the refractive index phase shift $\exp(\pi i t/\xi_o) \simeq 1 + i\pi t/\xi_o$ which may be regarded as producing a scattered wave of amplitude $\pi i t/\xi_o$. As a result of further phase shifts due to defocus and spherical aberration, the image of a small catalyst particle (where typically $\xi_o \simeq 100$ Å) is surrounded by Fresnel fringes (see Fig. 5(a),(b) and (c)). Precise structural details are difficult to sort out in these images and can also be further complicated by phase contrast effects from the support. Nevertheless it has been possible to image small Rh catalyst clusters on very thin carbon supports (Prestridge and Yates[31]).

Phase contrast can also result from the diffracted beams when these are not excluded by the objective aperture. In the case of small crystalline particles, these beams give rise to bright ghost images (see Fig. 5(c)) which are usually displaced from the central beam image of the particle because of defocus and spherical aberration. With the correct choice of defocus, however, the images can be made to coincide when lattice fringes may be observed. An example of such an image for a multiply-twinned icosahedral Ag particle is shown in Fig. 6.

The temptation to interpret lattice fringe images literally is frequently irresistible but is fraught with dangers, particularly when lattice defects are present and when the crystal thickness or diffraction conditions are varying from point to point. At the present time it is unclear whether lattice imaging will prove superior to high resolution dark field imaging for elucidating the internal structure (twin boundaries, dislocations, elastic strains etc.) or the surface structure (facets, notches, atomic ledges) of particles such as that shown in Fig. 6. These problems are so demanding that in all probability some combination of existing techniques may be required to solve them.

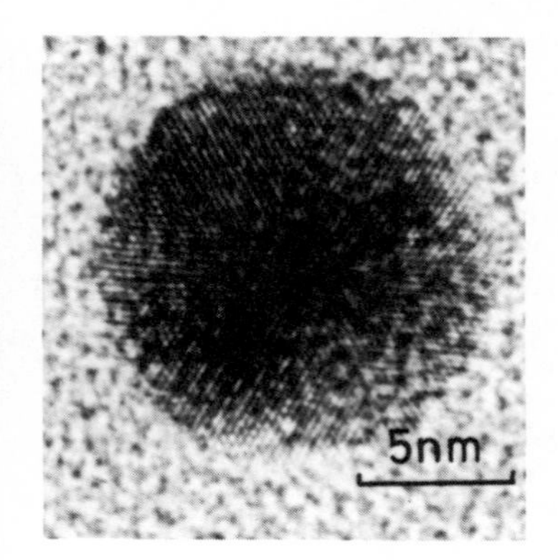

Fig. 6

High resolution lattice image of an icosahedral Ag particle. (Courtesy L.D. Marks and D.J. Smith.)

It is only in rather favourable cases that detailed information about surface structures of catalyst particles can be obtained from high resolution imaging techniques. If the crystal structure of the particles is fairly well known, it may be possible to deduce the crystallographic nature of the surfaces simply by observing various aspect ratios. This can be done for multiply-twinned particles (L.D. Marks, private communication) and has also been done[32] for $LaPO_4$ catalysts, where the nature of the acicular particle surface exposed could be correlated with catalytic activity. The formation of MoS_2 layers at the surface of hydrodesulphurisation catalyst particles has been observed by Sanders[33].

So far it has not been possible to use electron microscopy to detect reconstruction effects at surfaces of small particles. Such effects can be observed in the diffraction patterns from thin continuous films, and it should be possible to obtain images by dark field methods. It has, however, proved difficult to advance beyond the work of Cherns[34] who was able to use these techniques to image single atomic ledges in very thin single crystal gold films.

The assessment of porosity by electron microscopy is also an extremely difficult business if very high resolution is required. It has been possible[35] to follow the development of pores in single crystal boehmite under ideal conditions, but the problem is much more intractable in the typical highly disordered catalyst support material.

4. TECHNIQUES OF SCANNING TRANSMISSION ELECTRON MICROSCOPY

4.1 Image contrast

A great deal of the image contrast theory developed for the CTEM carries over to the STEM by virtue of the principle of reciprocity (Cowley[36]; Zeitler and Thompson[37]). The scattering probabilities are invariant under a reversal of the electron trajectories which converts for instance the CTEM bright field situation (Fig. 1b) with a narrow angle illuminating beam and a relatively wider angle imaging system into the STEM system with a convergent probe and a small angle axial detector (Fig. 1c). It can also be seen that the use of the annular detector in the STEM is equivalent to hollow cone dark field imaging in the CTEM. The range of hollow cone angles available in the STEM is much greater, however, and can be extended into the range where the diffuse scattering mechanism (eqn. 2) predominates and the coherent Bragg scattering is more or less absent. With increasing scattering angle, the electron scattering amplitude f(s) in eqn. (4) depends more strongly on atomic number Z (becoming proportional to Z) so that the high angle diffuse scattering is much more sensitive to the presence of heavy atoms.

The use of the annular detector signal to generate Z-contrast images in the STEM has been developed with spectacular results by Crewe et al.[10] to reveal isolated heavy atoms and very small heavy atom clusters on extremely thin amorphous carbon support films. Their work demonstrates that the visibility of the heavy atom images against the noise, arising from the disordered support structure as well as the electron beam, can be improved by displaying for instance an image formed from the ratio of the annular detector signal to the small angle energy loss signal collected with the spectrometer (Fig. 1c). These Z-contrast techniques have been applied with some success by Treacy et al.[38] to imaging of small Pd and Pt catalyst clusters on charcoal and γ-$A\ell_2O_3$ supports. An example is shown in Fig. 7. It is difficult to suppress entirely the image contrast from the support

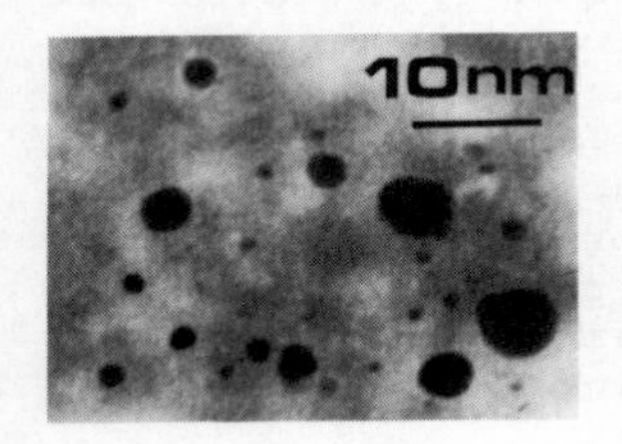

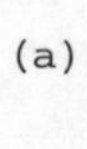

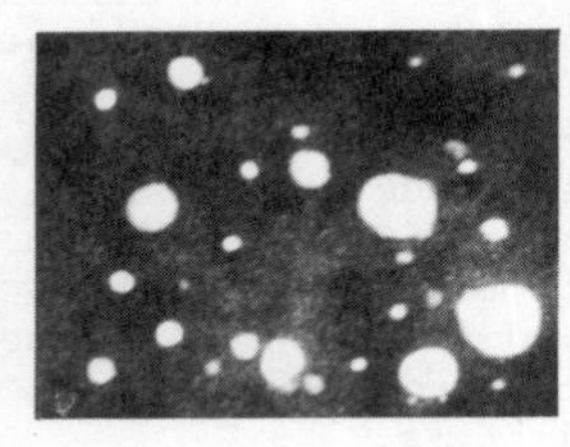

(b)

Fig. 7

Images of Pd catalyst particles on charcoal taken in the STEM. The bright field, zero-loss image is shown in (a); the image in (b), formed from the ratio of the annular detector signal and the small angle energy loss signal, shows the smallest particles more clearly.

particularly in the case of crystalline supports. Even if the annular detector used operates at angles large enough to escape all significant Bragg reflections from the support crystals, the theory[26] of diffuse scattering intensity from a crystal is more elaborate than eqn. (4) indicates and depends on crystal orientation because of channelling effects. Despite these complications, it seems likely that the Z-contrast method can be developed into the most effective technique for visualising small heavy atom clusters, particularly when they are held on the kind of support typical of catalyst specimens.

4.2 Microdiffraction patterns

The probe in the STEM can be stopped on an individual particle while a microdiffraction pattern is recorded either by scanning the scattered electron distribution over a detector or more simply by collecting it on a fluorescent screen. Although the spots in these patterns are necessarily rather coarse, because of the angular convergence of the probe they can provide a quick check to the structure of small particles (Brown[39]). An example in catalysis is provided by Fe 'graphimet' particles (a commercial product (and name) marketed by Ventron Corporation) where the structure has been identified from microdiffraction patterns as a mixture of graphite and Fe_3O_4 (Fisher et al.[40]).

4.3 Energy-loss spectra

The STEM also provides a powerful facility for collecting energy loss spectra from very small (~ 20 Å diameter) regions illuminated by the stationary probe. These spectra (see Fig.8) often contain a wealth of information about both composition and structure. So far the feature most utilised has been the characteristic edge structure associated with inner shell ionisation which can be used to identify the composition of small particles supported on very thin films. Fig. 8 shows the characteristic edges for O K shell and Fe L_{II-III} shell excitations

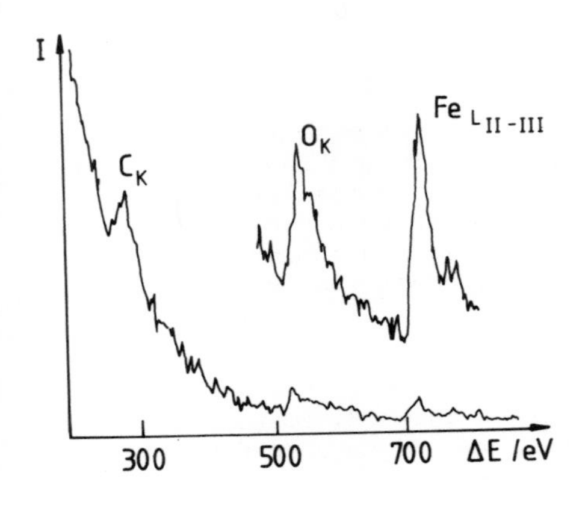

Fig. 8

Schematic energy loss spectra of Fe-graphimet particles after Fisher et al.[40]. The carbon, oxygen and iron edges are clearly seen. The magnified curve is typical of what can be obtained in a slow energy scan of several minutes.

observed by Fisher et al.[40] for the Fe 'graphimet' material described earlier. In the case of Ni 'graphimets' no oxygen was detected.

At energy losses ΔE well above the characteristic edge, corresponding to situations where the ionised electron is ejected with some kinetic energy, the energy loss spectrum sometimes shows weak oscillations as a function of ΔE which have roughly the same origin as the EXAFS effect (see articles by Joyner, Cox and by Pettifer in this volume). These features have been analysed to provide some structural information in the case of thin carbon films (Batson and Craven[41]) and it seems likely that they could also be of use in the study of small particles.

In the low-lying energy loss region $\Delta E \leq 30$ eV, the spectra often show pronounced peaks due to valence electron excitations which may be single electron excitations or collective excitations such as bulk plasmons and surface plasmons. It is doubtful at present whether the single electron excitations can ever be measured with sufficient precision by electron microscopists to give chemical information for individual small clusters. The surface plasmons, however, may well be sufficiently sensitive to details of surface structure and composition to provide useful if less directly interpretable information.

5. CONCLUSIONS

The electron microscope in its various forms has now become a very powerful instrument for the study of supported catalysts. Information about particle size distributions, particle dispositions, particle structures and compositions can be obtained for particles greater than about 15 Å in diameter. Care must be taken in the interpretation of the images, however, and this problem becomes more severe for smaller particles. The new techniques of scanning transmission electron microscopy, particularly Z contrast and energy loss spectroscopy, look promising as a way of answering some, if by no means all, of the questions posed by the catalyst chemist.

ACKNOWLEDGEMENTS

I am grateful to my colleagues T.G. Cunningham, R.M. Fisher, L.A. Freeman, L.D. Marks and M.M.J. Treacy who have taken part in the program of catalyst studies which underpins the writing of this article. The program was supported by S.R.C. as well as by I.C.I. and Johnson Matthey under the CASE scheme. I also express my thanks to the various colleagues who supplied figures for use in the text.

REFERENCES

1. D.B. Holt, M.D. Muir, P.R. Grant and I.M. Boswarva, Quantitative Scanning Electron Microscopy, Academic Press, London (1974).
2. J.I. Goldstein and H. Yakowitz (Eds) Practical Scanning Microscopy, Plenum Press, New York & London (1975).
3. R.M. Fisher and A. Szirmae "Electron Microscope Observations of Iron-Catalysed Gasification of Graphite" Electron Microscopy, 1, 452 (1978).

4. R.J. Lee and R.M. Fisher, 'Quantitative Characterisation of Particulates by Scanning and High Voltage Electron Microscopy' in Microanalysis of Particulates (in press) National Bureau of Standards, Washington (1979).
5. H.E. Bishop and D.M. Poole, Computer Control of the Electron Probe Microanalyser (in press) (1979).
6. P.B. Hirsch, A. Howie, R.B. Nicholson, D.W. Pashley and M.J. Whelan, Electron Microscopy of Thin Crystals, Krieger, New York (1977).
7. J.L. Hutchison and J.H. Lucas, 'Optimized Energy Dispersive X-ray Microanalysis', J. Microsc. 115, 295-297 (1979).
8. D.C. Joy and D.M. Maher, 'A Practical Electron Spectrometer for Chemical Analysis', J. Microsc. 114, 117-129 (1978).
9. C. Colliex and P. Trebia, 'Electron Energy Loss Spectroscopy in the Electron Microscope', in Electron Microscopy, 3, 268 (1978) (Microscopical Society of Canada - Toronto).
10. A.V. Crewe, J.P. Langmore and M.S. Isaacson, 'Resolution and Contrast in the STEM' in Physical Aspects of Electron Microscopy and Microbeam Analysis (Eds B.M. Siegel and D.R. Beaman) J. Wiley - New York, pp.47-62 (1975).
11. J.C. Mills and A.F. Moodie, 'Multipurpose High Resolution Stage for the Electron Microscope', Rev. Sci. Instr. 39, 962-969 (1968).
12. U. Valdre, E.A. Robinson, D.W. Pashley, M.J. Stowell and T.J. Law, 'A uhv Electron Microscope Specimen Chamber', J. Phys. E3, 501-506 (1970).
13. R.T.K. Baker and P.S. Harris, Controlled Atmosphere Electron Microscopy, J. Phys. E5, 793-797 (1972).
14. H.M. Flower, N.J. Tighe and P.R. Swann, 'Environmental Gas Reaction Cells' in High Voltage Electron Microscopy (Eds P.R. Swann, C.J. Humphreys and M.J. Goringe) pp.383-395 Academic Press, London (1974).
15. K. Heinemann, Rao D. Bhogeswara and D.L. Douglass, 'Oxide Nucleation on Thin Films of Copper, Oxid. Met. 9, 379-400 (1975).
16. R.T.K. Baker, P.S. Harris and S. Terry, 'Unique Form of Filamentous Carbon', Nature 253, 37-39 (1976).
17. R.T.K. Baker and R.J. Waite, 'Carbonaceous Deposits from Acetylene', J. Catal.37, 101-105 (1975).
18. A.F. Moodie and C.E. Warble, 'Transmission Electron Microscopy' in Sintering and Catalysis (Ed. G.C. Kuczynski) pp. 1-16 Plenum Press, New York (1975).
19. B.K. Ambrose, 'A Specimen Transfer Device', J. Phys., E9, 382-384 (1976).
20. B.K. Ambrose, D.A. Goulden and A. Howie, 'Electron Microscopy of Copper Oxidation' in Developments in Electron Microscopy and Analysis (Ed. J.A. Venables) pp.445-448, Academic Press, London (1976).
21. J.P.F. Levitt, The Structure and Composition of Metal Surfaces, Ph.D. Thesis, University of Cambridge, (1979).
22. D.A. Goulden, 'Electron Microscopy of Cuprous Oxide Island Growth' Phil. Mag. 33, 393-408 (1976).

23. E. Pope, W.L. Smith, M.J. Eastlake and R.L. Moss, 'The Structure and Activity of Supported Metal Catalysts', J. Catal. 22, 72-84 (1971).
24. L.A. Freeman, A. Howie and M.M.J. Treacy, 'Electron Microscopy of Pd Catalysts', J. Microsc. 111, 165-178 (1977).
25. P.C. Flynn, S.E. Wanke and P.S. Turner, 'TEM Studies of Supported Metal Catalysts', J. Catalysis 33, 233-248 (1974).
26. C.R. Hall and P.B. Hirsch, Thermal Diffuse Scattering of Electrons, Proc. Roy. Soc. A286, 158-177 (1965).
27. M.M.J. Treacy and A. Howie, 'Contrast Effects in Transmission Electron Microscopy of Supported Crystalline Catalyst Particles' (submitted to J. Catal.) (1979).
28. L.D. Marks and A. Howie, 'Multiply Twinned Catalyst Particles', Nature (to be published) (1979).
29. O.L. Krivanek and A. Howie, 'Kinematical Theory of Images', J. Appl. Cryst. 8, 213-219 (1975).
30. C.R. Hall and R.L. Hines, 'Electron Microscope Contrast of Small Atomic Clusters', Phil. Mag. 21, 1175-1186 (1970).
31. E.B. Prestridge and D.J.C. Yates, 'Imaging the Rhodium Atom with a Conventional High Resolution Electron Microscope', Nature 234, 345-347 (1971).
32. J.M. Cowley, J.C. Wheatley and W.O. Kehl, 'High Resolution Microscopy of $LaPO_4$ Catalysts', J. Catal. 56, 185- (1979).
33. J.V. Sanders, High Resolution Electron Microscopy of some Catalytic Particles, Chemica Scripta (in press) (1979).
34. D. Cherns, 'Direct Resolution of surface steps by transmission electron microscopy', Phil. Mag. 30, 549-556 (1974).
35. S.J. Wilson, 'Dehydration of Boehmite', Mineral. Mag. 43, 301-306 (1979).
36. J.M. Cowley, 'Image Contrast in STEM', Appl. Phys. Letts. 15, 58-59 (1969).
37. E. Zeitler and M.G.R. Thomson, 'Scanning Transmission Electron Microscopy', Optik 31, 258-280 (1970).
38. M.M.J. Treacy, A. Howie and C.J. Wilson, 'Z Contrast of Platinum and Palladium Catalysts', Phil. Mag. 38, 569-585 (1978).
39. L.M. Brown, 'Progress and Prospects for STEM in Materials Science' in Developments in Electron Microscopy and Analysis pp.141-148 (Ed. D.L. Misell), Institute of Physics, London, (1977).
40. R.M. Fisher, D.J. Smith, L.A. Freeman, S.J. Pennycook and A. Howie, 'Electron Optical Characterisation of Graphimets', Carbon Conference 14, 318 (1979).
41. P.E. Batson and A.K. Craven, 'Extended Fine Structure of the Carbon Core - Ionisation Edge obtained from Nanometer-sized Areas with Electron-Energy-Loss Spectroscopy', Phys. Rev. Letts. 42, 893-897 (1979).

CHARACTERISATION OF HETEROGENEOUS POLYOLEFIN CATALYSTS

A MICROSCOPE STUDY

By

R.T. Murray

1. INTRODUCTION

In this chapter we demonstrate the versatility and usefulness of optical and electron microscopy for the characterisation of reactions in which a heterogeneous catalyst is used to produce an insoluble polyolefin polymer of the kind described by Ballard et al.[1,2] elsewhere.

The problems presented by a major project such as the development of a novel industrial catalyst are manifold, and it is necessary for the microscopist to have access to a wide armoury of techniques. We will make reference to

1) Polarised light microscopy
2) Scanning electron microscopy
3) Energy dispersive electron microprobe analysis
4) Conventional transmission electron microscopy
5) Electron diffraction
6) High voltage electron microscopy
7) Ultramicrotomy.

Additional electron microscopic techniques are discussed in the preceding Chapter by Howie. It will be seen that, for the problem under consideration, several distinct advances were made as a result of predictions based on the microscopy. In addition, the understanding of the physics of polymer formation which was acquired during the course of the work prevented many abortive routes being pursued.

The objectives of the programme can best be understood by reference to the drawbacks of technology existing around 1970.

a) The conventional $TiCl_3$ based Ziegler catalysts produce polymer of a small particle size typically 0.5 mm in diameter which gives rise to handling problems. These are usually overcome by adding an expensive melt densification stage at the end of the polymerisation stream.
b) The catalyst residue, because of its corrosive nature, has to be washed out of the polymer.

We aimed to avoid both of these ancillary stages by developing a 'high mileage' supported catalyst achieving low residual levels and which produced larger polymer grains.

The catalyst family developed to meet these objectives consisted of a transition metal complex such as zirconium tetra benzyl $Zr(C_6H_5CH_2)_4$ bonded through surface hydroxyl groups to a nanocrystalline support such as alumina. Polymerisation, which takes place in a stirred tank, continuous process reactor, in the presence of a diluent which is a good solvent for the olefin, occurs by insertion of monomer units between the ligand and the zirconium[2]. Molecular weight is controlled by the addition of small quantities of hydrogen in the olefin feed. The catalyst is fed to the reactor as a slurry of 100 μ alumina particles which themselves are built up of 100 Å crystallites. The polymer is collected from the reactor in the form of 1 mm or larger spherical particles from which the diluent is removed by evaporation.

2. STRUCTURE OF THE POLYMER

When examined by a scanning electron microscope the polymer sphere is found to be an assembly of sausage-like units which curl and twist so as to intertwine and provide a physical source of strength (Plate 1). Additional linkages in the form of drawn polymer fibres are also found[3] as illustrated later (see Plate 6). The sausage-shaped objects are approximately 0.5 x 5 μ, though this will later be seen to be a process variable.

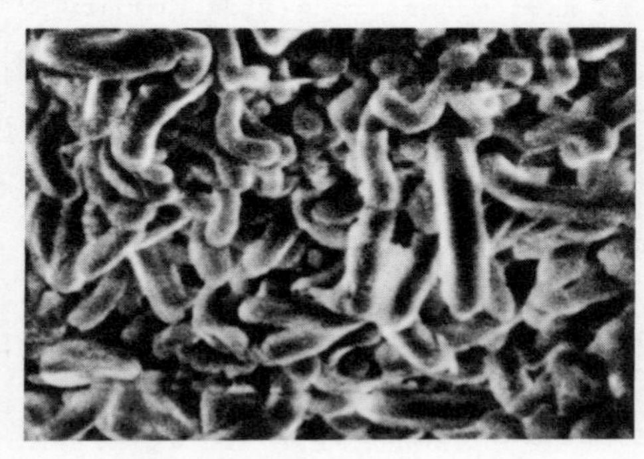

Plate 1

Sausage-like components which constitute the major portion of the polymer grains. (Mag. 4K.)

3. CATALYST RESIDUE AND MORPHOLOGY

We can now address ourselves to the problem which provided the impetus for this study, namely the catalyst residues and their location in the polymer. Plate 2, showing a dissected particle, soon put us on the wrong track for, as we see, the centre of a sphere is hollow and might well have contained a 100 μ particle of alumina catalyst which was lost during the process of 'surgery'. The missing cores could never be found so 1 μ sections were prepared from "Araldite" embedded polymer spheres and examined in the 1 Mev microscopes at Harwell and Oxford. Residues about 500-1000 Å diameter were found to be dispersed throughout the section in confirmation of the observations of Bulls and Higgins[4]. It then became appropriate to enquire into the relationship between the two building blocks, namely the sausage-like polymer and the activated microcrystals of alumina.

A single polymer particle of approximately 1 mm diameter is far too large to be studied in a TEM, so the problem was tackled by first grinding the spheres in liquid nitrogen and subjecting

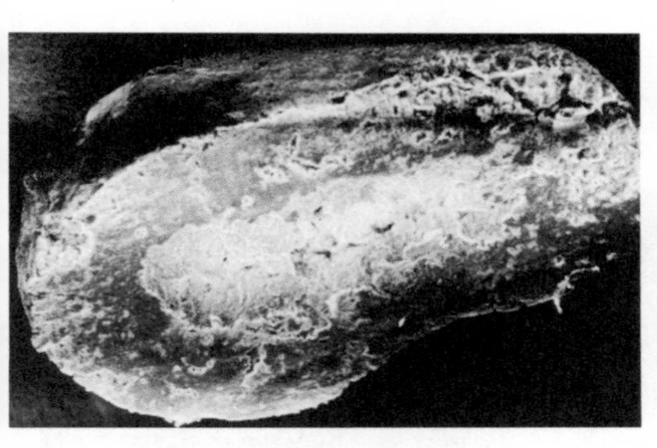

Plate 2

Cross-section through a polymer grain showing the hollow core caused by poor catalyst distribution across the support material (50x).

the resultant powder to ultrasonics in xylene or hexane for several hours. Very small assemblies of sausage-like polymer were thereby produced as shown in Plate 3, and on occasion individual segments were obtained. When examined in the high voltage EM a catalyst residue could clearly be distinguished at the distal end of each object, as shown in Plate 4. This material was polymerised on a catalyst of low activity so the residues are large and clearly visible, whereas Plate 5 is taken with a commercially more interesting catalyst giving rise to the desirable low residuals content. A single sausage-like object normally contains only one catalyst deposit, which is invariably found at its end away from branch points. It is reasonable to assume that this will be the growth point and that the fresh polymer is deposited behind the catalyst pushing it further from its starting point as polymerisation proceeds. The size of the residue is, however, greater than the size of an individual nanocrystal and contains maybe 100 to 1000 such units.

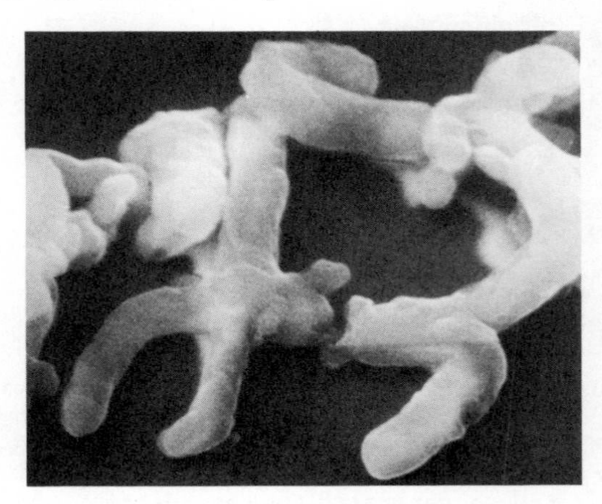

Plate 3

Sausage clusters after ultrasonic dispersion in xylene (15K).

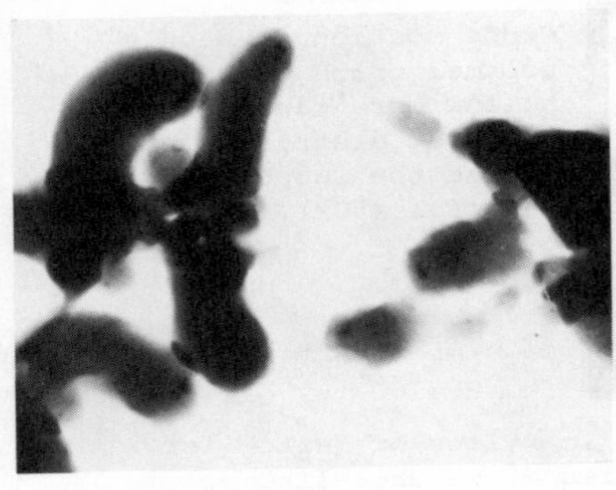

Plate 4

Sausages produced by a catalyst of low activity. Note the catalyst residue at one end of each sausage (17K).

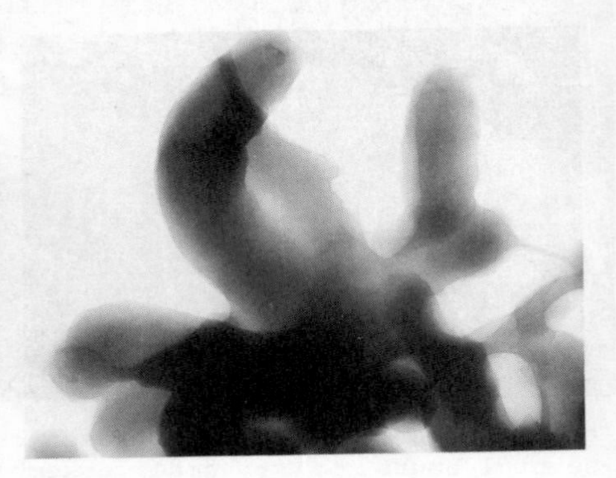

Plate 5

Sausages from an active catalyst: residues are smaller and of lower contrast (20K).

4. GROWTH PROCESS

The story as unfolded above has referred only to polymer/ catalyst at the end of a polymerisation run and curiosity prompts us to ask about the early stages of growth.

1) How does a sausage commence life?
2) How does the 100 μ alumina sphere break up?

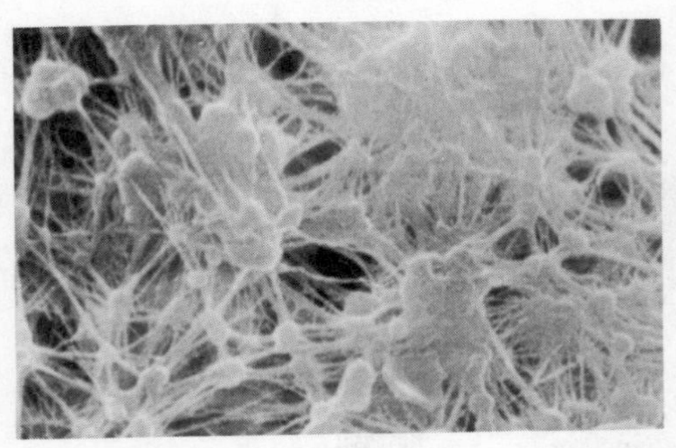

Plate 6

Cobweb texture produced by expansion at early growth stage. No sausages have been found at this stage (10K).

Plate 6 illustrates the nature of the growing particle after several minutes polymerisation. The sphere is breaking up into sizeable nodes which are held together by drawn threads of polymer. As polymerisation proceeds each node breaks up further under the swelling forces generated by the reaction and the space between nodes fills in (see Graff, Kortlev and Vonk[5]). At the final stage of disintegration of the catalyst each deposit or quantum is surrounded by its own microsphere of polymer, as shown in Plate 7. The next growth stage which I believe presents an intriguing problem for the more theoretically minded, results in these spheres converting to low aspect ratio 'sausages' for which further growth can be accommodated by axial extension, Plate 8.

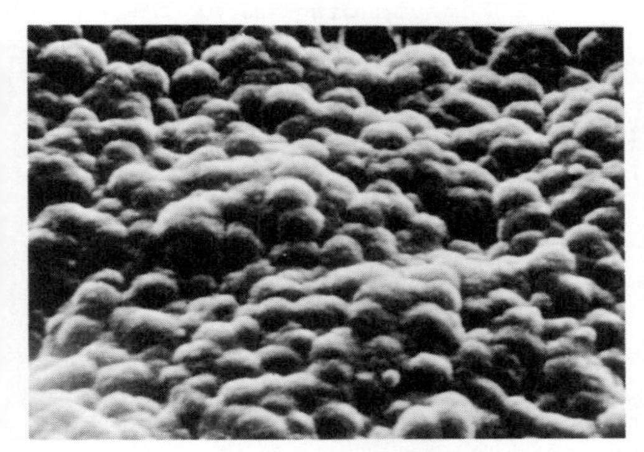

Plate 7

Polymer microspheres which later transform into sausages (10K).

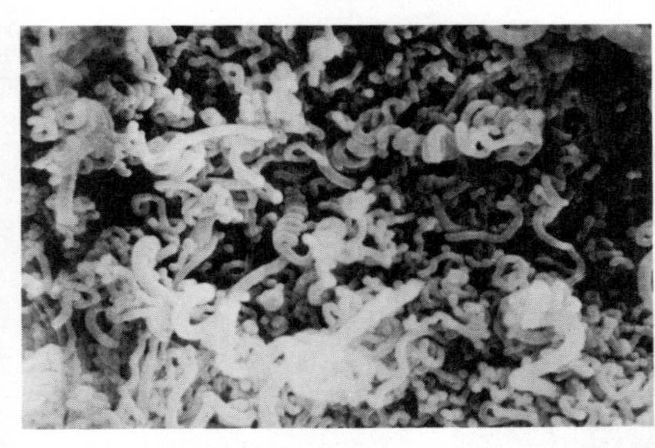

Plate 8

Sausages can be grown to high aspect ratios if long reactor dwell times are used (2K).

The growth process can thus be represented diagrammatically as in Fig. 1, the final element of which seeks to illustrate the crystalline texture of a 'sausage' as deduced from three significant observations :

1) when a sausage-like object is severely irradiated in the microscope it suffers degradation, and a stacked lamella structure is observed;
2) the individual sausages are birefringent under crossed polars, but this could be a shape effect; and
3) electron diffraction at 1 Mev on single sausages indicates chain orientation parallel to the sausage axis by arcing of the Debye Scherrer rings - Plate 9.

Following logically from Fig. 1, areas of support devoid or deficient in organometallic will give rise to voids in the polymer sphere. Reference to Plate 2 supports this view. Specimens were prepared for X-ray microprobe analysis by embedding the catalyst spheres in "Araldite" and facing the block on a good glass knife. The sample was made conducting by an evap-

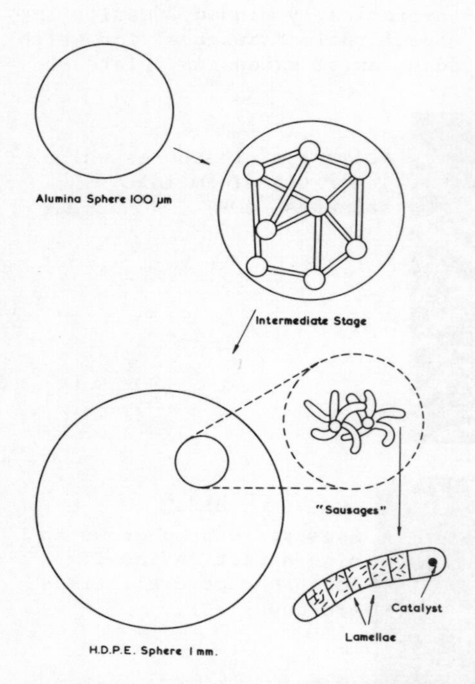

Fig. 1

Schematic diagram of the growth stages.
Note the partial orientation of the polymer chains in the final sausage stage.

Plate 9

(110) and (200) rings obtained by selected area diffraction from a single sausage. Note the arcing due to orientation of the polymer chains.

orated carbon layer and analysed by stepping the probe across a near diametrical cross-section. The resultant Zr/Aℓ ratio demonstrated high catalyst loading over the outer 20 μ and very low levels at the centre. Improved methods of impregnation led to the elimination of this defect.

We have already referred to the improved economics of a process in which no densification stage is required. The polymer containing voids when fed directly to an extruder, traps sufficient air to lead to bubbles in the extrudate, which ruin the mechanical properties of the product. Elimination of the voids together with further increases in powder density by increasing reactor dwell time (predicted on the basis of the microscopic model) removed this barrier to commercial success.

The description so far has been of a workable process in which all the steps make technological sense. However, there are many pitfalls for the unwary, and it is perhaps worth illustrating the results of using an inappropriate grade of alumina. As indicated by Fig. 1, the mechanical characteristics of the support can determine the stages of growth; if a readily dispersible grade is chosen, polymerisation which proceeds rapidly at first leads to the disintegration of the spherical catalyst, and combined with the action of the stirring blade leads to the formation of stirrer crystals (Pennings and Keil[6]) and Shishkebabs - see Plate 10. These rapidly entangle, raise the effective viscosity of the medium and eventually clog up the reactor which can only be restored to useful operation by vast expenditure of human labour.

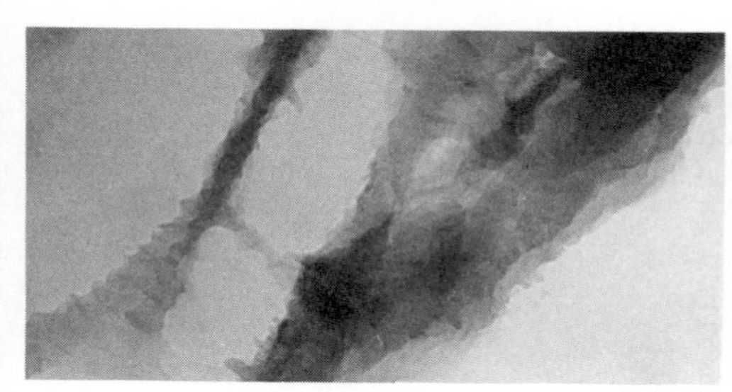

Plate 10

Stirrer crystals produced when a weak catalyst support is used.

5. ZIEGLER-NATTA SYSTEMS

Previous studies in this area have concentrated on the use of replica techniques to overcome the problem of introducing air-sensitive materials into the microscope. A quite simple anaerobic cell[7] has been designed to fit the normal specimen insertion device of our Philips EM300, and its effectiveness demonstrated by the preservation as shown by electron diffraction of the structure of a Stauffer Ziegler Natta catalyst. Commercial interest is now directed at materials of much higher surface area, as illustrated in Plate 11. The lattice fringes indicate a high degree of preservation and define clearly the extent of a single crystal from which a surface area can be calculated[8]. Good agreement with X-ray and N_2 absorptiometry has been found.

A great deal of interest has been shown in determining the location of active sites for polymerisation on the titanium trichloride systems (Guttman and Gullet[9]), but no previous observations of the earliest stages of growth under realistic conditions are known to the author. Polymer and catalyst killed off after very short reaction times has been examined using the anaerobic cell, but no signs of preferred sites have been detected (Plate 11). Thus, either the complete surface of the catalyst particle is active or, more likely, the exothermic nature of the reaction has produced melting followed by encapsulation of the catalyst crystallites.

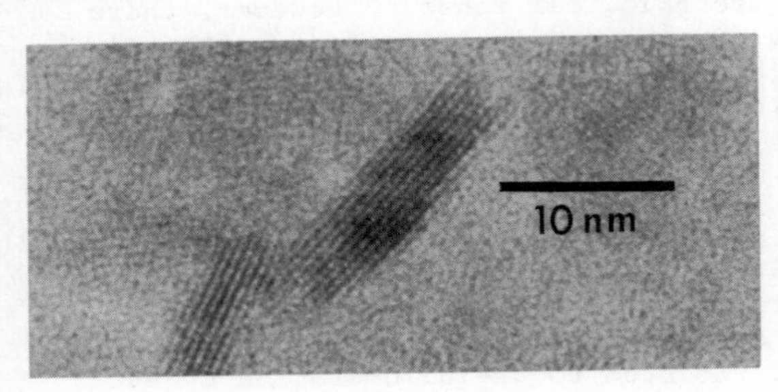

Plate 11

Lattice resolution image of a high surface area $TiC\ell_3$ catalyst showing crystallite size.

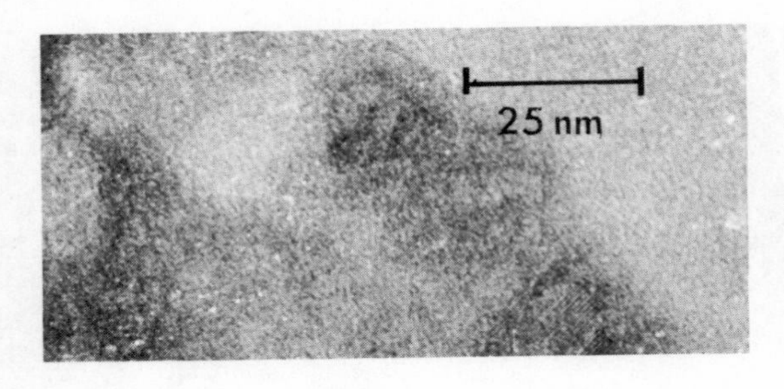

Plate 12

The same catalyst surrounded by initial polymer. No clear sites can be distinguished.

6. CONCLUSION

The experiments we have described are not the first essay by polymer scientists to understand the physics of polymer growth. Their value lies in the ease with which we have been able to observe simultaneously the morphology of the polymer and the catalyst, and in particular to do so under near commercial conditions of production.

The origin of the growth of sausage-like objects remains a mystery. Observations on polymer growth from $VC\ell_3$ supported on planar materials led Marchessault and co-workers[10] to postulate a mechanism based on overcrowding within the plane. The isotropic growth of the polymer precludes the existence of a directional constraint, so it is possible that the sausage shape is attained by a surface tension mechanism reinforced by diffusion control in which the energy required to expand the solidified skin of the microsphere is no longer available from the exotherm-ic polymerisation. A point of weakness will be required to permit further growth, but once this has been found polymerisa-tion can continue indefinitely at a constant rate. Inadequate data are available to permit a calculation of the maximum sphere diameter.

REFERENCES

1. British Patent No. 1314828 (1973).
2. D.G.H. Ballard, E. Jones, R.J. Wyatt, R.T. Murray and P.A. Robinson, Highly active polymerisation catalysts of long life devised from σ- and π-bonded transition metal alkyl compounds, Polymer, 15, 169 (1974).
3. T. Davidson, Microstructure of Particle foam polyethylene, J. Pol. Sci., Pol. Letters, 8, 855 (1970).
4. V.W. Bulls and T.H. Higgins, A particle growth theory for heterogeneous Ziegler polymerisation, J. Poly. Sci., A1, 8, 1037 (1970).
5. R.L.F. Graff, G. Kortleve and C.G. Vonk, On the size of the primary particles in Ziegler catalysts, J. Pol. Sci., Pol. Letters, 8, 735 (1970).
6. A.J. Pennings and A.M. Kiel, Fractionation of polymers by crystallisation from solution III, Kolloid Zschr. Z. Polymer, 205, 160 (1965).
7. R.T. Murray, D.G.H. Ballard and D. Platt, The polymerisa-tion of α-olefins: a microscope study. 6th European Conf. on EM, Jerusalem, 1976, 393.
8. R.T. Murray, R. Pearce and D. Platt, Observations of the fundamental particles in a titanium trichloride based polymerisation catalyst by electron microscopy, J. Poly. Sci., Pol. Letters, 16, 303 (1978).
9. J.Y. Guttman and J.E. Guillet, Mechanism of propylene polymerisation on single crystals of α-titanium trichloride, Macromolecules, 3, 470 (1970).
10. R.H. Marchessault, B. Fisa and H.D. Chanzy, Nascent morphology of polyolefins, Critical Reviews in Macromol. Sci., 315 (1972).

USE OF MÖSSBAUER SPECTROSCOPY FOR CATALYST CHARACTERISATION

By

W. Jones

1. INTRODUCTION

The technique of Mössbauer (or nuclear gamma resonance) spectroscopy has, since its discovery[1] in 1958, found considerable application within the area of materials science. Those parameters, such as isomer shift, quadrupole splitting and magnetic hyperfine splitting, which may be extracted from a Mössbauer spectrum allow a detailed analysis of the chemical state of certain atoms within a material. Furthermore, for some isotopes it is possible to determine these parameters under conditions which are much closer to actual catalyst operating conditions than is permitted with many of the other techniques available for the study of catalysts (i.e. at high pressures and temperatures and various gaseous environments). The use of specimens in the form of microcrystallites, supported or otherwise, ensures that the information obtained bears considerable significance to the chemical state of the surface of the catalyst. In addition to these favourable points, the recent development of conversion electron Mössbauer spectroscopy enhances the surface sensitivity of the Mössbauer technique, enabling systems which are not microcrystalline to be studied and also allowing the possibility of obtaining spectra from various depths within the solid.

We begin by summarising the essential features of a Mössbauer spectrum, and in the second section illustrate some recent applications of Mössbauer spectroscopy in heterogeneous catalysis. The approach is one of indicating the general manner in which the technique has been applied and the total Mössbauer/catalysis literature is not comprehensively summarised. In such a way we shall draw upon Mössbauer experiments which have been performed in other areas of materials science but which are well-suited to catalysis.

2. THE MÖSSBAUER EFFECT

The discovery that the incorporation of the emitting and absorbing nuclei within a solid matrix enables resonant absorption and emission of γ-rays forms the basis of the technique. Its importance lies in the very narrow linewidth of the emitting photon resulting from the relatively long lifetime of the excited nuclear state - typically the mean life of the excited state is 10^{-8}s, corresponding to a natural linewidth of the order of 10^{-8} eV - and the consequent ability to probe the variations in nuclear energy levels resulting from any discrete changes in the chemical state and/or environment of the Mössbauer nucleus.

Such changes in nuclear energy levels are measured by modifying the energy of the probing γ-ray by applying a Doppler shift, and a Mössbauer spectrum therefore consists of a plot of counts against applied Doppler velocity (positive and negative), the velocities corresponding to the addition and subtraction of Doppler energy shifts to the γ-ray energy. Mössbauer spectroscopists, consequently, refer to energy changes using units of velocity.

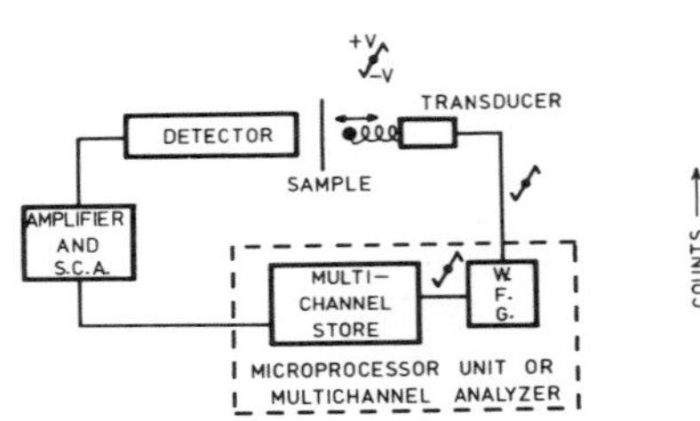

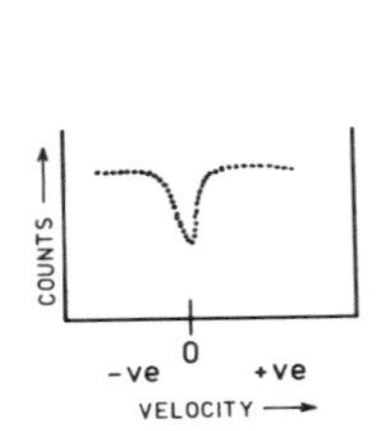

Fig. 1

Diagram illustrating the components of a Mössbauer spectrometer.

Experimentally the technique is relatively simple and the instrumentation required is shown schematically in Fig. 1. The source of γ-rays is attached to a transducer, the velocity of which is determined by a ramp wave which is simultaneously fed into a multichannel analyzer, allowing each channel to correspond to a particular instantaneous velocity. The γ-rays may be monitored by a variety of detectors according to their energy. Scintillation-crystal and gas-filled proportional counters are used within the range 0-100 keV, the gas-filled counters giving better resolution below 30 keV. Lithium drifted germanium detectors, which possess better resolution than both of those previously mentioned, are also used but suffer from the necessity of keeping the detector at liquid nitrogen temperatures. They are particularly useful when the separation of emissions of fairly similar energy is required. Separation of the required signal from X-rays and electronic background noise is obtained by use of pulse height analysis and a signal channel analyzer. Microprocessor-based systems for Mössbauer work are now available at a much lower cost than systems based on multichannel analyzers.

(a) Isomer shift

A Doppler shift required to produce resonance between source and absorber when they are not in chemically identical environments, the chemical isomer shift (δ), arises because the nuclear energy levels are sensitive to changes in electron density at the nucleus. Such changes, the extent of which govern the magnitude of δ, may arise from differences in local topology as well as changes in valence state of the Mössbauer isotope. We may express (see Fig. 2) to a good approximation the isomer shift as

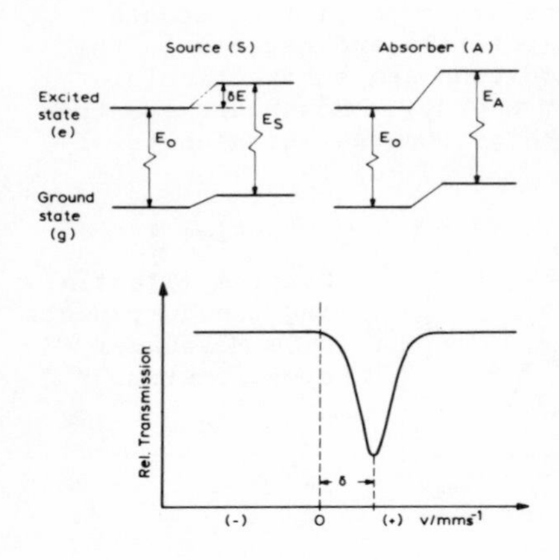

Fig. 2

The origin of isomer shift. The measured value of δ depends on the value E_S-E_A and consequently it is necessary to quote isomer shift values relative to some particular reference. For ^{57} Mössbauer an iron foil is frequently used. (Reproduced by permission of Prof. P. Gutlich and Springer-Verlag, see Bibliography.)

$$\delta = (\Delta E)_A - (\Delta E)_S = \frac{2}{3}\pi Ze^2[|\psi(0)|_A^2 - |\psi(0)|_S^2].[\langle R^2\rangle_e - \langle R^2\rangle_g]$$

where e = electronic charge, Z = nuclear charge, R_e, R_g are the nuclear radii in the excited and ground state respectively, and $|\psi(0)|_A^2$ and $|\psi(0)|_S^2$ are the total electron densities at the nucleus of the absorber and source respectively. The term $\langle R^2\rangle_e - \langle R^2\rangle_g$ is a nuclear constant for a particular isotope. Screening effects from orbitals which do not have a finite electron density at the nucleus will alter the value of $|\psi(0)|_A^2$ and $|\psi(0)|_S^2$ and will therefore modify the value of the isomer shift. Consequently, changes in valence charge resulting from the addition or removal of, say, d electrons will also modify the value of the isomer shift. It is, therefore, possible to determine changes in oxidation state, or gain an indication of the strength of bonding between the Mössbauer isotope and the surrounding atoms or ions. This ability to monitor bond strengths is particularly useful for investigating species adsorbed on bimetallic catalysts since it permits, for example, an investigation of the degree of interaction between the adsorbed species. We should note at this stage that the value of δ (which must be considered relative to the natural linewidth of the γ-ray) depends directly upon the difference in radii of the ground and excited nuclear states. We shall return to this point in section 2(d).

(b) Quadrupole splitting

This arises with any nucleus with spin quantum numbers greater than I = 1/2 and which, therefore, possess a non-spherical charge distribution. The magnitude of this non-sphericity is given by the nuclear quadrupole moment Q, the sign of Q depending on the shape of the deformation, i.e. whether the charge distribution is oblate or prolate about the nuclear spin axis. If the symmetry of the surrounding electronic charge is not spherical (for the atomic electrons) or cubic (for the surrounding ions) the degeneracy of the nuclear spin will be lifted and two or more absorption lines will be seen, the number depending on the values of I in the ground and excited states.

The magnitude of the field gradient may be expressed as

$$q = (1 - R)q_{val} + (1 - \gamma_{\infty})q_{lat}$$

where $q_{lattice}$ is the electric field gradient contributed by the secondary ligand charges and $q_{valence}$ results from the valence electrons (filled core levels presenting a spherically symmetric field). γ_{∞} and R are the so-called Sternheimer antishielding factors.

Defining an asymmetry parameter as

$$\eta = \frac{\left(V_{xx} - V_{yy}\right)}{V_{zz}}$$

where V_{xx}, V_{yy} and V_{zz} are the electric field gradients along the respective directions $\left(|V_{xx}| > |V_{yy}| > |V_{zz}|\right)$.

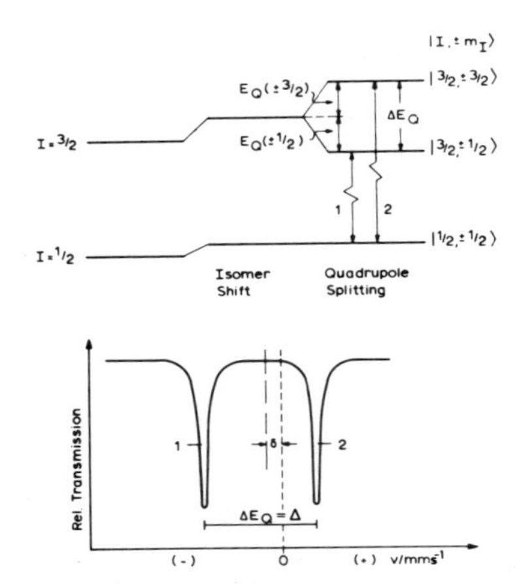

Fig. 3

The quadrupole splitting present when I = 3/2 in the excited state and I = 1/2 in the ground state (e.g. in ^{57}Fe and ^{119}Sn Mössbauer). The associated Mössbauer spectrum gives the value of Δ and δ.
(Reproduced by permission of Prof. P. Gutlich and Springer-Verlag - see Bibliography.)

The magnitude of the produced quadrupole moment, assuming axial symmetry of the field gradient, i.e. $V_{xx} = V_{yy}$ and $\eta = 0$ will be given by

$$Q.S. \quad = \quad 1/2\ e^2 qQ(1 + \frac{1}{3}\eta^2)^{1/2}$$

for an I = 1/2 ground state and an I = 3/2 excited state (e.g. for ^{57}Fe and ^{119}Sn, see Fig. 3). We note, therefore, that the magnitude of the quadrupole splitting is a function of the two parameters eq and eQ; eQ being a nuclear constant for a given isotope and eq a parameter depending upon the intra- and inter-electronic configuration around the isotope. For more complicated spin states (e.g. $I_{gd} = 3/2$, $I_{ex} = 5/2$) the spectra become increasingly complex.

(c) Magnetic hyperfine interaction

Any interaction between the nuclear dipole moment with a magnetic field will lift the degeneracy of the magnetic sub-levels and give rise to 2I + 1 levels (I being the nuclear spin quantum number). For iron, where I = 1/2 and 3/2, this results, after application of the selection rule that $m_I = \pm 1/2$, in a characteristic 6-line pattern (Fig. 4). Further interactions

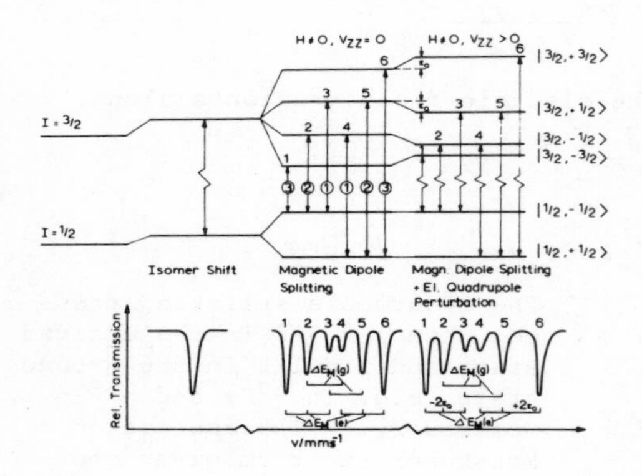

Fig. 4

Schematic illustration of the magnetic dipole splitting seen in ^{57}Fe, (a) with and (b) without electric quadrupole perturbation. (Reproduced by permission of Prof. P. Gutlich and Springer-Verlag - see Bibliography.)

giving rise to isomer shifts and quadrupole splittings will also modify the spectrum. The measurement of the magnetic hyperfine splittings permits a determination of extremely useful information about the magnetic properties of the compounds investigated, thereby affording information about directions of easy magnetization and size of magnetic interactions. Particle size determination may also be made since the temperature at which magnetic ordering occurs depends critically on the size of the particle.

(d) Additional parameters

Whether isomer shift and quadrupole splitting data may be obtained depends upon the natural width of the resonant line (determined by the mean lifetime of the excited state) and also by the value of the nuclear contraction constant ($\delta R/R$) which directly influences the magnitude of the isomer shift. The narrower the resonance line the smaller need ($\delta R/R$) be for significant information. ^{57}Fe and ^{119}Sn have comparatively small values for half-width and large values for $\delta R/R$, whereas for ^{121}Sb the half-width (2.10 mms^{-1}) is quite large although the value of $\delta R/R$ of $- 8.5 \times 10^{-4}$ means that data with respect to oxidation state determination may be obtained. Full details on Mössbauer isotopes are given by Greenwood and Gibb and in the Mössbauer Effect Data Index (see Bibliography and the Periodic Table, Fig. 5).

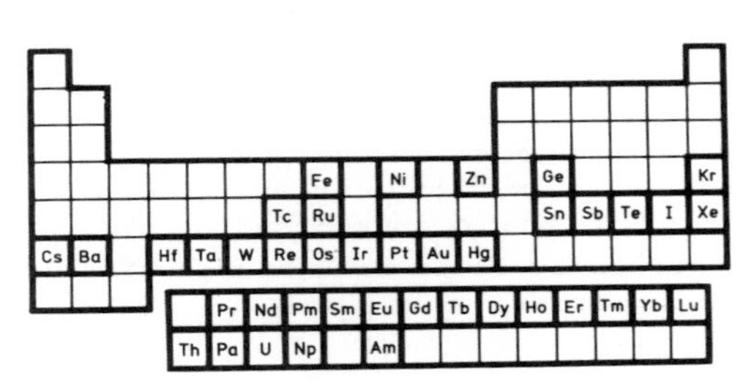

Fig. 5

Periodic Table of Mössbauer active elements. (See Mössbauer Effect Data Index in Bibliography.)

Significant recoil-free-fractions occur only for gamma energies less than 150 keV, and, whilst some isotopes have quite low energy gamma rays, as the energy of the gamma ray increases so also does the necessity of collecting data with the sample and/or source at low temperatures. Thus, for ^{121}Sb Mössbauer it is customary to hold both the source and sample at temperatures close to liquid nitrogen temperatures (or lower). This limitation obviously means that high temperature Mössbauer work (see next section) is possible only for a limited number of elements, (e.g. ^{57}Fe, ^{119}Sn, ^{151}Eu).

The combination of natural abundance of the isotope (sometimes leading to the necessity of working with isotope-enriched samples) coupled with the half-life of the source also limits to some extent the feasibility of a study. ^{181}W, for example, the source for ^{181}Ta Mössbauer, has a half-life of 140 days and a natural abundance of 99.99%. For nickel Mössbauer (^{61}Ni isotope) the source ^{61}Co has a half-life of 99 minutes and a natural abundance of only 1.25%.

Gutlich, Link and Trautman (see Bibliography) have recently summarised the literature of Mössbauer spectroscopy and transition metal chemistry.

(e) In-situ studies

The very nature of catalysts, and the sometimes complex combination of temperature and pressure under which they operate, demands techniques which ideally permit the catalyst to be studied under such extreme conditions. Because of the nature of the Mössbauer process and the relatively high energy of the gamma ray, it is possible to design cells which allow samples to be investigated at high pressures (up to several hundred kilobar), high temperatures (well above 1000 K) and under various gaseous atmospheres (or vacuum), and in most cases the limitations are imposed by the suitability of the cell, e.g. the thickness of the beryllium windows or the resistance of the cell to corrosive gases.

Delgass et al.[2] have designed several cells, with one of which they collected europium Mössbauer data at temperatures up to 800 K in mixtures of reacting gases up to pressures of 1.4 atmosphere. In a cell designed by Clausen et al.[3] Mössbauer spectra were obtained in the temperature range 78 - 725 K with the sample either in ultra-high vacuum or in the presence of various gases, Fig. 6, and as an illustration of the possible use of such a cell the reaction of α-Fe_2O_3 supported on silica gel was investigated. The necessity of in-situ studies resulted from the susceptibility to oxidation to Fe^{3+} of the small metal particles of iron produced on reduction of the supported α-Fe_2O_3.

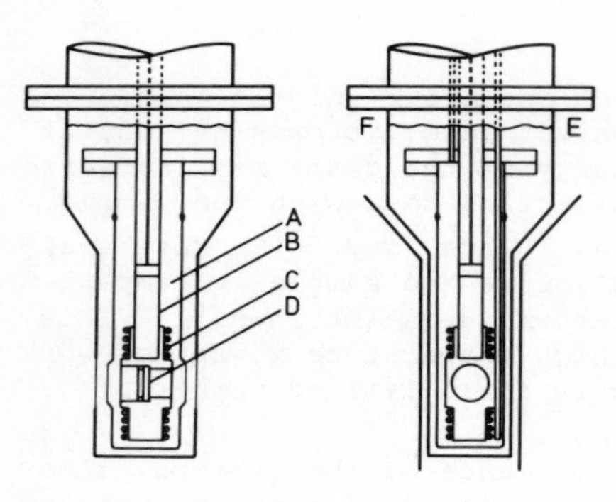

Fig. 6

Diagram of an in-situ Mössbauer cell. A, in-situ chamber. B, Liquid-N_2 tube. C, Heating coil. D, Mössbauer sample wafer. E, Gas inlet tube. F, Gas outlet tube. The design permits studies in the presence of applied magnetic fields. (From reference 3. Reproduced by permission of The Institute of Physics.)

For the 6.2 keV Mössbauer transition of tantalum (^{181}Ta) Saloman et al.[4] have observed resonance at temperatures up to 2400 K using a W-181 source at room temperature. In an early paper Preston et al.[5] studied the magnetic hyperfine interaction within iron foils up to 1300 K.

Drickamer and Frank[6] have successfully used Mössbauer spectroscopy to investigate the electronic changes resulting from the application of high pressures to magnetite and have obtained

Mössbauer spectra whilst the sample was under several hundred kilobar pressure.

Clearly there is considerable scope for designing Mössbauer cells working at high temperatures and pressures to investigate a variety of catalyst systems.

3. EXAMPLES OF CATALYST SYSTEMS INVESTIGATED BY MÖSSBAUER SPECTROSCOPY.

(a) Oxide Catalysts

The selective heterogeneous oxidation and ammoxidation of olefins over mixed oxide catalysts has been studied by Mössbauer spectroscopy. For oxidation catalytically active oxides containing Mossbauer isotopes include Fe_2O_3, Fe_3O_4, $Fe_4Bi_2O_9$, $FeAsO_4$ (the latter two being selective), with the overall activity of these oxides, i.e. the nature of the active site, believed to be the combination of two adsorption sites. For the formation of butadiene from but-1-ene the A-site is where strong adsorption of butene occurs, and the B-site where weak dissociative adsorption occurs (Gates et al.[7]). Adsorption sites suggest that a strong correlation between heat of adsorption and catalytic activity as well as selectivity exists. Thus Fe_3O_4 and $Fe_4Bi_2O_9$ strongly and irreversibly adsorb butadiene and but-1-ene, whereas $FeSbO_4$ and $FeAsO_4$ both weakly and strongly adsorb butadiene only weakly but-1-ene. Matsuura[8] has shown, using ^{57}Fe Mössbauer, that all catalysts active in oxidation and showing reversible adsorption contain Fe^{3+} in a bulk octahedral environment, and furthermore, a strong correlation is found between heat of adsorption with isomer shift and quadrupole splitting.

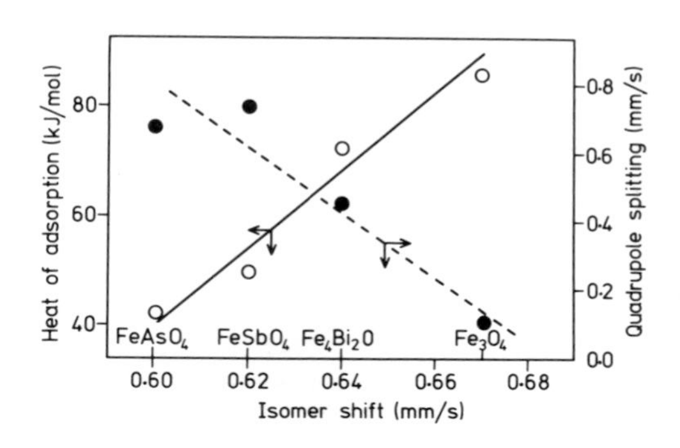

Fig. 7

The relationship between heat of adsorption of but-1-ene and the measured isomer shift and quadrupole splitting. Isomer shifts are given relative to the reference material $Na_2[Fe(CN)_5NO].2H_2O$. (From reference 8. Reproduced by permission of The Chemical Society.)

Thus in Fig. 7 we see increasing distortion of the oxygen octahedra as we move along the series Fe_3O_4, $Fe_4Bi_2O_9$, $FeSbO_4$ and

$FeAsO_4$, and it may be concluded, therefore, that B sites are connected with bulk octahedral Fe^{3+}. Comparison of these observations with those made on bismuth molybdate (assuming equivalent roles of iron and bismuth) is good[9].

Jeitschko et al.[10] have investigated scheelite mixed oxide catalyst $Bi_3(FeO_4)(MoO_4)_2$ using Mössbauer spectroscopy and X-ray structural analysis in both ordered and disordered phases. In the ordered phase the Mössbauer spectrum shows only 1 type of iron species present - the high spin trivalent species expected from the composition and structure (Fig. 8). However, the

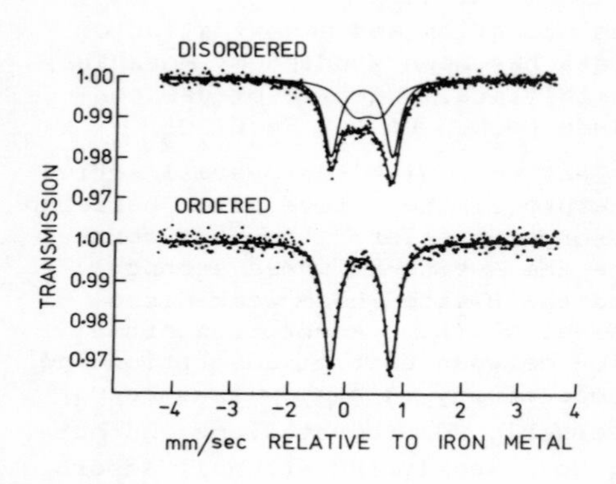

Fig. 8

^{57}Fe Mössbauer spectra of the ordered and disordered phase of $Bi_3(FeO_4)(MoO_4)_2$. (From reference 10. Reproduced with permission.)

results once again clearly demonstrate the relationship between the observed Mössbauer parameters and oxidation state and bond length. Thus the isomer shift for the iron species (δ = 0.28 mms^{-1} relative to iron) is characteristic of high spin Fe^{3+} with the quadrupole splitting of 1.04 mms^{-1} reflecting a significant deviation from cubic symmetry. In Fig. 9 the variation in isomer shift against average Fe-O distance is plotted for several

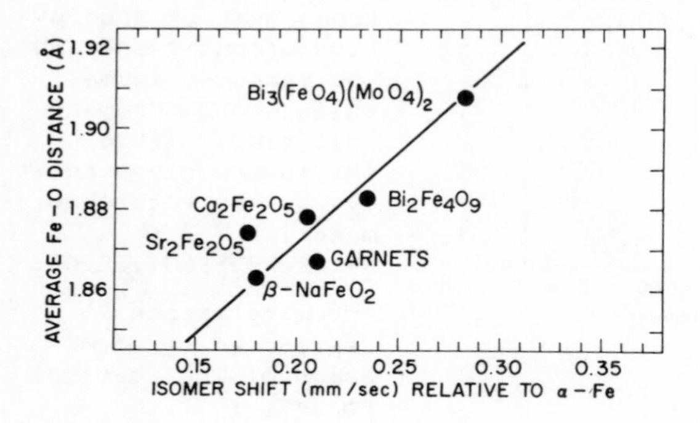

Fig. 9

The variation of isomer shift with average Fe-O distance for several high-spin Fe^{3+} oxides. (From Ref. 10. Reproduced with permission.)

oxides. The value of δ and Fe-O distance, 1.90 Å in this particular scheelite, clearly falls within an established pattern.

In the disordered material the several kinds of Mössbauer parameters expected (the Figure shows an arbitrary fit to two pairs) are clearly evidenced, indicating a random distribution of Fe and Mo on the tetrahedral cation sites. The small quadrupole splittings suggest that the iron in the disordered phase has a higher coordination than in the ordered phase.

In the catalyst system, $SnSbO_4$, Berry and Maddock [11] have followed the variation of isomer shift and quadrupole splitting with increasing antimony concentration by calcining mixtures of SbO_2 and SnO_2. Fig. 10 reproduces their results using ^{121}Sb Mössbauer. Interestingly, changes in isomer shift and quadrupole splitting occur up to concentrations of 10% Sb, indicating the formation of a biphasic material at this stage. The selectivity of a catalyst for the oxidation of propylene to acrolein is also enhanced with Sb concentrations up to 10% excess.

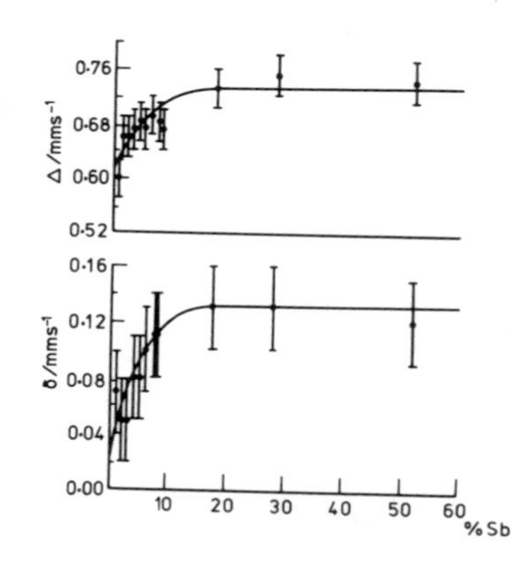

Fig. 10

Increasing distortion of the environment around the tin nucleus (rejected in the value of Δ) is seen as the antimony concentration increases - isomer shifts relative to an iron foil calibration. (From reference 11. Reproduced by permission of the authors and Elsevier Sequoia.)

A further example is given by Evans[12] using ^{121}Sb Mössbauer to investigate the system U-Sb-O : (Evans, 1976) : the sensitivity of the ^{121}Sb isomer-shift is utilised to investigate the bond strength and topologies and the relative activities of $USbO_5$, USb_3O_{10} and a commercial catalyst, for the synthesis of acrylonitrile. He suggests that within the series $KSbF_6$, $NaSbF_6$, Sb_2O_5 and $SbCl_5$ (each with an Sb ion valence of +5) the change in isomer shift from -3.7 mms^{-1} to +3.7 mms^{-1} reflects good agreement with the increasing covalency and decreasing electronegativity difference in the series. Fig. 11 illustrates ^{121}Sb Mössbauer spectra of pure U-Sb-O phases and of a supported active catalyst. Clearly from this figure we see that only Sb^{5+} is present in both USb_3O_{10} and $USbO_5$. In the commercial catalyst Sb^{3+} is indicated, and comparison with spectra reported previously is confirmed as Sb_2O_4. Area analysis showed the catalyst to correspond to 42% USb_3O_{10} and 58% Sb_2O_4.

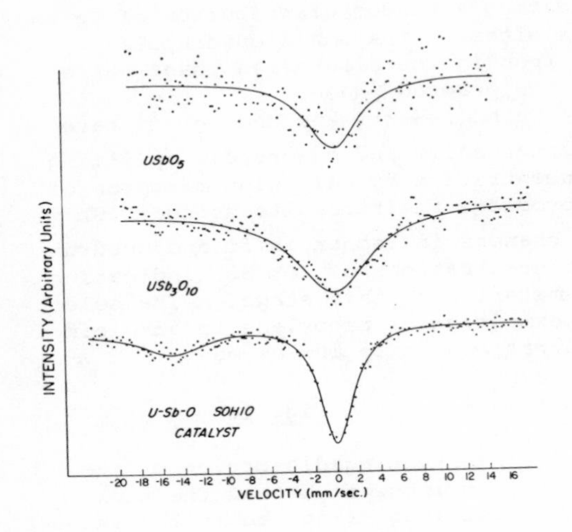

Fig. 11

^{121}Sb Mössbauer spectra of pure U-Sb-O phases and of a commercial supported catalyst preparation. Sb^{5+} and Sb^{3+} are present in the commercial catalyst. ($BaSnO_3$ source) (From Ref. 12. Reproduced by permission of Academic Press.)

Furthermore, Evans concludes that the similarity of isomer shifts for Sb in USb_3O_{10} and $USbO_5$ suggests that differences in their catalytic activity cannot be due to differences in covalence of the Sb-O bonds. The differences in linewidth between USb_3O_{10} and $USbO_5$, however, may explain the better selectivity of USb_3O_{10} - the larger linewidth of this material agreeing with the proposal that there are two Sb sites in USb_3O_{10}, the increased width being consistent with the superposition of two absorptions from slightly equivalent sites.

(b) Bimetallic catalysts

The interest in bimetallic catalysts (e.g. FePt, FeIr, PdAu) stems from both the catalytic properties they display and also from the possibility of using such clusters to investigate the electronic factors governing the activity of transition metals in a variety of catalyst systems[13]. In order to characterise fully such clusters it is necessary to know (a) the degree of interaction of the two metals both with themselves and also with the support, and (b) the factors which influence the degree of dispersion and uniformity of the catalyst and the catalyst activity. When one (or both) of the metals is a Mössbauer active element, information relative to both these questions may be obtained. Fig. 12 (taken from the work of Garten[14]) shows some of these results for PdFe on η-Al_2O_3. In (a) the physical mixture of Fe and Pd (made by grinding 0.2% Fe/η-Al_2O_3 with 9.5% Pd/η-Al_2O_3 and reducing at 673 K) shows essentially the same ferrous species

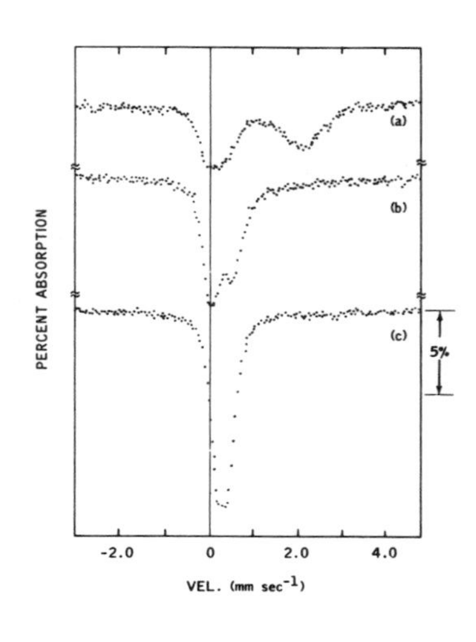

Fig. 12

^{57}Fe Mössbauer spectra of 0.2% Fe, 9.5% Pd/η-Al_2O_3; (a) physical mixture, reduced 1 hr. at 673°K; (b) co-impregnated sample, reduced 1 hr. at 673°K; and (c) co-impregnated sample, reduced at 973°K.
(From Ref. 14.
Reproduced with permission of New England Nuclear Corporation.)

which is obtained when preparing supported iron catalysts. In co-impregnated samples with the same composition, however, complete reduction at 673 K to iron is obtained with the isomer shift (δ = 0.28 mms^{-1}, Δ = 0.47 mms^{-1}) corresponding closely to the value obtained for dilute concentrations of iron in bulk PdFe alloys (b). Reduction of the co-impregnated sample at 973 K results in the collapse of this spectrum to give a doublet with an isomer shift again similar to that of the bulk alloys (c). The decrease in the quadrupole moment (Δ = 0.24 mms^{-1}) indicates agglomeration of the PdFe clusters.

Exposure to oxygen at 25°C of this bimetallic cluster results in oxidation of the iron to Fe^{3+}, whilst exposure to hydrogen, again at 25°C, clearly reduces the ferric species to a mixture of ferrous and metallic iron. This is shown in Fig. 13 and provides direct support for the formation of PdFe bimetallic clusters. In addition, Garten has indicated how, in other bimetallics (e.g. where both metals are from Group VIII), small amounts of iron may be added to act as a probe for cluster formation. There appears to be considerable scope for Mössbauer spectroscopy in this area of catalysis.

(c) Promoted iron catalysts

In order to increase the activity and the surface area of the reduced catalyst resulting from the avoidance of sintering of the small iron particles, it is customary to add various promoter oxides such as Al_2O_3, K_2O and Na_2O to iron-based synthetic ammonia catalysts, the addition of small amounts of oxides (ca 1%)

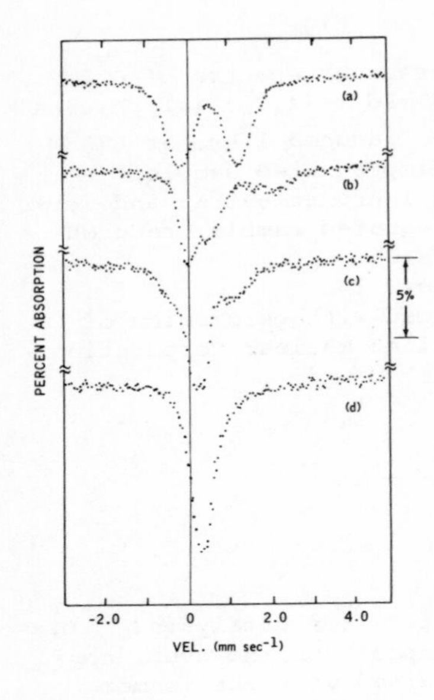

Fig. 13

0.2% Fe, 9.5% Pd/η-Al_2O_3.
(a) Reduction at 673°K and exposure to oxygen at 298°K. (b) Following treatment (a) evacuation and exposure to hydrogen at 298°K.
(c) Reduction at 973°K and exposure to oxygen at 298°K. (d) Following treatment (c) exposure to hydrogen at 298°K.
(From reference 14. Reproduced with permission of New England Nuclear Corporation.)

being believed to result in up to 70% of the iron surface being covered with promoter. It has, however, been suggested, as a result of X-ray (powder) line broadening on 3% Al_2O_3 promoted iron catalysts, that the aluminium is present within the bulk of the α-Fe as $FeAl_2O_4$. Mössbauer spectroscopy provides a viable means of investigating the exact role that such promoters play. Topsoe et al.[15] have shown how the inclusion of randomly distributed $FeAl_2O_4$ would be expected to change the quadrupole interaction of neighbouring iron atoms compared with these within the pure lattice. The magnetic field present at the iron nucleus would also be modified by the presence of $FeAl_2O_4$ molecules. Expressing the change in magnetic field relative to that of pure α-iron, for any given site, by H we may write :

$$H = \sum_{n=1}^{n=5} M_n \Delta H_n$$

where M_n is the number of solute atoms in the nth neighbour shell and ΔH_n is the internal field shift due to solute atoms in the nth shell. n is set at 5, since in dilute FeAl and FeSi alloys changes in magnetic field are detected as far away as the 5th neighbour shell. A variation in hyperfine splitting is, however, not observed, and it is concluded that the amount of aluminium in the catalyst present as randomly distributed $FeAl_2O_4$

domains must be less than 2%. Furthermore, the Fe^{2+} signal expected from $FeAl_2O_4$ would, if from an inverse spinel structure, appear in between these absorptions due to iron. No such peaks were observed.

An enhanced area of one of the iron absorptions due to the singlet absorption of the cubic iron symmetry within a normal spinel was also not observed. From their Mössbauer spectra Topsoe et al. concluded that less than 15% of the aluminium is present in a form which would give rise to a ferrous absorption characteristic of bulk $FeAl_2O_4$. Analysis of the curvature of the background component of the spectra removes the possibility of sub 30A diameter particles. The results suggest that for the completely reduced catalyst the aluminium is present as Al_2O_3 inclusions.

(d) Zeolites and aluminosilicates

Although the three-dimensional arrangement of the aluminosilicate framework of zeolites has been determined using single crystal X-ray diffraction, the positions of the exchangeable cations (and their movement on dehydration of the zeolite) required to compensate the negative charge on each aluminium tetrahedra are, for many systems, not well-established. Certain cations (e.g. cobalt, copper and nickel) may readily be studied using e.p.r., but for some the spectra resulting from the cations in sites of low symmetry are so complicated that determination from e.p.r. spectra is, if not impossible, extremely difficult. Iron constitutes an example of where e.p.r. is not particularly advantageous, and consequently iron Mössbauer measurements on such systems have proved worthwhile.

Delgass et al.[16], using Mössbauer spectroscopy to study ferrous exchanged Y-zeolites (see Fig. 14), and coupled with other measurements (e.g. infra-red and X-ray diffraction), obtained information about the location of these cations in both the hydrated and dehydrated forms. For the dehydrated zeolite two doublets, with parameters $\delta_1 = 1.37$ mms^{-1}, $\Delta_1 = 2.44$ mms^{-1}, $\delta_2 = 1.04$ mms^{-1}, $\Delta_2 = 0.62$ mms^{-1} (with respect to a ^{57}Co-Cr source) in the ratio of approximately 6:1 were observed and were identified with ferrous ions sitting at sites I and II' respectively. They also obtained evidence for Fe^{2+} having tetrahedral coordination with water fragments within the β-cages. Clear evidence for the interaction of adsorbed gases with one of the ferrous ions was also obtained, but the site accessible to the gases (site II' within the large cavity) giving rise to too small a Mössbauer absorption (less than 15% of total) to permit detailed analysis. For zeolite-A, however, the situation is improved, since the hexagonal prism between the sodalite cages (site I) does not occur, whereas the accessible sodalite window site (site II') is available. Dickson and Rees[17], therefore, studied zeolite-A after partial ferrous exchange (composition $Fe_{2.66}Na_{6.68}Al_{12}Si_{12}O_{48}0.9NaAlO_2$). Fig. 15 illustrates the Mössbauer spectra of a ferrous A-zeolite at various stages of hydration.

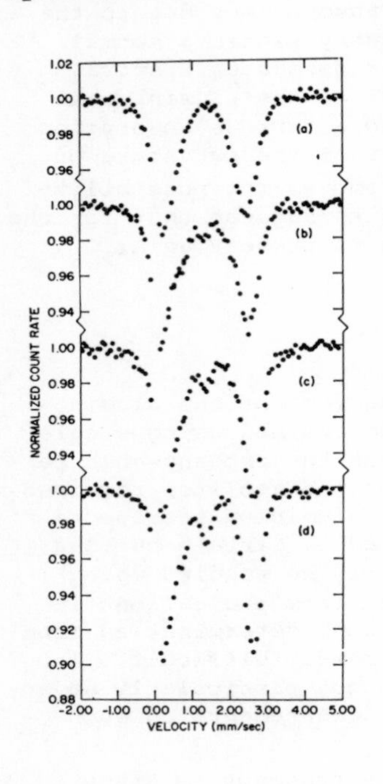

Fig. 14

Room temperature spectra showing the dehydration of a ferrous Y-zeolite. Samples evacuated at 298°, 381°, 509° and 673°K. (Isomer shifts relative to ^{57}Co-Cr source.) (See reference 16. Reproduced by permission of The American Chemical Society.)

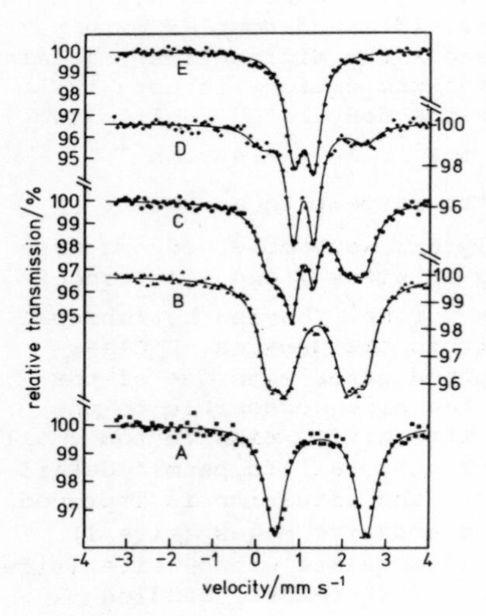

Fig. 15

Room temperature spectra of A, hydrated in ferrous A-zeolite; B, C, and D, the zeolite at various stages of dehydration; and E, fully dehydrated. Dehydration conditions are: B, heat *in vacuo* 30 mins at 333°K and 1 hr. at 363°K; C, further 2 hrs at 363°K; D, heat *in vacuo* for 14 hrs at 280°K; E, 24 hrs *in vacuo* at 633°K. (Isomer shifts relative to $Na_2[Fe(CN)_5NO].2H_2O$.) (see reference 16. Reproduced by permission of The Chemical Society.)

In the hydrated form one doublet is seen ($\delta = 1.50$ mms^{-1}, $\Delta = 2.10$ mms^{-1}, parameters with respect to sodium nitroprusside) which upon dehydration changes to give a doublet with parameters $\delta = 1.08$ mms^{-1}, $\Delta = 0.46$ mms^{-1}. Furthermore, adsorption of gases (e.g. methanol, acetonitrile and ethylene) which are restricted to the large cavities of the zeolite suggested that the ferrous ions were located on the sodalite window site, the value of the quadrupole splitting indicating three-fold coordination to lattice oxygens. Area analysis of the spectra (Fig. 16) enabled them to show that two water molecules are added for each Fe^{2+} ion. Comparison of the above spectra with those

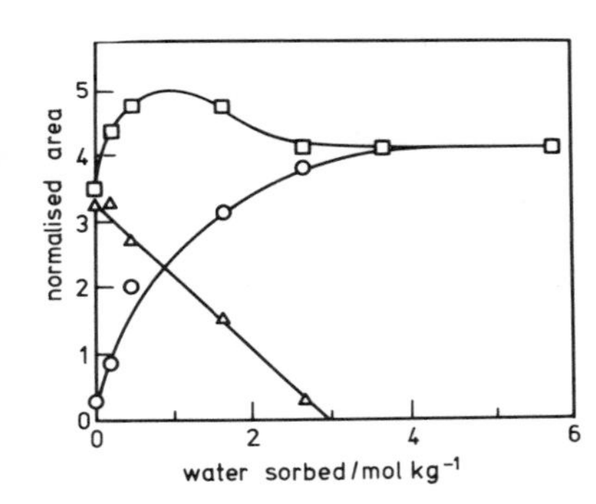

Fig. 16

Area under inner doublet Δ, outer doublet O, and the total area of absorption from spectra of a ferrous A-zeolite at various stages of dehydration. (see reference 17. Reproduced by permission of The Chemical Society.)

obtained from ^{57}Co A-zeolite emission work - an X-ray study of the cobalt A-zeolite having shown that all the cobalt ions are situated in the sodalite window-site - enabled Dickson and Rees to determine the coordination of the ferrous ion as $(Z\text{-}O)_3\ Fe^{2+}\ (H_2O)_2$ where (Z-O) indicates a zeolite oxygen anion.
Closer examination of Fig. 15 shows that two types of iron (related by a temperature-dependent process) are present. Further addition of water is coordinated to the sodium ions and to the lattice, until, after 10 molecules per large cavity are adsorbed, the spectra of the two types of ferrous ion begin to merge, forming $(Z\text{-}O)_3\ Fe^{2+}\ (H_2O)_3$. The use of total area measurements (which relates to strength of bonding to lattice) indicates that the ferrous ions do not become completely hydrated but remain bound to the zeolite framework.

In a further extension of this study, Dickson and Rees[18] studied the adsorption of ethanol in dehydrated zeolite-A. The conclusions are that two molecules of ethanol are now coordinated to each ferrous ion which again have 5-fold coordination. The similarity of the spectra from water and ethanol adsorbed samples suggests that the ferrous ions are again accommodated on the large cavity side rather than on each side of the window. Unlike the case of water, however, there appears to be insufficient room within the cavity for 6-fold coordination to occur. At saturation 6.5 molecules/unit cell are adsorbed, compared with the 5.2 molecules per unit cell required, on average, to complex completely all the ions.

Mössbauer spectroscopy readily yields for a range of other aluminosilicates, such as natural and synthetic montmorillonite, isomer shifts and quadrupole splitting values. This in turn gives a straightforward method of determining the oxidation state and coordination number of adsorbed and intercalated species. For example, in the study of the interaction between montmorillonite and organic molecules (such as benzidene) valence changes associated with intercalation may readily be seen. In the reversible colouration between montmorillonite and benzidine, Tennakoon et al.[19] observed an original ferrous/ferric ratio of 0.16:1. When the montmorillonite is intercalated in the presence of water, however, the amount of ferric iron present, in the now blue material, is reduced. Upon dehydration the material becomes yellow and the ferrous/ferric ratio returns to its hydrogen peroxide the blue colouration is enhanced, and it is concluded that the ferric ion acts as an electron acceptor site following the reaction

$$\text{benzidine} + Fe^{3+} \rightleftarrows \text{benzidine}^{+} + Fe^{2+}$$

(e) <u>Supported iron catalysts</u>

Mössbauer spectroscopy has been used by Boudart et al.[20] to investigate the conditions required for the preparation of small, well-dispersed, particles of iron on various supports. Their results indicate that it is possible, with MgO supports, to prepare iron particles (within the range 1.5 → 3.0 nm) which show little tendency to sinter in either hydrogen or under conditions of ammonia synthesis. Fig. 17 illustrates the spectra obtained from samples containing from 1-40% (by weight) of iron on MgO. The spectra may be analysed into three components: a magnetically split six-line pattern and a central singlet due to superparamagnetic iron, and a central doublet resulting from a Fe^{2+} species.

Furthermore, the parameters of the Fe^{2+} component are those of Fe^{2+} in MgO, the magnitude of the quadrupole splitting suggesting that the local environment of the Fe^{2+} ion is at least 20 times richer in iron than is likely from a random distribution of Fe^{2+} in MgO. This supports electron microscopic work which suggests that the iron particles are located in regions of MgO enriched in iron.

In a further study[21] of such catalyst preparations it is found that the ratio of magnetically split to superparamagnetic absorption (a central singlet) increases when hydrogen is passed over those samples which contain iron particles of diameter <u>ca</u>

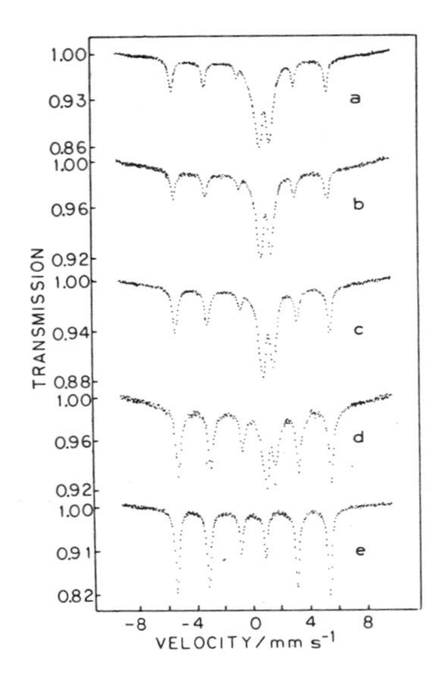

Fig. 17

Mössbauer spectra of reduced Fe/MgO samples. (a) 1% Fe/MgO, (b) 5% Fe/MgO, (c) 16% Fe/MgO, (d) 40% Fe/MgO and (e) iron calibration. (From reference 20. Reproduced by permission of Academic Press.)

1.5 nm. An explanation for this apparent decrease in the magnetisation anisotropy barrier is a change in surface structure of the particle and consequently change in crystal shape (for particles of this size 50% of the atoms are at the surface). No changes in the ratio of magnetic/superparamagnetic absorption were seen for larger particles of iron (*ca* 8 nm diameter) where only 10% of the atoms are at the surface.

(f) Conversion Electron Mössbauer Spectroscopy

Unlike transmission Mössbauer spectroscopy, which requires high surface-area solids, conversion electron Mössbauer spectroscopy (CEMS) is an approach suitable where the surface to bulk ratio is low (for example, the study of foils). Following excitation of the Mössbauer nucleus (^{57}Fe and ^{119}Sn are particularly suited to CEMS work) de-excitation occurs with a high yield of emitted conversion electrons. These electrons (for K shell emission) have initially kinetic energies of the order of *ca* 7 keV and therefore act as surface sensitive probes for the top 2000 Å of the sample. Experimentally the electrons are detected using a He/CH_4 proportional counter or in a high vacuum system using an electron multiplier. The technique is particularly suited to the oxidation and corrosion of foils. (For a recent review, see Tricker[22].)

Fig. 18(a) shows the CEM spectrum of an oxidised iron foil (transmission Mössbauer showed no indication of oxide) clearly indicating the extent of surface oxidation. More recently energy analysed CEMS (where the relationship between the final energy of the emitted electron after modification, following

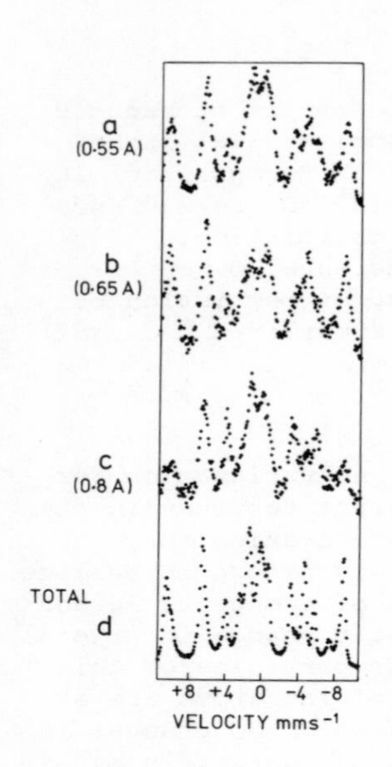

Fig. 18

Depth analysed ^{57}Fe CEMS spectra of an iron foil oxidised at $723^{o}K$ in air for 20 minutes. (From reference 23. Reproduced by permission of North-Holland Publishing Company.)

inelastic collisions, as a function of escape depth is exploited) has shown several possible applications to surface studies[23]. Fig. 18(b) & (c) compares the spectra obtained using the K and L conversion electrons, respectively. (The L conversion electrons having much greater bulk sensitivity because of their higher initial kinetic energy.)

4. CONCLUDING REMARKS

The principles of Mössbauer spectroscopy with particular emphasis for catalysis work have been described, and it has been shown that, allied with other techniques, this approach provides a useful means of directly probing the nature of certain catalytic species. The additional bonus of being able to obtain data for certain elements at high temperature and pressure will probably be exploited more in the future, as will the approach of using the Mössbauer isotope as a probe to study other non-Mössbauer active elements. The fact that the technique is not of itself surface sensitive does not, from the very nature of most catalysts, present a great problem, since high surface area/bulk ratios frequently prevail in catalyst systems. In those cases where the surface/bulk ratio is low, conversion electron Mössbauer spectroscopy proves to be an advantage.

REFERENCES

1. R.L. Mössbauer, Z. Physik., 151, 124 (1958).
2. W.N. Delgass, L.-Y. Chen and G. Vogel, Rev. Sci. Inst., 47, 136 (1976).
3. B.S. Clausen, S. Morup, P. Nielsen, N. Thrane and H. Topsoe, J. Phys. E., 12, 439 (1979).
4. D. Salomon, W. Wallner and P.J. West, Mössbauer Effect Methodology, 10, 291 (1976).
5. R.S. Preston, S.S. Hanna and J. Heberle, Phys. Rev., 128, 2207 (1962).
6. H.G. Drickamer and C.W. Frank, Electronic Transitions and the High Pressure Chemistry and Physics of Solids, Chapman and Hall, London (1973).
7. B.C. Gates, J.R. Katzer and G.C.A. Schuit, Chemistry of Catalytic Processes, McGraw-Hill, London (1979).
8. I. Matsuura, Proc. 6th Int. Congress on Catalysis, 2, p.819, Chemical Society, London (1977).
9. J. Haber, Proc. 6th Int. Congress on Catalysis, 2, p.825, Chemical Society, London (1977).
10. W. Jeitscuko, A.W. Sleight, W.R. McClellan and J.F. Weiher, Acta Cryst., B32, 1163 (1976).
11. F.J. Berry and A.G. Maddock, Inorg. Chem. Acta, 31, 181 (1978).
12. B.J. Evans, J. Catalysis, 41, 271 (1976).
13. J.H. Sinfelt, Accts. Chem. Res., 10, 15 (1977).
14. R.L. Garten, Mössbauer Effect Methodology, 10, 69 (1976).
15. H. Topsoe, J.A. Dumesic and M. Boudart, J. Catalysis, 28, 477 (1973).
16. W.N. Delgass, R.L. Garten and M. Boudart, J. Phys. Chem., 73, 2970 (1969).
17. B.L. Dickson and L.V. Rees, J.C.S. Faraday I, 70, 2038, 2051 (1974a).
18. B.L. Dickson and L.V. Rees, J.C.S. Faraday I, 70, 2060, (1974b).
19. D.T.B. Tennakoon, J.M. Thomas and M.J. Tricker, J.C.S. Dalton, 2211 (1974).
20. M. Boudart, A. Delbouille, J.A. Dumesic, S. Khammouma and H. Topsoe, J. Catalysis, 37, 486 (1975).
21. M. Boudart, J.A. Dumesic and H. Topsoe, Proc. Natl. Acad. Sci., 74, 806 (1977).
22. M.J. Tricker, Surface and Defect Props. Solids, Chem.Soc. Spec. Per. Report, I, 106 (1977).
23. W. Jones, J.M. Thomas, R.K. Thorpe and M.J. Tricker, Appl. Surface Sci., 1, 388 (1978).

BIBLIOGRAPHY

Bancroft, G.M. (1973). Mössbauer Spectroscopy, McGraw-Hill, London.

Delgass, W.N. (1976). 'Mössbauer Spectroscopy in Heterogeneous Catalysis,' in Mössbauer Effect Methodology, 10, 1.

Dumesic, J.A. and Topsoe, H. (1977). 'Mössbauer Spectroscopy Applications to Heterogeneous Catalysis' in Adv. in Catalysis, 26, 121.

Gager, H.M. and Hobson, M.C. (1975). 'Mössbauer Spectroscopy' in Catalysis Rev.-Sci. Eng., 11, 117.

Greenwood, N.N. and Gibb, T.C. (1971). Mössbauer Spectroscopy, Chapman and Hall, London.

Gutlich, P. (1975). 'Topics in Applied Physics 5, Mössbauer Spectroscopy', Editor U. Gonser, Springer-Verlag.

Stevens, J.G. and Stevens, V.E. (1965-1975). Mössbauer Effect Data Index, Adam Hilger, London.

Stevens, J.G. and Stevens, V.E. (1976-). Mössbauer Effect Data Index, Plenum, New York.

Wertheim, G.K. (1964). Mössbauer Effect: Principles and Applications, Academic Press, New York.

CATALYST CHARACTERISATION BY TEMPERATURE AND VOLTAGE PROGRAMMING

By

Brian D. McNicol

1. INTRODUCTION

In the field of heterogeneous catalysis, despite the many advances that have been made in characterisation of catalyst surfaces, the design of a successful practical catalyst remains an art rather than a science. The practical catalyst generally evolves through the following of established procedures such as precipitation, impregnation, drying, calcination, reduction etc. In many cases the reasons for carrying out such procedures are unclear but the steps are necessary if success is to be achieved. The reason that so little progress has been made can, in part, be put down to the fact that the materials that require study to answer the unresolved questions are not readily amenable to investigation by many of the techniques available for the study of solids and surfaces. The catalysts usually consist of a high surface area support, e.g. silica, alumina, carbon, which in many cases is amorphous and contains a catalyst, which may be a metal, present in very small concentration and which is very finely dispersed. In addition there may be other additives present to promote the catalyst activity or to function as co-catalysts. Some additives may be present to modify the metal function of the catalyst, e.g. as a second metal; other additives serve the function of modifying the properties of the support. An example highlighting this situation is the case of catalytic reforming where highly complex multimetallic supported catalysts are used. For instance, one of the most successful reforming catalysts is

$$Pt/Re/S/A\ell_2O_3/C\ell$$

where the platinum metal function is modified by alloying with rhenium, the metallic activity is tempered by sulphiding, and the acidity of the $A\ell_2O_3$ support is modified by incorporation of chloride. When catalyst systems are so complex it is not surprising that there is little understanding of the processes occurring during their preparation or the origin of their special catalytic properties.

In the field of electrocatalysis, where the problems encountered are very similar to those of heterogeneous catalysis, as progress in catalyst development has taken place a similar lack of understanding prevails about how supported catalysts function. The support most commonly used in this field is carbon, owing to its electrical properties; but again the problems of the complex nature of catalyst preparation arise and success relies on adherence to established procedures.

It is appropriate to discuss together the success that has recently been achieved in the understanding of catalyst preparation and function using the relatively new technique of temperature-programmed reduction (TPR)[1] and the associated desorption technique (TPD)[2] and the more established electrochemical technique of cyclic voltammetry (CV)[3]. Both techniques, though not completely analogous, provide similar information on the nature of the catalyst. The cyclic voltammetric technique is a surface technique, whereas TPR is a bulk technique, though when carried out hand in hand with TPD it gives information on both surface and bulk.

We illustrate the principles of each of the above techniques and demonstrate their usefulness in catalyst characterisation by appropriate examples of the advances in understanding of catalysts that have been made through their application. We shall also draw comparisons between the techniques where possible, i.e. where results have been obtained on the same catalysts using each technique. Such comparisons, of course, can only be limited owing to the inherent restriction of cyclic voltammetry, namely that the catalyst under study be electrically conducting.

2. THE TECHNIQUE

2.1 Temperature-Programmed Reduction (TPR) and Associated Techniques

The term TPR was first used in a paper by Robertson et al.[4] but the basic idea of characterising catalysts by monitoring their reducibility was first suggested by Holm and Clark[5]. Later the TPR technique as we know it today, employing thermal conductivity cells, linear temperature programmers, etc., was developed in the Shell Laboratories in Houston and Amsterdam[1].

The technique involves the adaptation of the temperature-programmed gas chromatograph to the purpose of measuring reduction. The requirements are a gas handling system, a thermal conductivity cell and associated electronics, linear temperature programmer, recorder, sample holder, furnace and cold traps; altogether a relatively inexpensive set-up. A typical TPR apparatus is shown in Fig. 1. The catalyst sample is contained in a quartz tube surrounded by a small tubular electric furnace whose temperature is controlled by a linear temperature programmer. The catalyst sample can be pretreated in different gas streams prior to the actual TPR experiment. When the sample is ready for TPR measurement the gas stream is switched to a gas of composition 5%v H_2/95%v N_2 and this passes through one arm of the thermal conductivity cell then through the reactor and via a series of traps (to remove reduction products) through the other arm of the thermal conductivity cell where the change in hydrogen concentration of the gas stream brought about by any reduction process is monitored via the thermal conductivity change (H_2 and N_2 have widely differing thermal conductivities). The change in hydrogen concentration is displayed on a recorder. Since the gas flow is constant the change in hydrogen concentration is proportional to the rate of catalyst reduction. Distinct reducible

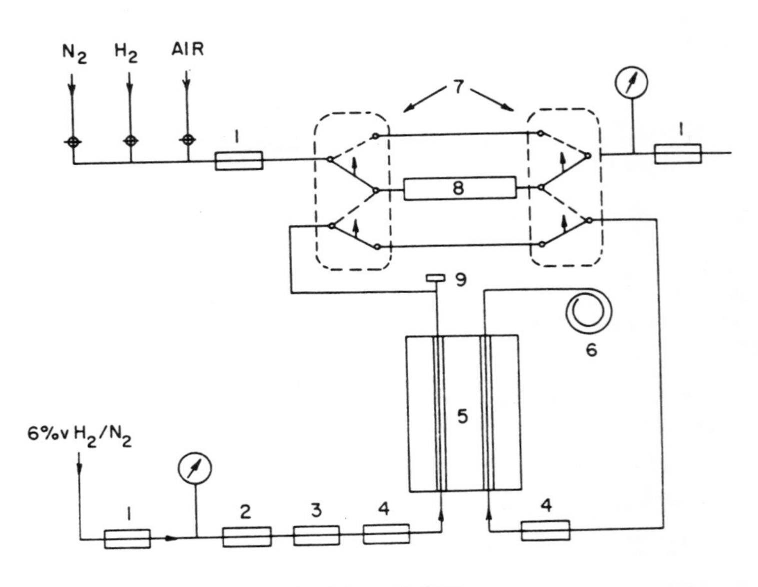

1. Zamba Negretti Reduction Valve.
2. $Pt/A\ell_2O_3$ Catalyst.
3. Molecular Sieves (Linde 5A).
4. Dewar Vessel (-80°C).
5. Thermal Conductivity Cell.
6. Brake Capillary.
7. Gas Flow Switch.
8. Reactor (Quartz U-Tube).
9. Injection Point for Calibration.

<u>Fig. 1</u>

Block diagram of apparatus used for TPR.

species in the catalyst show up as peaks in the TPR spectrum on the recorder. Basically the apparatus is similar to that of Benesi et al.[6] for hydrogen chemisorption. In fact, at the end of a TPR experiment the sample can be cooled to 196 K and the subsequent temperature-programmed desorption (TPD) profile is measured. In this case the hydrogen desorption is monitored by the change in thermal conductivity of a more dilute 0.1%v H_2/ 99.9%v N_2 gas stream. From the TPD profile information relating to the platinum surface area and also the nature of the platinum surface can be obtained. This latter procedure is similar to the TPD procedure described by Cvetanovic and Amenomiya[2] and Aben et al.[7]

There are several experimental parameters that can be varied in a TPR experiment and these are listed in Table 1. Gentry,

Table 1

Variable parameters in TPR/TPD

Flow rate of hydrogen
Concentration of hydrogen
Linear heating rate
Mass of solid sample

Hurst and Jones[8] have recently carried out a TPR study of copper ions in zeolites in which they derived kinetic equations for TPR. Following the procedure of Cvetanovic and Amenomiya[2] they derived the equation:

$$2 \log_e T_m - \log_e \beta + \log_e [H_2]_m = E/_{RTm} + \text{constant} \qquad (1)$$

where β is the rate of linear heating

T_m is the temperature at the peak maximum

E is the activation energy.

By plotting the left-hand side of the above equation versus $^1/_{T_m}$ a straight line is obtained of slope $^E/_R$. A similar equation has been derived for TPD. Values of E obtained for reduction of copper ions in Zeolite Y using this method[8] were in good agreement with results found using an isothermal kinetic method[9].

The mathematical treatment for TPD assumes that no readsorption of gas takes place during desorption and that the molecules are adsorbed on a homogeneous surface without interactions between adsorbed molecules. Surface inhomogeneity, slow heating and large surface areas and resulting readsorption make the mathematical treatment more complicated[2].

The experimental variation in T_m with hydrogen concentration and flow rate agrees closely with the predictions of the equation. Gentry et al.[8] found an effect of mass of zeolite on the TPR profile. This they attributed to the temperature differences between the sensor and the catalyst when large masses were used. Also significant hydrogen concentration gradients occurred within the solid bed leading to non-homogeneous reduction. Practice at Shell Research for measuring TPR of catalysts such as metals on $A\ell_2O_3$, SiO_2, carbon etc. has been to keep the mass of sample so far as possible constant and small: by so doing, problems such as the above do not arise.

Basically, since the technique depends on the wide variation in thermal conductivity between the reactant gas and its inert carrier, any reaction can be studied by temperature programming provided a convenient gas mixture can be obtained. For instance temperature-programmed oxidation can readily be studied by using He/O_2 gas mixtures.

The sensitivity of temperature-programmed reduction is high, reductions involving hydrogen consumptions of 1 μmole being readily detectable. In the TPD mode surface areas of Pt of 0.005 m^2 g^{-1} are measurable. About 0.5 g of catalyst containing $<$ 0.5%w reducible species is sufficient for adequate resolution. Smaller amounts, e.g. a few milligrams, of unsupported catalysts may be used. Increased sensitivity is attained by using a more dilute reducing gas stream. The catalyst sample under study need not be crystalline: the sole requirement is that it be in an oxidised state. Information on reduced samples can often be obtained by subjecting them to prior mild oxidation.

Quite apart from the application of temperature programming to catalyst characterisation, there are potential applications in the field of corrosion chemistry, mineral processing and as an analytical technique to detect reducible or oxidisable impurities. In fact the technique is so sensitive that the presence of trace impurities can complicate profile interpretation. In the field of heterogeneous catalysis and electrocatalysis it has proved invaluable in the detection of alloying between metallic components, the interaction between catalysts and supports and the evaluation of metal crystallite size (see later).

2.2 Cyclic voltammetry: Background

Cyclic voltammetry (CV) is an electrochemical technique which has been widely used in the study of electrode reactions and the characterisation of electrode surfaces in electrocatalysis[3]. It is a surface technique in that the first few atomic layers are analysed. In many respects it may be regarded as the electrochemical analogue of the temperature-programmed technique of reduction (TPR), desorption (TPD) and oxidation (TPO) but where electrochemical potential rather than temperature is the parameter that is linearly varied. It provides similar information on catalyst condition as do the gas phase techniques. Cyclic voltammetry monitors the change in free energy of equilibrium processes occurring at the electrocatalyst surface via a linear control of potential with time. Because it is an electrochemical technique and requires the catalyst to be electrically conducting, CV has not attracted much interest in the field of heterogeneous catalysis, though there are many areas where it could usefully be applied. Here we draw comparisons between the voltage and temperature-programming techniques: it will emerge that, in certain cases, identical information concerning the nature of the respective catalysts can be obtained with both techniques.

CV monitors the state of the electrode surface under conditions identical to those under which the electrode functions during electrochemical operation, and in this respect is preferable to ex-situ techniques for surface examination such as X-Ray Photoelectron Spectroscopy, Auger Electron Spectroscopy, Low Energy Electron Diffraction etc.

The technique involves the periodic variation of the potential of the electrode under study and as such can be classified as a linear perturbation technique. Since it reveals changes in the nature of the surface due to the passage of electric charge, it is necessary to count the charge for the Faradaic process involved in the adsorption and desorption of surface species. The relationship between the passage of charge and the applied potential gives information regarding the processes occurring as a function of their relative free energy changes. Information can also be obtained on the state and extent of double layer charging in regions where no Faradaic surface processes are occurring. Since such double layer transient currents have relaxation times of 10^{-6}-10^{-7}s they can be separated from the Faradaic currents by selection of appropriate rates of voltage sweep. The main practical limitation of the technique is the requirement that the electrode catalyst be conducting and as a result it is restricted to the study of metal surfaces, certain metal sulphides, carbon or graphite supported catalysts, certain mixed oxide catalysts and semiconducting supported catalysts.

For an electrode process of the kind

$$M + A \underset{k_R}{\overset{k_f}{\rightleftarrows}} MA_{ads}. + e \tag{2}$$

with surface coverage of M with A_{ads} represented as θ_A the currents in the forward and reverse reactions are as follows:

$$i_f = nFk_f\ (1-\theta_A)\ C_{A^-}\ \exp\beta VF/RT \tag{3}$$

$$i_R = nFk_r\ \theta_A\ \exp -(1-\beta)\ VF/RT \tag{4}$$

where V is the metal/solution potential difference, $\beta = 1/2$ is the Bronsted symmetry factor for the charge transfer process, C_{A^-} is the pseudocapacitance. For processes where desorption or deposition of A occurs to or from a monolayer limit, i_f and i_r are not continuous and are observed only by a transient technique.

At a sweep rate $S = dV/dt$ (i.e. $V = V_{init.} \pm St$) the time-dependent current i_t yields the coverage at any point from:

$$\theta_A = \int \frac{i_t dt}{Q} \tag{5}$$

where Q is the charge required to form a monolayer. Equation (4) can be written as :

$$\theta_A = \int \frac{i_t dV}{QS} \tag{6}$$

for adsorption, and

$$\theta_A = \theta_{initial,A} - \int \frac{i_t dV}{QS} \tag{7}$$

for desorption. Now since $i = dq/dt$ where dq is the charge passed in the time dt, i can be written:

$$i = dq/dt = \frac{dq}{dV}\frac{dV}{dt} \qquad (8)$$

$$= \frac{dq}{dV}.S \qquad (9)$$

dq/dV or C is the capacitance associated with the electrode process.

$$\therefore i = CS \qquad (10)$$

and C consists of a double layer charging component C_{dl} plus a component arising from the potential dependence of the coverage θ_A of the surface by species A.

This latter component C_A is known as the adsorption pseudo-capacitance for species A.

$$\therefore C = C_{dl} + C_A \qquad (11)$$

and

$$i = C_{dl}S + QS.\frac{d\theta_A}{dV} \qquad (12)$$

$$= C_{dl}S + C_AS \qquad (13)$$

for any coverage ≤ 1. The profile of i with respect to V at sweep rate S gives $d\theta/dV$, which is the differential coefficient of the electrochemical isotherm for adsorption of A according to equation (2). The double layer charging contribution in equation (13) can be subtracted from i by calculation or by an empirical procedure.

We can express equation (3) above in terms of S under sweep conditions as follows:

$$i_t = dq/dt = nFk_f[1 - \frac{i_f dt}{Q}]\, C_{A^-} \exp\beta(V_{init.} \pm St)\, F/RT \qquad (14)$$

and this can be evaluated numerically to give the current time profile for the surface process as θ_A goes from 0 to 1 or vice versa.

If reaction (2) is in equilibrium we observe reversible behaviour in the CV, i.e. adsorption and desorption occur at the same potentials in both anodic and cathodic sweeps. This can be represented by equating i_f and i_r in equations (3) and (4) which for the Langmuir case gives:

$$\frac{\theta_A}{1 - \theta_A} = K_1\, C_{A^-} \exp(VF/RT) \qquad (15)$$

where $K_1 = k_r/k_f$. For more complex adsorption behaviour where, for instance, phenomena such as surface heterogeneity and interaction of adsorbed species exist, appropriate treatments have been developed[10,11].

For irreversible reactions the θ_A versus f(V) relationship together with the corresponding C_A have to be recalculated from equations (3) and (4) separately. Current/voltage profiles are found which are asymmetric and depend on sweep speed. By variation of sweep speed over a wide range, e.g. 2-3 decades, a change from reversible to irreversible behaviour can occur and valuable kinetic information relating to the surface process can then be obtained.

The use of cyclic voltammetry to derive kinetic information in the manner described above is generally applicable for the study of reactions on smooth electrodes, and since this review is primarily concerned with catalyst characterisation and the analogies between gas phase temperature programming and CV, we shall not dwell further on its use in obtaining kinetic information. We shall, in the main, be concerned with measurements carried out on catalysts at comparable voltage sweep rates for CV and similar linear temperature changes for TPR/TPD. However, the area of characterisation of smooth low-surface-area surfaces will be covered. For further detailed information on CV see reviews in refs 3, 12, 13.

2.3 Experimental Set-Up for Cyclic Voltammetry

The reaction cell for a CV experiment consists of three compartments containing respectively the test electrode, counter electrode and reference electrode (Fig. 2). The ohmic potential drop between test and reference electrodes is kept low by using a Luggin capillary placed close to the test electrode surface and connected to the reference compartment. The electrolyte may be acid, base, buffer or salt solution or some non-aqueous electrolyte, though in this review we will be concerned only with acid electrolyte. The temperature and atmosphere within the cell are controlled and materials used for the electrode must have high electrical conductivity, e.g. metal sheets, foils, compressed powders, metals dispersed on conducting substrates such as carbon, or semiconductors.

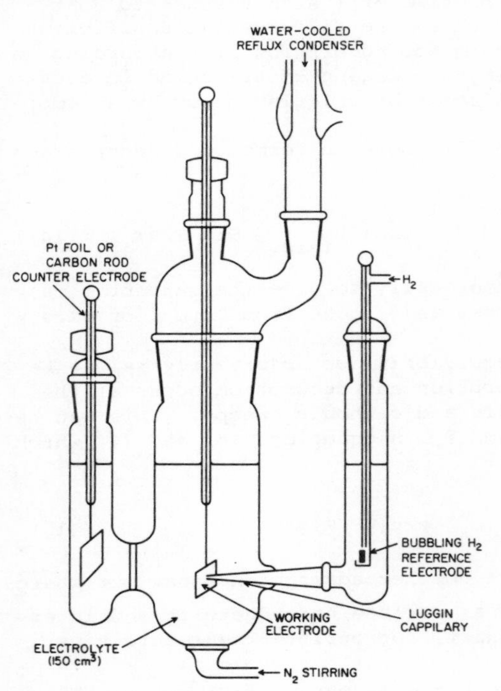

Fig. 2
Test cell for cyclic voltammetry

Of critical importance when dealing with the characterisation of very-low-surface-area catalysts such as metal sheets is the purity of the electrolyte. This must be free of trace organic impurities since these may deposit on the electrode during CV leading to blocking or modification of catalytic sites on the surface. Generally in such systems the water used to make up the electrolyte must be purified by pyrodistillation[14] and mixed with ultra high purity (e.g. Aristar grade) chemicals. For high-surface-area catalysts the requirements are less critical and usually electrolytes are made up from Analar grade chemicals and doubly distilled water.

The electronic equipment consists generally of a programmable potentiostat linked to a waveform generator and the potentiostat delivers a cycling voltage of varying magnitude at a constant sweep rate (ca 20-200 mv/s). This equipment, together with devices for measuring and displaying the variation of current and voltage, constitutes the CV electronic equipment. For further details of the experimental set-up see the papers by Breiter[15] and Brown[16].

3. APPLICATIONS OF THE TECHNIQUES IN CATALYST CHARACTERISATION

3.1 Smooth Noble Metals and Alloys

No work has been carried out on characterisation of smooth low-surface-area catalysts using TPR/TPD employing the thermal conductivity detection technique, but a good deal of attention has been paid to such materials using CV; and several examples will serve to highlight the application of the technique. Where possible, comparisons will be drawn between results found with CV and those from non-electrochemical measurements.

Most work on smooth catalytic electrodes has been carried out on noble metals and, in particular, platinum. In acid or base electrolyte these metals adsorb/desorb hydrogen and oxygen by Faradaic charge transfer processes and these processes are revealed as distinctive features on a CV profile. The profile for smooth metallic platinum (Fig. 3) is divided into several regions. Between 0 and 0.4 V hydrogen is adsorbed on the cathodic sweep and is desorbed on the anodic sweep. It will be noticed that at least three chemisorption types are observed and that the process is reversible. It has been proposed that these peaks arise from H adsorbed on different crystal faces of platinum[17] and there is much recent evidence to support this view. However, Conway et al.[12] have observed multipeaks even on well-characterised single crystal faces of platinum and the authors speculated that the behaviour observed arose from induced heterogeneity associated with a number of possible effects that are itemised in their paper[12]. The gas phase studies of smooth platinum have revealed two types of chemisorbed hydrogen species[18].

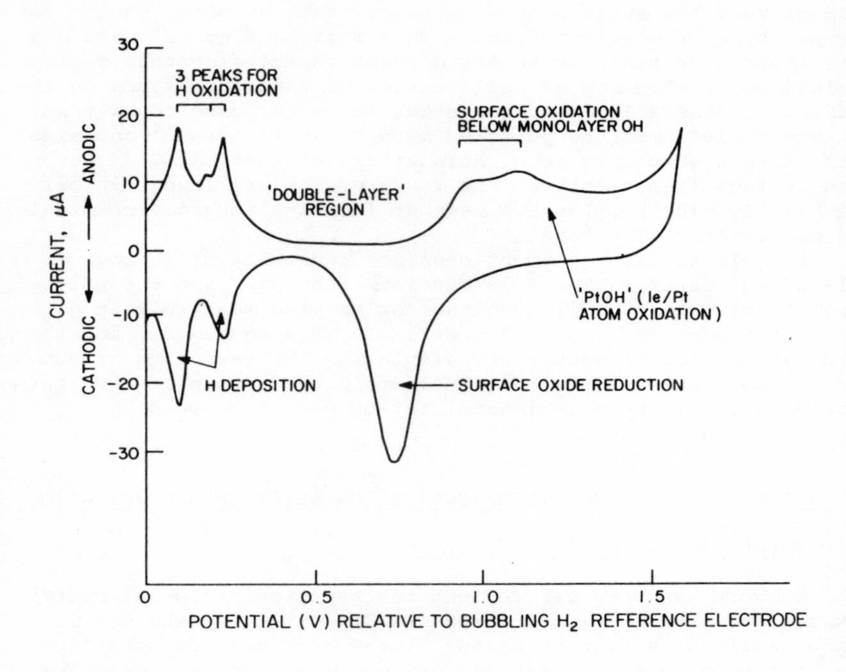

Fig. 3

Typical cyclic voltammogram for a smooth platinum electrode.

From the charge passed during H adsorption or desorption and assuming a certain packing density of Pt atoms of say 1.31×10^{19} atoms/m^{-2} then it can be calculated that for 1 cm^2 of Pt a charge of 210 μC is required[19]. Thus, by measuring the charge, we can estimate the available surface area of the platinum.

In the region between 0.4 and 0.8 V the surface is free from adsorbed species and this is the double layer region. Above 0.8V oxygen chemisorption commences on the anodic sweep and three peaks can be distinguished below monolayer coverage with OH species according to the reaction:

$$Pt + H_2O \rightarrow Pt\text{-}OH + H^+ + e^- \qquad (16)$$

The oxygen chemisorption, like the hydrogen chemisorption, is characteristic of the metal involved, and though initially reversible becomes increasingly irreversible as the potential is raised. It has been suggested[20] that this is due to a rearrangement of the chemisorbed oxide film on the surface to a two-dimensional oxide lattice amongst the surface metal atoms which is thermodynamically more stable than the adsorbed layer. At still higher potentials surface PtO_2 is formed and oxygen evolution begins at about 1.5 V. The reduction of the platinum oxide manifests itself as a peak at about 0.75 V on the cathodic sweep, i.e. at a lower potential than oxide formation, thus marking the irreversibility of the process. The potential of oxide formation is highly characteristic of the metal involved ranging from 0.33 V for rhodium to 1.20 V for gold. As we shall see later, this is an important property in terms of catalyst characterisation. The nature of the structure of the oxide formed on such metal surfaces is still in dispute, LEED studies[21] indicating a chemisorbed layer and studies in oxygen atom penetration[22] into platinum indicating oxide formation.

For noble metal alloys a relevant question to pose is whether the second alloying component actually participates in the catalytic reaction or simply modifies the catalytic metal. CV can give useful information on the role of alloy catalysts and their surface composition.

Woods[23] studied the surface of Pt/Au and his results indicated that, whilst a homogeneous Pt/Au alloy existed in the bulk, the surface behaved as separate Pt and Au phases. The CV profile of a 64% Pt/36% Au alloy is shown in Fig. 4, and it can be seen that the profile is almost identical to the sum of the profiles of the individual metals. Even if cycled continuously to

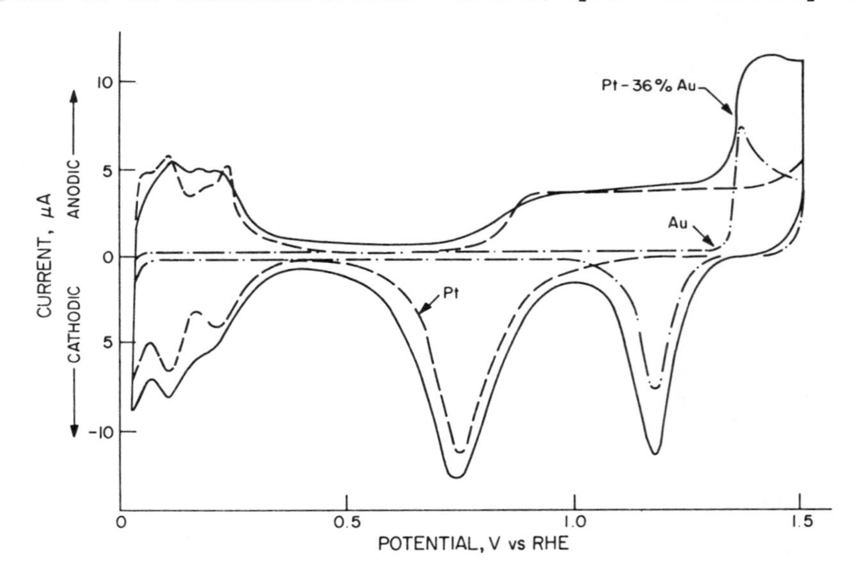

Fig. 4

Cyclic voltammograms for platinum, gold, and a platinum/36%-gold alloy in 1 M H_2SO_4 at 25°C. Triangular potential sweeps at 40 mV s^{-1}. (Reproduced from reference 23).

dissolve some of the surface metal, the catalyst profile remained unchanged, indicating that the new surface exposed was still behaving identically to individual Pt and Au. Thus Pt/Au alloys separate at the surface into equilibrium phases and this is typical of alloys whose phase diagrams show a wide miscibility gap. In the gas phase it has been noted that the alloying of Pt with Au has the same effect as decreasing the Pt crystallite size[24]. The dilution of the Pt surface with Au, which does not adsorb hydrogen, leads to a dramatic decrease in hydrogen chemisorption[18]. This effect would not be noticed in CV since the chemisorption of hydrogen on platinum

$$Pt + H^{+} + e \rightarrow Pt\text{-}H \tag{17}$$

is a different process from the dissociative gas-phase adsorption of hydrogen which is thought to require at least two Pt sites. Thus we might expect differences in the conclusions from both studies. What is clear from both CV and gas-phase H_2 chemisorption is that there is no major electronic or ligand effect of Au on Pt.

On the contrary Pt/Rh, Pd/Rh and Pd/Au smooth alloys behave completely differently, displaying surface characteristics in CV different from those of the individual metals[25]. The oxide reduction peaks in CV for example appear at potentials intermediate between those characteristic of the individual metals and the position of the peak can be a probe of alloy composition. Rand and Woods[25] studying Pt/Rh (Fig. 5) showed how cycling between potential limits led to a more rapid dissolution of Rh leading to a Pt-rich phase, and they developed a relationship between surface composition and oxide reduction peak potential.

$$\varphi_p = C_A \varphi_p^A + C_B \varphi_p^B \tag{18}$$

where φ_p is the alloy oxide reduction peak potential and φ_p^A is the oxide reduction peak potential of metal A and φ_p^B that of metal B. C_A and C_B are the concentrations in atom fractions of metals A and B respectively. They were able to show how surface enrichment in Rh occurred when the alloy was heated at high temperatures in air. Surface compositions of similar magnitude to these obtained by CV were also obtained in Auger Electron Spectroscopy (AES) though the CV values were somewhat greater than the values found for the outermost layer of the alloy by AES[26]. The authors were also able to show by CV that the surface involved in chemisorption is confined to the first few atomic layers.

The modification of the surface of Pt by Rh can be put to useful purpose, e.g. the oxidation of methanol is accelerated by alloying Pt with Rh[27]. The greater ease of oxygen chemisorption in the Pt/Rh alloy favours removal of the adsorbed intermediates in the reaction. A similar effect was noticed in studies of smooth alloys of Pt with Ru[28]. In these cases we are dealing with a ligand effect in the alloy where the adsorption properties of the one metallic component are altered via an electronic interaction with the second component.

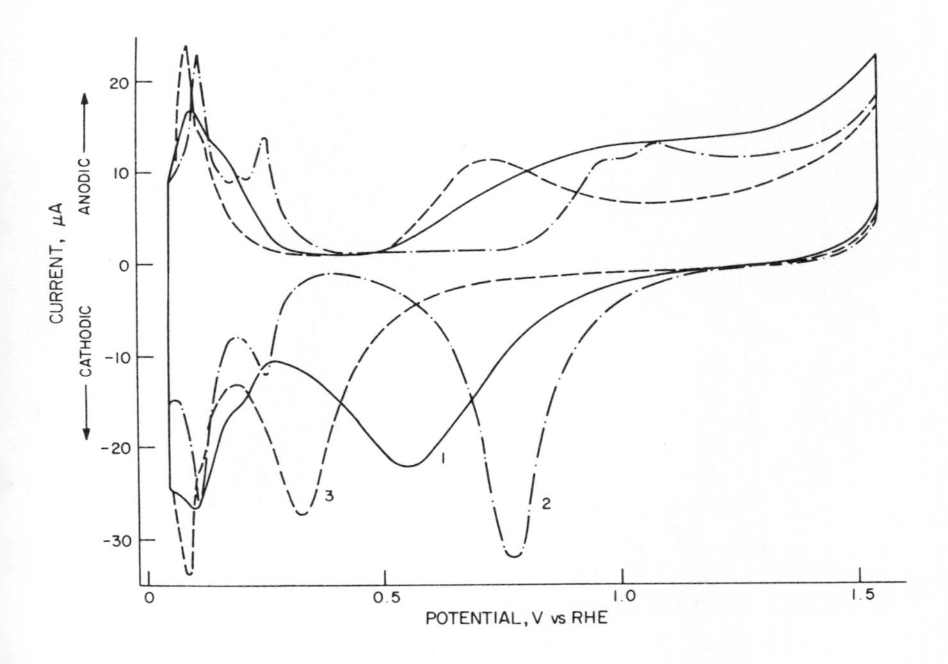

Fig. 5

Cyclic voltammograms for (1) a platinum/rhodium alloy with 1:1 surface composition, (2) platinum, (3) rhodium, in 1 M H_2SO_4 at 25°C. Triangular potential sweeps at 40 mV s^{-1}. (Reproduced from reference 25)

3.2 Finely Divided Platinum Catalysts

The CV of unsupported finely divided platinum, e.g. Pt Adams catalyst and Pt blacks, resembles closely that of the smooth metal though, of course, the currents involved in the various processes are much higher because of the high metal surface area involved. At least three types of hydrogen chemisorption are found. This is in agreement with gas-phase TPD studies on unsupported Pt black[29] and Pt supported on $\gamma A\ell_2O_3$[7] where multi-adsorbed species of hydrogen were also observed (Fig. 6). Also the free energy differences between the most weakly and strongly adsorbed hydrogen was about 11.7 kJ mol^{-1} in both gas phase and electrochemical measurements[12]. Although, as pointed out previously, we are dealing with the electrosorption of hydrogen atoms on the one hand and the dissociative adsorption of hydrogen mole-

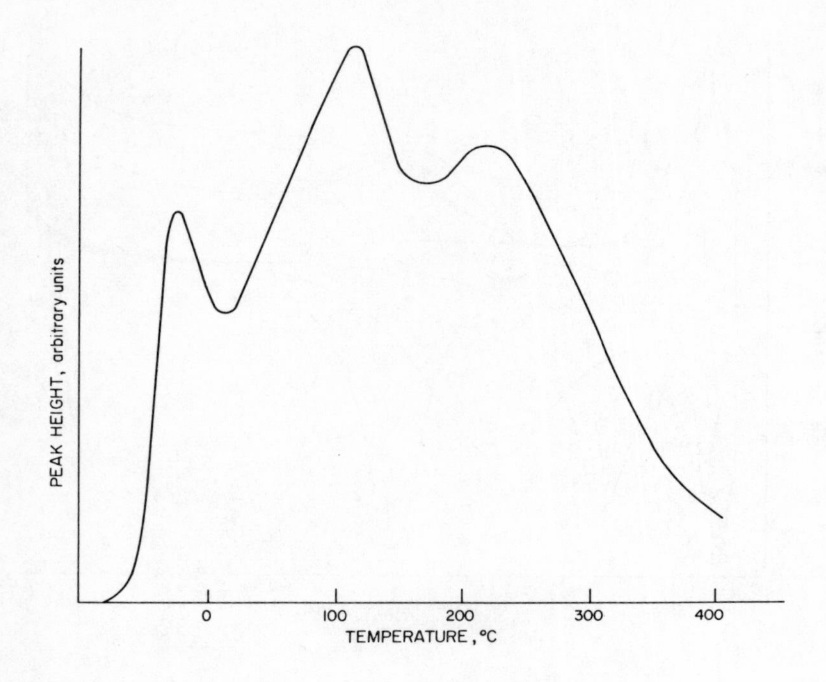

Fig. 6

TPD of hydrogen from 0.46%w Pt/Al_2O_3 (H/Pt = 1.0). (Reproduced from reference 7)

cules on the other, the site energies on the Pt surface should be the same, and so the agreement is not fortuitous.

We have recently compared the electrochemical technique of CV with the gas-phase technique of TPD in the evaluation of hydrogen chemisorption on Adams Pt catalyst[30]. Assuming a packing density of 1.31×10^{19} atoms Pt m^{-2}, this can be done by an integration of the hydrogen desorption area of a CV profile and the area under the hydrogen TPD profile. The results of fresh and sintered Adams Pt catalyst are given in Table 2 where, for comparison, values calculated from X-ray powder diffraction are also given. The results are in very good agreement with each other, thereby affording mutual corroboration.

Table 2

Pt Surface Areas of Pt Catalysts by TPD, CV and X-ray Diffraction (XRD)

Catalyst	Technique	Pt Surface Area ($m^2\ g^{-1}$)
Fresh	TPD	45.5
	CV	36.6
	XRD	40.0
Sintered	TPD	1.73
	CV	1.9
	XRD	1.75

We have also been involved in the development of Pt supported on carbon fibre paper as catalysts for methanol electro-oxidation. These catalysts are prepared by ion exchange of a pre-oxidised pyrographite-coated carbon fibre paper with $Pt(NH_3)_4(OH)_2$ aqueous solutions followed by drying and activation in air at 300°C. CV and TPR have been applied to the study of such catalysts and to less active ones prepared by simple impregnation of untreated carbon fibre paper[31]. The latter have very low Pt metal areas (of the order 5-20 m^2g^{-1}) whereas the ion-exchanged catalysts have Pt metal areas of up to 130 m^2g^{-1}. These results highlight the similarities in the characterising information obtained using the two techniques. We have measured the reducibility of the air-activated catalysts using both CV and TPR. With CV this was possible by commencing the voltage sweep at 0.7V, proceeding anodically to 1.6V and then cathodically down to 0.0V. The results for the two catalysts are shown in Fig. 7(a,b). It is clear that the ion-exchanged variant shows much higher currents for all the processes occurring on the catalyst surface, reflecting the much higher Pt metal area. The Pt 'oxide' species formed upon 300°C air activation is reduced at 0.20V for the ion-exchanged catalyst and 0.3V for the impregnated catalyst. Clearly the oxide formed under this condition is radically different from that formed upon the surface of reduced Pt crystallites during cycling. Since CV is a surface process, the reduction peaks at 0.2V and 0.3V reflect the reduction of only the surface of the oxide. However the charge involved in this process for the ion-exchanged catalyst is a much larger fraction of the charge expected for the total Pt inventory in the catalyst than for the impregnated catalyst, a reflection of the higher surface area of the ion-exchanged catalyst. In the TPR experiments no reduction peaks were observed above room temperature. Consequently the TPR system was adapted to measure reduction profiles down to -160°C.

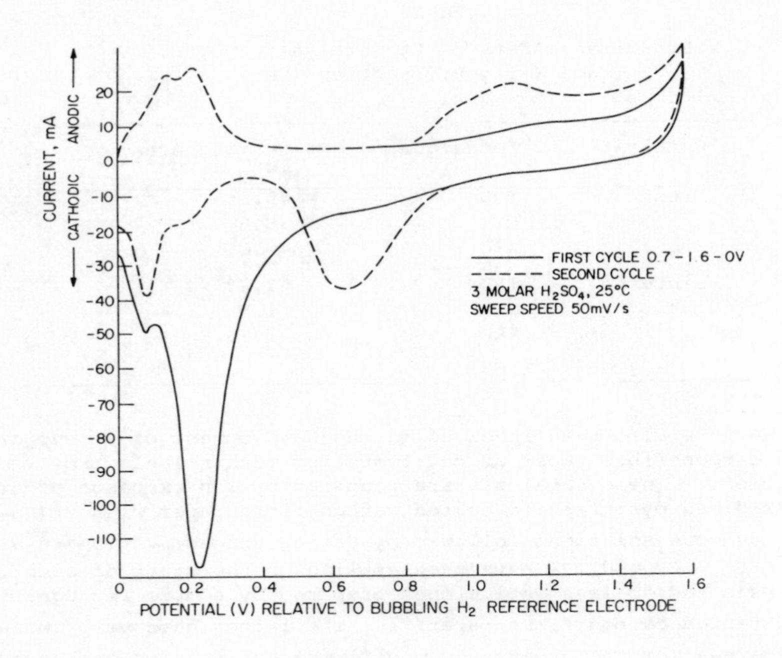

Fig. 7(a)

Cyclic voltammogram of Pt ion-exchanged onto oxidised carbon fibre paper-air activated 300°C.

The results are shown in Fig. 8 where it can be seen that the ion-exchanged catalyst is the more difficult to reduce. There is clearly agreement with the CV study, which showed that the ion-exchanged catalyst reduced at the lowest potential. The H_2 consumption during the reduction of the ion-exchanged catalyst was greater than that of the impregnated catalyst. The reason for this lies in the fact that 300°C and actuation of the impregnated catalyst results in the formation of some Pt metal as confirmed by X-ray powder diffraction. This Pt metal has a very low surface area and corresponds to the species formed during air activation of unsupported $Pt(NH_3)_4(OH)_2$. In the impregnated catalyst the main reduction peak occurs at -97°C with a smaller peak at -78°C. The ion-exchanged catalyst shows a main peak at -57°C with only a very weak shoulder at high temperatures. The increased difficulty of reduction of the ion-exchanged catalyst reflects the stronger binding of the Pt species to the oxide groups on the carbon surface. The H_2 consumption of the ion-exchanged catalyst was equivalent to that required for reduction

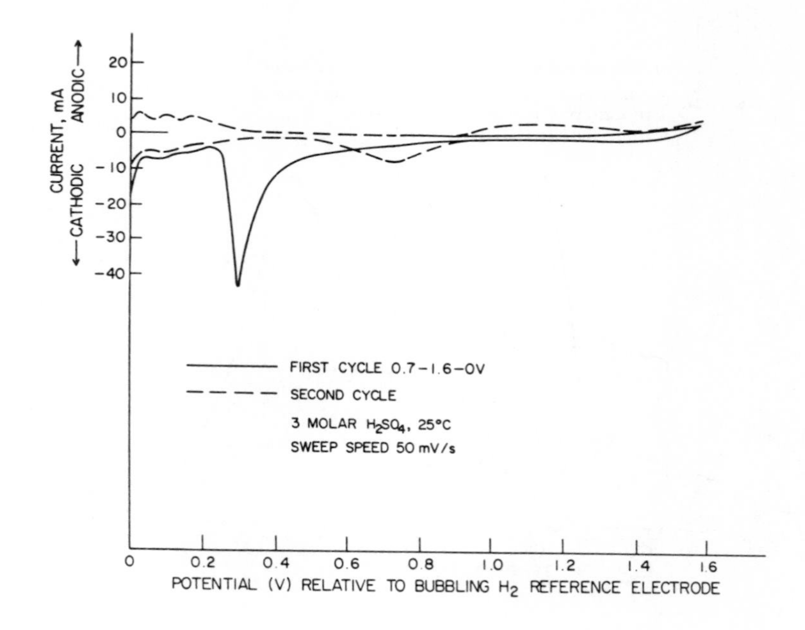

Fig. 7(b)

Cyclic voltammogram of Pt impregnated onto untreated carbon fibre paper - air activated 300°C.

of Pt^{2+} oxide. This is in line with the X-ray photoelectron spectroscopic (XPS) studies of the catalyst[32], where it was shown that about 75% of the Pt in the surface was present as Pt^{2+}, the remainder being Pt^{4+}. These results demonstrate the similarities in information obtainable with, and also the fine differences that exist between, the CV and TPR techniques.

In the search for alternative conducting supports to carbon for electrocatalysts we recently carried out an evaluation of semiconducting tin oxide and Sb-doped tin oxide as supports for platinum[33]. It was hoped that via the ion exchange interactions that occur between SnO_2 and platinum species that highly stable well-dispersed Pt could be obtained. Pt on Sb-doped SnO_2 had been claimed to be an effective electrocatalyst for O_2 electroreduction[34], and there are many examples of the use of SnO_2 as a support in gas-phase heterogeneous catalysis[35]. The catalysts were made by either impregnation or ion exchange of the SnO_2 substrate with solutions of H_2PtCl_6 followed by drying

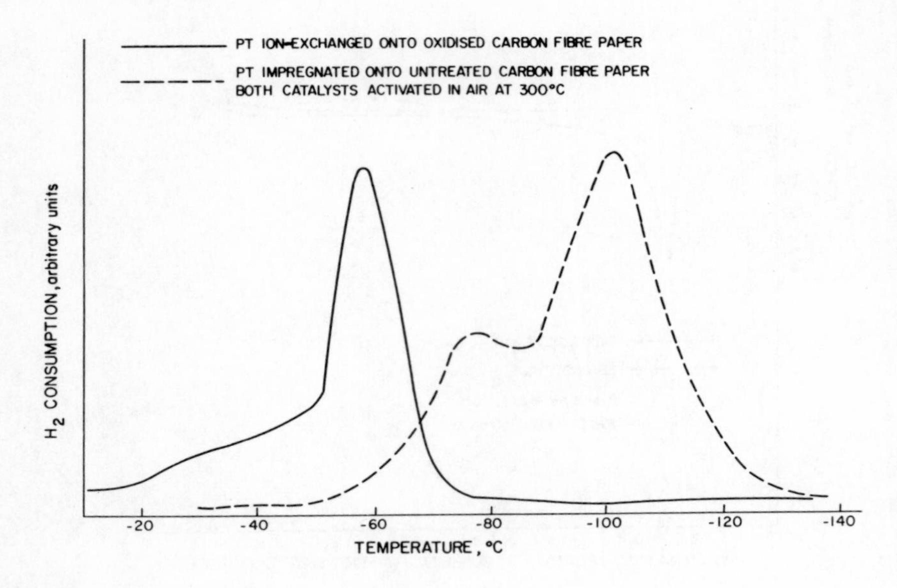

Fig. 8

TPR profiles of air activated Pt supported on carbon fibre paper.

and calcination at 300°C or 500°C in air. This is the normal procedure carried out in the preparation of supported catalysts. The TPR results on some of these catalysts are shown in Figs 9 and 10. It can be seen that the catalysts dried at 120°C show reduction peaks in the temperature range 100-150°C close to where unsupported $H_2PtC\ell_6$ is reduced: a broader peak is present above 200°C corresponding to a platinum species involved in a stronger interaction with the SnO_2 support. The amount of H_2 consumed is close to the stoichiometric quantity required for a $Pt^{4+} \rightarrow Pt^{o}$ transformation. The SnO_2 support is reduced in a single peak of maximum temperature 650°C, some 100°C lower than that of pure SnO_2, reflecting some catalytic influence of Pt on the reduction of SnO_2. The calcined samples show either a reduction in intensity or absence of any low temperature reduction of Pt. The high temperature reduction peak of SnO_2 increases in intensity as the low temperature peaks disappear, indicating that the platinum species is probably being reduced in this region. Clearly the calcination step renders the platinum species difficult to reduce, perhaps because of a Pt-support interaction. Even with $\gamma A\ell_2O_3$-supported Pt, however, where the metal-support interaction is thought to be strong, the reduction

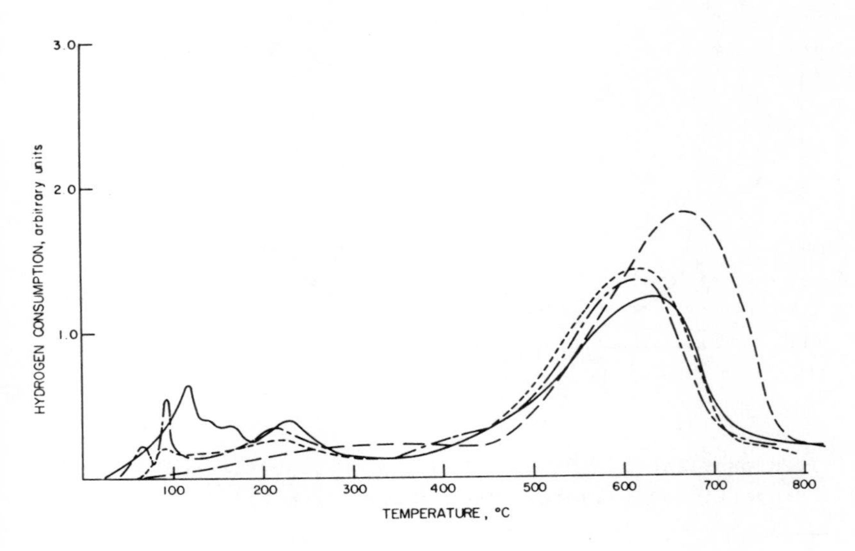

Fig. 9

TPR profiles of SnO_2 (NH_4OH-precipitated) supported catalysts. (Reproduced from reference 33)

— — — Pure SnO_2.

———— 10%w Pt on SnO_2 (H_2PtCl_6-impregnated, dried at 120°C)

—·—·— Dried Pt/SnO_2 catalyst, calcined at 300°C in air.

-- -- -- Dried Pt/SnO_2 catalyst, calcined at 500°C in air.

Heating rate 6°C min^{-1}.

of platinum species in a calcined catalyst occurs at the comparatively low temperature of 280°C. The SnO_2-supported catalysts were also examined by CV. Here the calcined catalysts were electrochemically reduced at room temperature; the CV profiles are shown in Figs 11 and 12, where no features that could conclusively be attributed to Pt were observed, although it is clear that the Pt influences electrochemical surface processes occurring on the SnO_2 (note Fig. 11 especially). Thus the electrochemical reduction is insufficient to reduce the platinum formed on calcination; and the CV results therefore convincingly confirm the conclusions from the TPR results. As a check on the features to be expected from Pt on SnO_2 using CV we electrodeposited Pt onto a Sb-doped SnO_2 disc and the CV profile of such a catalyst is shown in Fig. 13 where the typical hydrogen and oxygen adsorption and desorption regions of Pt can be seen

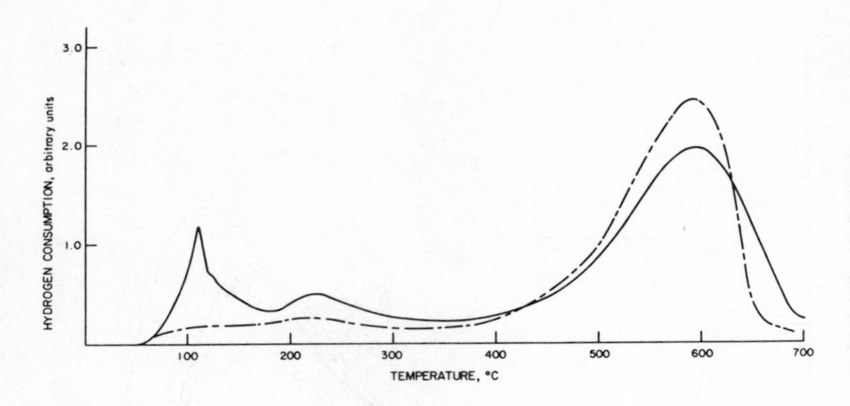

Fig. 10

TPR profiles of Sb-doped SnO_2 (NH_4OH-precipitated) supported catalysts. (Reproduced from Ref.33.)

——— 10%w Pt on Sb-doped SnO_2 (H_2PtCl_6-impregnated, dried at 120^oC).

- - - - -Dried Pt on Sb-doped SnO_2, calcined at 300^oC in air. Heating rate $6^oC\ min^{-1}$.

superimposed upon the background arising from the SnO_2 support.

The Pt on SnO_2 catalysts that had been calcined were, as might be expected on the basis of the foregoing, inactive for methanol electro-oxidation. If the calcination step is omitted and the dried catalysts directly reduced, then catalytic activity is indeed found, though some corrosion occurs, as seen from dissolution of reduced surface Sn species. The platinum species formed upon direct reduction has a very low surface area as might be expected from the TPR profiles, which showed that the predominant platinum species resembled unsupported H_2PtCl_6. TPR and CV demonstrate in this case a negative metal-support interaction brought about by the calcination step. The results also indicate that the higher the temperature of the calcination step the more effective is the metal-support interaction.

Often in electrocatalysis, as in heterogeneous catalysis, poisoning of the catalyst surface by species such as H_2S, CO tends to occur. In electrocatalysis the poisons originate from impurities in the electrolyte, and this can be critical for smooth low-surface-area catalysts. In the case of Pt, poisoning by adlayers of S may occur, though in some instances, e.g. in the oxidation of formic acid and carbon monoxide[36], the adlayers may serve as promoters. Using CV we can observe such effects since on sweeping to high potentials the poison may be oxidised and its removal is reflected in the appearance of an additional

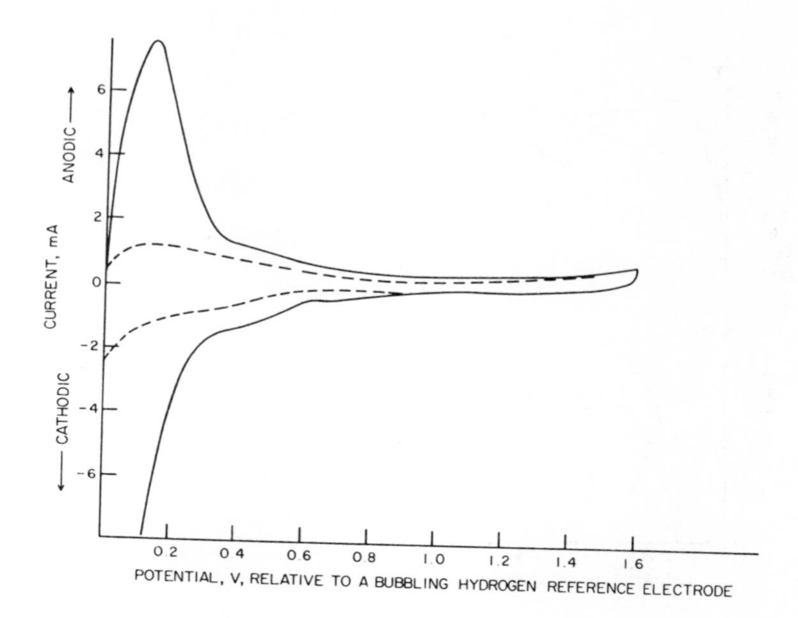

Fig. 11

Cyclic voltammetry of SnO_2 and of Pt on SnO_2. (Reproduced from Ref. 33.)

——— 10%w Pt as H_2PtCl_6 on SnO_2 dried at 120°C, calcined at 300°C in air.

– – – Pure SnO_2

3-molar sulphuric acid at 25°C. Sweep speed 50 mV s^{-1}.

feature in the CV profile. When Pt is partly covered by S in H_2SO_4 electrolyte the H adsorption/desorption area is markedly reduced in intensity, and, above 0.65V, extra peaks in the profile arise from the oxidation of S to SO_2 [36]. Much work has been carried out on Pt catalysts in the presence of small concentrations of H_2S [37-39]. It is thought that a platinum sulphide is formed on the surface, which is then oxidised as follows :

$$PtS + 4H_2O \rightarrow PtO_2 + SO_2 + 8H^+ + 8e \qquad (19)$$

giving rise to a peak in the voltammogram at 1.3-1.4V. In an evaluation of trifluoromethanesulphonic acid as an electrolyte for methanol-air fuel cells a poisoning effect by S on Pt was revealed[40] by the reduction in H desorption peak intensities and the appearance of a new peak at 1.3-1.4V in the CV profile when the poisoned catalyst was cycled in H_2SO_4 electrolyte (Fig.14).

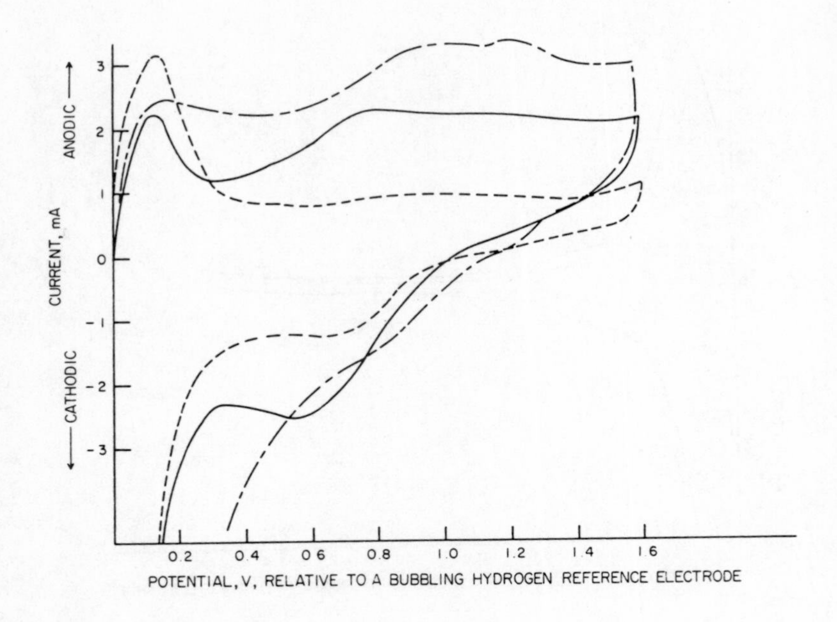

Fig. 12

Cyclic voltammetry of Sb-doped SnO_2 and of Pt on Sb-doped SnO_2 (Reproduced from reference 33.)

— - - — Sb-doped SnO_2.

——— 10%w Pt as H_2PtCl_6, on Sb-doped SnO_2, dried at 120^oC, calcined at 300^oC in air - first cycle.

- - - - - Above after continued cycling

3-molar sulphuric acid at 25^oC

Sweep speed 50 mV s^{-1}.

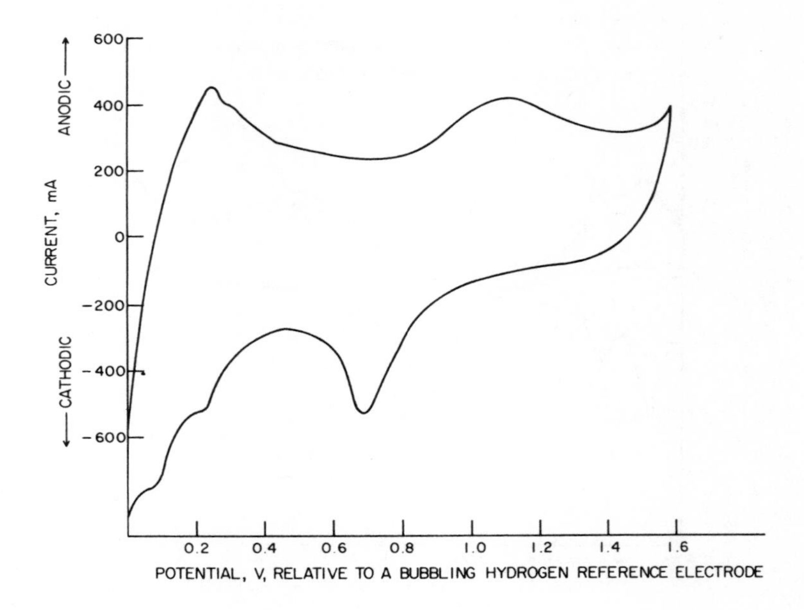

Fig. 13

Cyclic voltammetry of Pt, electro-deposited onto Sb-doped SnO_2-

10 mg Pt on Sb-doped SnO_2 disc.

(Reproduced from reference 33.)

3-molar sulphuric acid at 25°C

Sweep speed 50 mV s^{-1}.

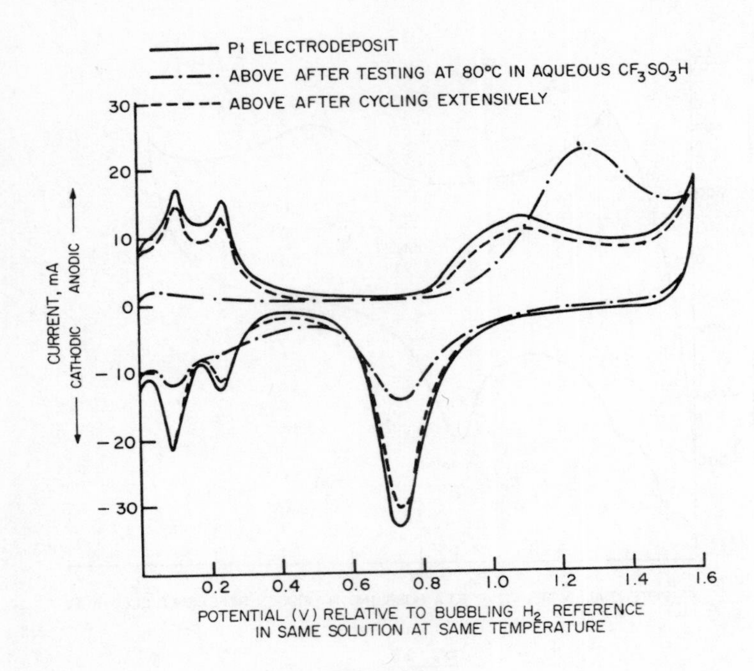

Fig. 14

Poisoning of Pt electrodeposited catalyst in aqueous CF_3SO_3H.

(Reproduced from reference 40.)

The sulphur species arises from the decomposition of the $CF_3SO_3^-$ acid radical. Similarly Janssen and Moolhuysen[41] observed the effect of S poisoning on finely divided Pt for methanol electro-oxidation. The sulphur species can be removed by successive cycles to high potential (Fig. 14). The effect of S through presulphiding of supported Pt reforming catalysts is also readily observed using TPR, the sulphided platinum species reducing at different temperatures from platinum oxide species[42]. Successive TPR runs result in the removal of the sulphur species as H_2S and with intermediate re-oxidation between runs in TPR, profiles identical with that of an unsulphided catalyst can be obtained. The analogy between TPR and CV is again apparent.

3.3 Finely Divided Bimetallic Catalysts

TPR and CV are likely to find their most extensive application in the study of finely divided bimetallic catalysts, which, in certain cases, display greater activity, selectivity and stability for many electrocatalytic and gas-phase catalytic reactions. Moreover, information on the nature of the metal-metal interactions responsible for the effects is often incomplete. Frequently such catalysts contain very small amounts of metal in finely divided form, and conventional techniques of catalyst characterisation are inadequate. TPR and CV studies of comminuted catalysts and electrocatalysts have provided valuable information on the nature of the catalyst and are now regularly used in such studies. The work of Robertson et al.[4], on Ni/Cu catalysts supported on SiO_2, demonstrated the application of the TPR technique to the detection of alloying between the two metals. The Cu component catalysed the reduction of the Ni species but the interaction between the two metals was shown to take place during the reduction step of the catalyst preparation. TPR has also proved useful in the area of characterisation of bimetallic noble-metal-based reforming catalysts where the metal species are present in concentrations $<$ 1%w and are not amenable to detection by X-ray powder diffraction, X-ray photoelectron spectroscopy or even transmission electron microscopy. TPR and TPD studies of Pt, Pt/Ge and Pt/Re on γAl_2O_3 catalysts have increased our understanding of these systems. It has been shown conclusively[43] that the Re component of an Al_2O_3-supported Pt/Re reforming catalyst, when prepared via an impregnation, drying, calcination and reduction procedure, is completely reduced and that alloying between Pt and Re occurs during the reduction step (compare analogous to the behaviour of the Cu/Ni system). A similar situation exists for Pt/Ge on Al_2O_3[44] and the reducibility profiles are shown in Fig. 15. Note that, in the calcined catalyst, the platinum species reduces at 280°C, some 150°C higher than the unsupported H_2PtCl_6 impregnating salt, and the germanium species at about 600°C. The H_2 consumption corresponds to the reduction of Pt^{4+} to Pt^o and of Ge^{4+} to Ge^{2+}. No evidence of a Pt-Ge interaction is found at this stage of the preparation. If, however, the catalyst is monitored for hydrogen chemisorption at various stages of the TPR profile[45] then, as soon as Ge starts to reduce, the H/Pt ratio of the catalyst decreases sharply, implying that Ge is building into the surface of the Pt and impeding chemisorption of hydrogen. Upon completion of the reduction, a mild re-oxidation followed by a further TPR experiment confirms that Ge has indeed built into the Pt and now reduces in the same temperature region as the platinum (Fig. 16). It seems that a fraction of the Ge^{4+} present in the calcined catalyst is reduced completely to Ge^o and is alloyed with Pt. The remainder of the Ge inventory probably remains in the 4+ valence state involved in interaction with the Al_2O_3 support.

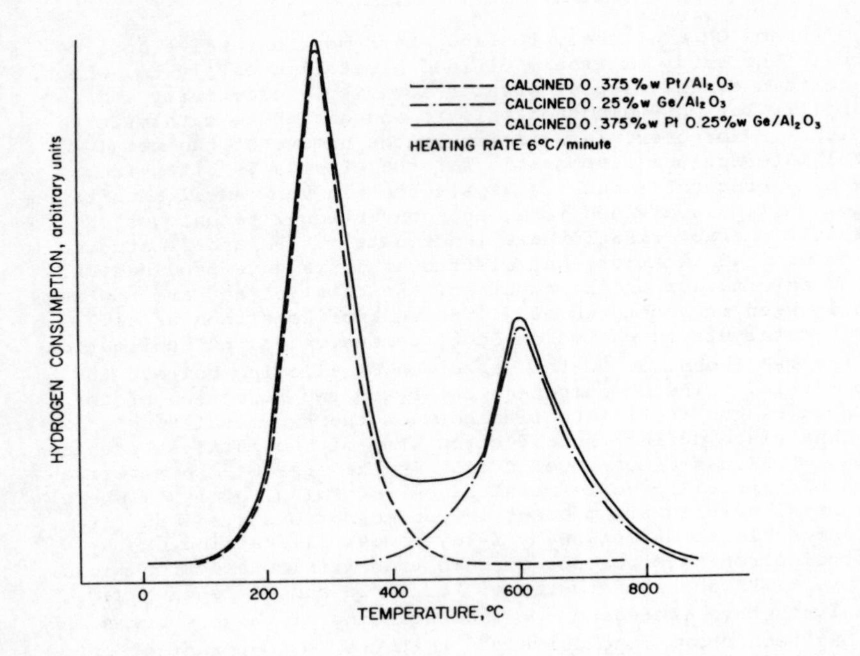

Fig. 15 Reducibility of calcined reforming catalysts.

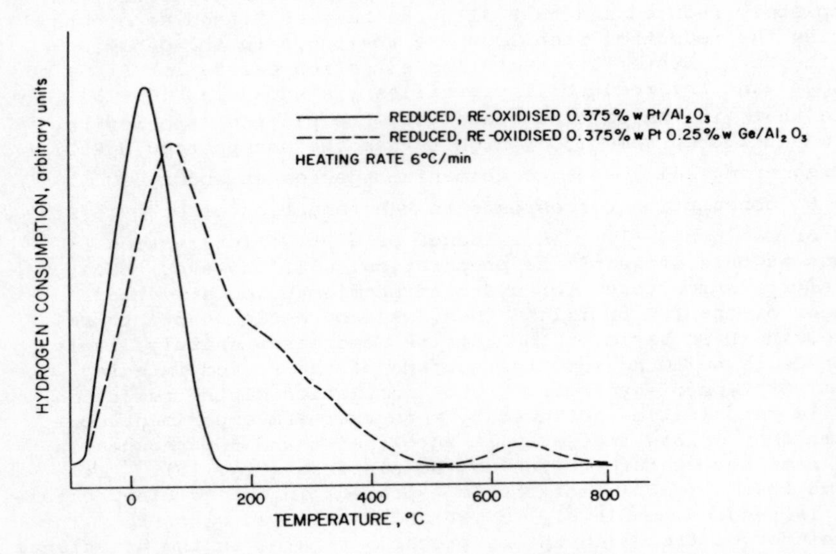

Fig. 16 Reducibility of re-oxidised reforming catalysts.

Systems that are readily studied by both CV and TPR are the Adams-type catalysts, which are finely divided (up to 100 m^2 g^{-1} in area), unsupported and metallic conductors. They are prepared as oxides by high temperature melting of the salts with $NaNO_3$ followed by leaching of the $NaNO_3$, washing and drying of the finely divided oxide catalyst. This catalyst can then be reduced and used as appropriate. A TPR and CV study of PtO_2, RuO_2 and mixed PtO_2/RuO_2 Adams catalysts has recently been carried out[46], from which (Figs 17(a-d) the TPR profiles of the separate and mixed oxides may be seen. It is clear that RuO_2

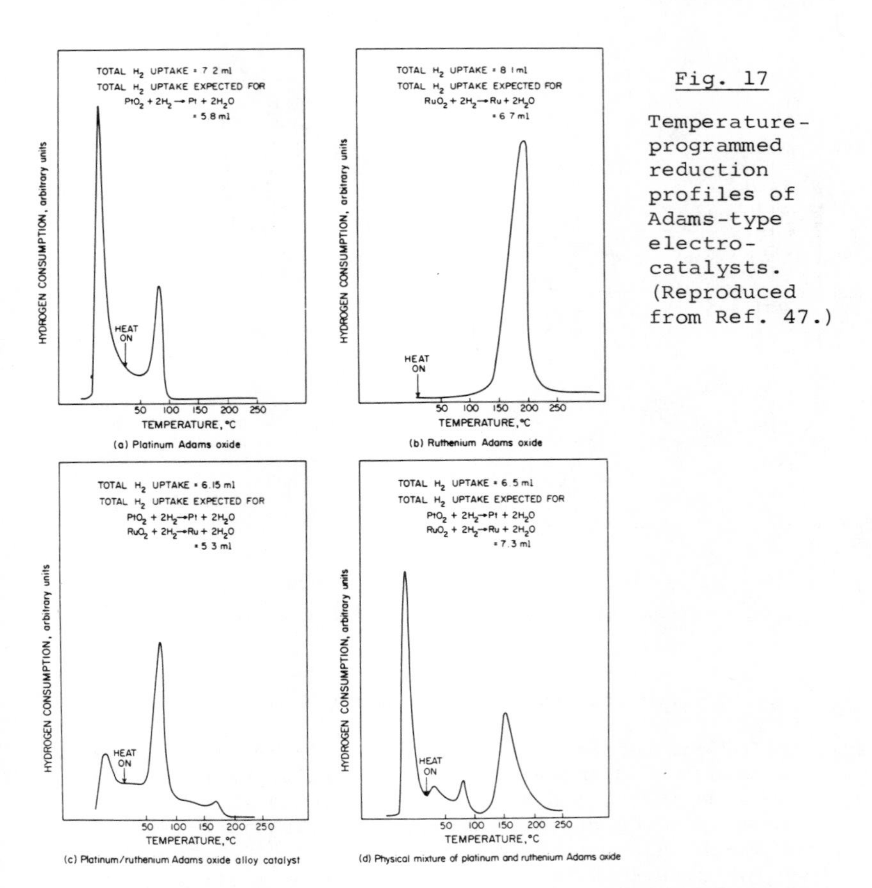

Fig. 17

Temperature-programmed reduction profiles of Adams-type electro-catalysts. (Reproduced from Ref. 47.)

is much more difficult to reduce than PtO_2, which is largely reduced at room temperature.

A physical mixture of PtO_2 and RuO_2 shows essentially the reducibility characteristics of the separate phases (Fig.17d) whilst the mixed oxide of PtO_2/RuO_2 shows reducibility characteristics intermediate between those of the two free oxides. A single reduction peak is found at 75°C compared to 170°C for the free RuO_2 and room temperature and 80°C for free PtO_2. The existence of small peaks at room temperature and 170°C indicates the presence of some free PtO_2 and RuO_2 in this sample (Fig. 17c). An inspection of Fig. 17 reveals that, in all cases, the hydrogen uptake corresponds closely to that expected for the reduction of the 4+ valence state of the metals to the zerovalent state. Similar effects can be seen using CV to study the reduced phases (Fig. 18). The adsorption and desorption charac-

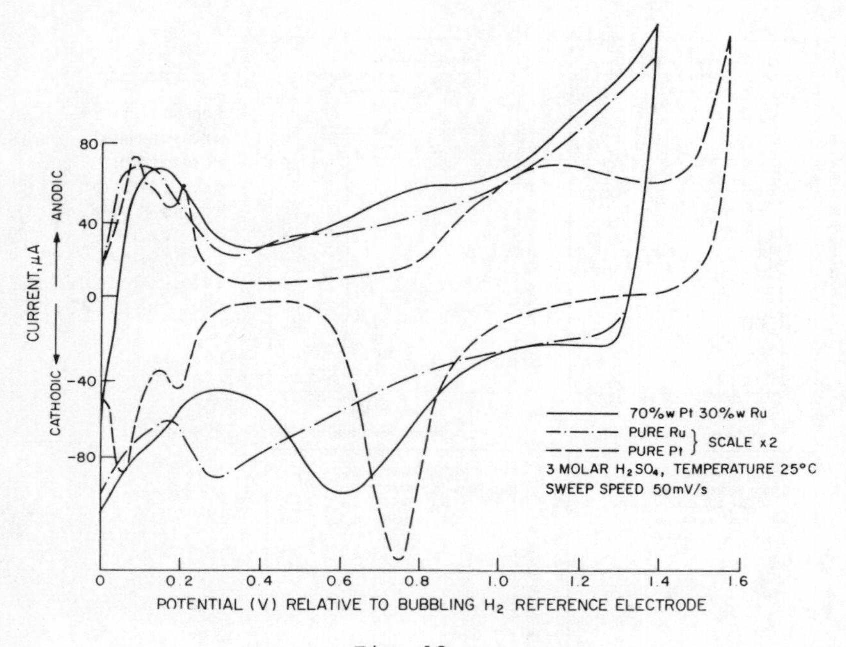

Fig. 18

Cyclic voltammograms of Adams catalysts.

teristics of the reduced mixed oxide are altered from those of the free metals. Clearly in the mixed oxide phase an interaction between the Pt and Ru components occurs which leads to an alloy phase upon reduction. Catalytic evaluation[46] revealed that the mixed oxide after reduction was some 30 times more active than either Pt or the physical mixture catalyst for methanol electro-oxidation, thus confirming the modification of the Pt catalytic properties by the ruthenium species.

As with gas-phase techniques such as TPD and Auger electron spectroscopy CV provides information on the surface composition of finely divided alloys and although not so quantitative as some of the techniques discussed elsewhere in this volume, it can provide valuable guides to the mode of action of the catalyst. For instance, in a recent study of Pt/Ru on carbon fibre paper catalysts for methanol electro-oxidation[47], it was shown how activation of the catalyst in air led to catalysts more active by some 30-fold than catalysts directly activated in H_2.

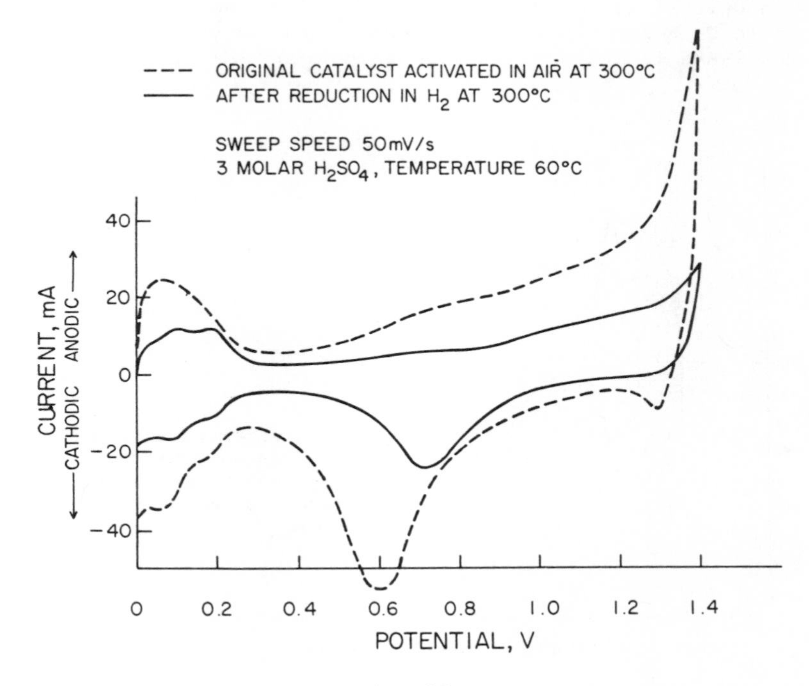

Fig. 19

Cyclic voltammograms of Pt/Ru on pyrographite-coated carbon fibre paper. (Reproduced from Ref. 46.)

An inspection of cyclic voltammograms (Fig. 19) revealed that hydrogen activation had led to a catalyst whose surface resembled that of pure Pt, i.e. surface enrichment in Pt had taken place leading to a surface enriched in the component forming the strongest bond with the gas used for the activation. In air, there is an increase in the Ru concentration. These results partly explain the wide divergence in reported optimum compositions of finely divided Pt/Ru alloys for methanol electro-oxidation.

It was discussed earlier how CV revealed the existence of a ligand effect in Pt/Rh alloys. A similar effect is thought to exist, on the basis of gas-phase TPD measurements[48], with Pt/Sn catalysts. CV of finely divided Pt/Sn prepared by electrodeposition reveals that the hydrogen and oxygen adsorption/

desorption characteristics are radically altered from that of pure Pt[49]. In particular the Pt oxide reduction peak shifts markedly in a cathodic direction (Fig. 20). As Sn is removed from the surface of the catalyst by cycling the peak shifts gradually back to the position of pure Pt (Fig. 20). The shift in the peak is analogous to that found for smooth Pt/Rh alloys and is indicative of a ligand effect of Sn on Pt and is thought to be responsible for the exhanced catalytic activity of such catalysts in methanol electro-oxidation[50,51]. The incor-

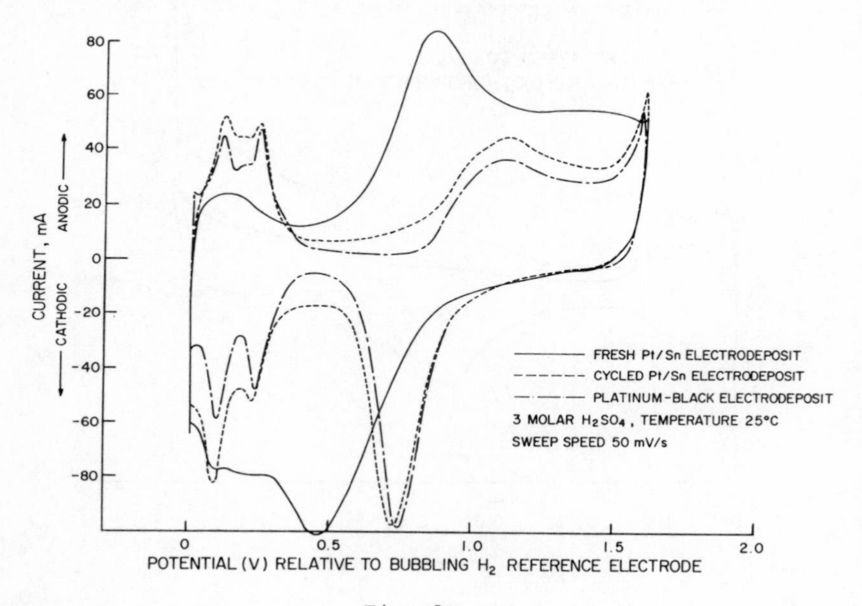

Fig. 20

Cyclic voltammograms for platinum-black
and Pt/Sn electrodeposits
(Reproduced from Reference 49.)

poration of excess Sn in the catalyst can lead however to a blocking of the platinum activity[51]: this is illustrated in Fig. 21. Such a catalyst has zero catalytic activity until the less strongly bound blocking species is removed by cycling. Likewise Pt/Sn catalysts always contain small amounts of SnO_2 species on the surface, and these can readily complex with hydroxylated anions such as silicate or phosphate to produce a polymeric species that blocks the catalyst surface[52]. This effect by silicate is illustrated in Fig. 22, where it can be seen that all the processes that normally occur on the surface of Pt during CV are absent.

Similar promoting effects on Pt were demonstrated for the finely divided Pt/Ti system using CV[41]. Again modifications of the adsorption properties of Pt by Ti were found and substantial increases in catalytic activity observed.

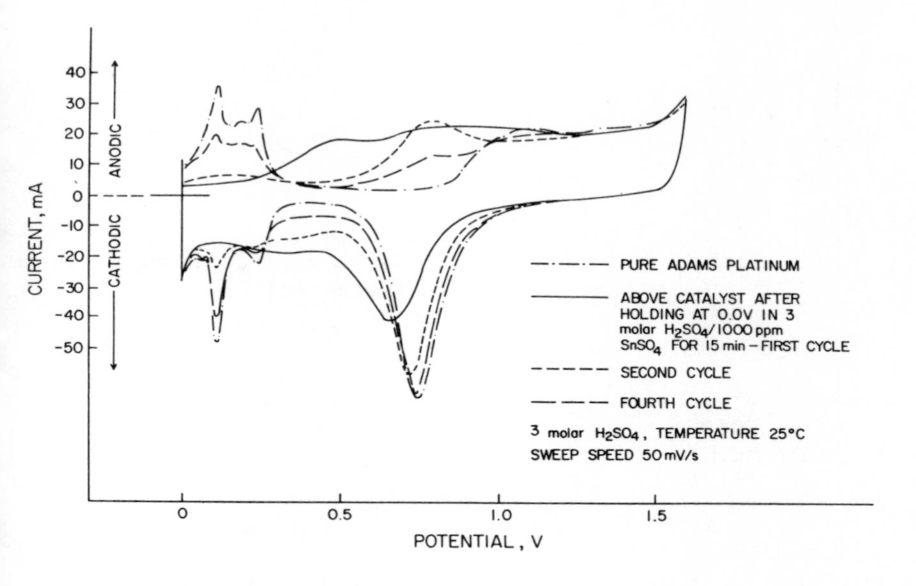

Fig. 21

Cyclic voltammetry of pure Adams platinum and Adams Pt/Sn catalysts between 0 and 1.6 volts (Reproduced from Reference 51.)

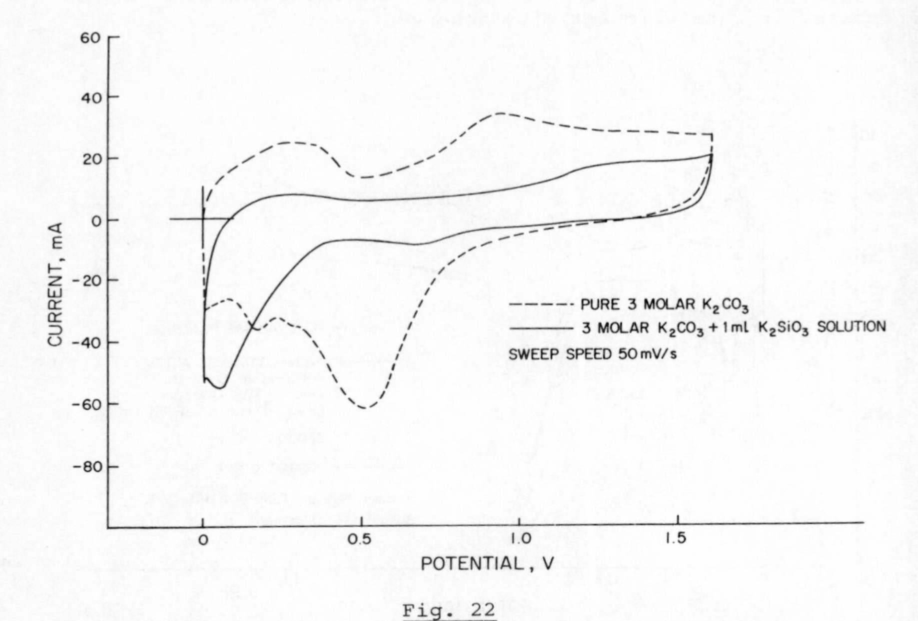

Fig. 22

Cyclic voltammograms of Pt/Sn electrodeposit in 3 molar K_2CO_3 and 3 molar K_2CO_3 + 1 mℓ K_2SiO_3 solution
(Reproduced from Reference 52.)

REFERENCES

1. J.W. Jenkins, B.D. McNicol and S.D. Robertson, Chemtech., 7, 316 (1977).
2. R.J. Cvetanovic and Y. Amenomiya, Adv. Catal., 17, 103 (1967).
3. D.R. Lowde, J.O. Williams and B.D. McNicol, Applications of Surface Science, 1, 215 (1978).
4. S.D. Robertson, B.D. McNicol, J.H. de Baas, S.C. Kloet and J.W. Jenkins, J. Catal., 37, 424 (1975).
5. V.C.F. Holm and A. Clark, J. Catal., 11, 305 (1968).
6. H.A. Benesi, L.T. Atkins and R.B. Mosely, J. Catal., 23, 211 (1971).
7. P.C. Aben, H. Van Der Eijk and J.M. Oelderik, Proc. 5th Int. Congr. Catal., 1, 717, (1973). Ed. J. Hightower (North-Holland, Amsterdam).
8. S.J. Gentry, N.W. Hurst and A. Jones, J. Chem. Soc. Faraday I (in press).
9. P.A. Jacobs, M. Tieren, J.P. Linart, J.B. Uytterhoeven and H. Beyer, J. Chem. Soc. Faraday I, 72, 2793 (1976).

10. S. Srinivasan and E. Gileadi, Electrochim. Acta, 11, 321 (1966).
11. P. Stonehart, H. Angerstein-Kozlowska and B.E. Conway, Proc. Roy. Soc., A310, 541 (1969).
12. B.E. Conway and H. Angerstein-Kozlowska, Proceedings of the Workshop on Electrocatalysis on Non-Metallic Surfaces, National Bureau of Standards Gaithersburg (1975). Ed. A.D. Franklin. p.107.
13. P. Stonehart and P.N. Ross, Catal. Rev. - Sci. Eng., 12, 1 (1975).
14. B.E. Conway, H. Angerstein-Kozlowska, W.B.A. Sharp and E. Criddle, Anal. Chem., 45, 1331 (1973).
15. M.W. Breiter, J. Electrochem. Soc., 112, 845 (1965).
16. O.R. Brown, Electrochim. Acta, 13, 317 (1968).
17. F. Will, J. Electrochem. Soc., 112, 481 (1966).
18. F.J. Kuijers, R.P. Dessing and W.M.H. Sachtler, J. Catal., 33, 316 (1974).
19. T. Biegler, D.A.J. Rand and R. Woods, J. Electroanal. Chem., 29 269 (1971).
20. B.E. Conway, E. Gileadi and M. Dzieciuch, Electrochim. Acta, 8, 143 (1963).
21. C.W. Tucker, J. Appl. Phys., 35, 1897 (1964).
22. S. Shuldiner, J. Electrochem. Soc., 112, 212 (1965).
23. R. Woods, Electrochim. Acta, 16, 655 (1971).
24. W.M.H. Sachtler, Catal. Rev. - Sci. Eng., 14, 193 (1976).
25. D.A.J. Rand and R. Woods, J. Electroanal. Chem., 36, 57 (1972).
26. B.G. Baker, D.A.J. Rand and R. Woods, J. Electroanal. Chem., 97, 189 (1979).
27. D.A.J. Rand and R. Woods, Proc. Symp. on Electrocatalysis, Electrochem. Soc. San Francisco. Ed. M.W. Breiter, p.140 (1973).
28. H. Binder, A. Kohling and G. Sandstede. "From Electrocatalysis to Fuel Cells" Ed. G. Sandstede (Univ. of Washington Press, Seattle, 1972), p.43.
29. S. Tsuchiya, Y. Amenomiya and R.J. Cvetanovic, J. Catal., 19, 245 (1970).
30. P.A. Attwood and B.D. McNicol, unpublished results.
31. P.A. Attwood, R.J. Bird, B.D. McNicol and R.T. Short, Paper presented at Euchem. Conf. on Dispersed and Colloidal Metal Particles in Catalysis, Namur, Belgium, 1977.
32. D.R. Lowde, J.O. Williams, P.A. Attwood, R.J. Bird, B.D. McNicol and R.T. Short, J. Chem. Soc. Faraday I (in press).
33. V.B. Hughes and B.D. McNicol, J. Chem. Soc. Faraday I, (in press).
34. A.C.C. Tseung and S.C. Dhara, Electrochim. Acta, 19, 845 (1974).
35. See for example M.J. Fuller and M.E. Warwick, J. Catal. 20, 441 (1973).

36. H. Binder, A. Kohling and G. Sandstede, "From Electrocatalysis to Fuel Cells" Ed. G. Sandstede, (University of Washington Press, Seattle, 1972). p.59.
37. E. Najdeker and E. Bishop, J. Electroanal. Chem., 41, 79 (1973).
38. T. Loucka, J. Electroanal. Chem., 31, 319 (1971).
39. N. Ramasubramanian, J. Electroanal. Chem., 64, 21 (1975).
40. V.B. Hughes, B.D. McNicol, M.R. Andrew and R.T. Short, J. Appl. Electrochem., 7, 161 (1977).
41. M.M.P. Janssen and J. Moolhuysen, Electrochim. Acta, 81, 869 (1976).
42. B.D. McNicol, unpublished results.
43. B.D. McNicol, J. Catal., 46, 438 (1977).
44. B.D. McNicol, Paper presented at Surface Reactivity and Catalysis Discussion Group Meeting, Nottingham 1976.
45. J.W. Jenkins, unpublished results.
46. B.D. McNicol and R.T. Short, J. Electroanal. Chem., 81, 249 (1977).
47. B.D. McNicol and R.T. Short, J. Electroanal. Chem., 92, 115 (1978).
48. A.S. Belyi, Reaction Kinetics and Catalysis Letters, 7, 461 (1977).
49. M.R. Andrew, J.S. Drury, B.D. McNicol, C. Pinnington and R.T. Short, J. Appl. Electrochem., 6, 99 (1976).
50. M.M.P. Janssen and J. Moolhuysen, J. Catal., 46, 289 (1977).
51. B.D. McNicol, R.T. Short and A.G. Chapman, J. Chem. Soc. Faraday I, 72, 2735 (1976).
52. B.D. McNicol, A.G. Chapman and R.T. Short, J. Appl. Electrochem., 6, 221 (1976).

PARTICLE METHODS OF CHARACTERISING CATALYSTS

PART I NEUTRONS

By

C.J. Wright

1. INTRODUCTION

In recent years neutron sources have become widely available both within the U.K. and in many other parts of the world, with the result that scientists can now look upon them as part of their portfolio of instruments for solving a particular problem in a way similar to that in which they may regard Raman or Mössbauer Spectroscopy.

It is the intention of this chapter to explain and illustrate the areas in catalysis research where advantages will accrue from using neutron techniques to illuminate a particular problem. The introduction will include an outline of the different classes of neutron experiment and a description of how the neutron nucleus interaction expressed through the cross-sections leads to many of the individual applications. After this the technique will be described area by area, with examples chosen from the literature to underscore points of importance. A recent review of neutron scattering from adsorbed molecules surfaces and intercalates emphasise certain aspects of the field in more detail, and may prove useful to the reader[1].

As a result of scattering, neutrons can change both their energies, $\hbar\omega$, and momenta, $\hbar\varkappa$, and the magnitudes of these changes, the energy or momentum transfer are defined as follows:

$$\hbar\omega = \hbar\omega_{initial} - \hbar\omega_{final}$$

and

$$\hbar Q = \hbar(\underset{\sim}{\varkappa}_{initial} - \underset{\sim}{\varkappa}_{final}) = \hbar(\varkappa_i^2 + \varkappa_f^2 - 2\varkappa_i\varkappa_f\cos\theta)^{\frac{1}{2}}$$

where θ is the scattering angle.

In terms of these variables, the neutron scattering field can be divided up into four experimental areas with the approximate boundaries:

(1) $\hbar\omega = 0$, $\hbar Q \leq 0.2$ Å^{-1}. The area of <u>small angle scattering</u> where information on the textural properties of materials such as pore structure and particle size can be determined.

(2) $\hbar\omega = 0$, $1 \leq Q \leq 10$ Å^{-1}. <u>Diffraction</u> for structure analysis.

(3) $\hbar\omega \leq 4$ meV, $Q < 4$ Å^{-1}. <u>Quasi-elastic scattering</u> for characterising the spatial and temporal aspects of atomic and molecular diffusion.

(4) $\hbar\omega > 4$ meV, $Q > 4$ Å^{-1}. <u>Inelastic scattering</u> for studying vibrational transitions of atoms and molecules.

The intensity of neutron scattering from a particular element is determined by the scattering lengths, b, of its constituent isotopes, and although the magnitudes of these scattering lengths have a weak dependence upon atomic number, proportional to $z^{1/3}$, there are sizeable fluctuations superimposed upon this from nucleus to nucleus. If the element comprising the scattering sample has just one isotope, of zero nuclear spin, then the scattering lengths of all the nuclei of that element will be the same, and coherent scattering will be the result.

In any other case where there is a mixture of isotopes, and/or nuclei with non-zero nuclear spin, then the scattering consists of a coherent component, the cross-section of which, σ, is related to the sum of the mean scattering lengths of the nuclei, $\sigma_{coh} = 4\pi\bar{b}^2$, and an incoherent component whose cross-section is related to the deviations of the scattering lengths from this mean, $\sigma_{inc} = 4\pi\,(\overline{b^2} - \bar{b}^2)$. The values of the cross-sections for particular atoms of importance in catalysis are shown in Table 1 together with neutron, X and gamma ray total cross-sections (absorption plus scattering)[2,3]. The X-ray absorption cross-sections are those for copper Kα radiation and the gamma rays are those for ^{57}Fe.

The intensities of the scattering in the four areas mentioned above depend on different aspects of the total scattering. The small angle scattering depends upon scattering length density differences between different regions of a material, whilst the inelastic and quasi-elastic scattering that have been predominantly used in catalysis research depend upon the magnitude of the incoherent cross-sections of the atoms investigated. Finally, this contrasts with diffraction where the incoherent scattering is an unwanted background and all the information is obtained in the scattering arising from the coherent cross-sections.

On inspecting this table several important generalisations emerge. Firstly, the very low values of the neutron absorption cross-sections, for heavy elements, relative to those for X and gamma rays make it much easier not only to investigate bulk samples and consequently representative samples of material, but also to explore their properties under the extreme conditions that are found in furnaces, pressure vessels and cryostats. Secondly, one can see that the incoherent cross-section of hydrogen is considerably greater than that from any other atom, which means that the technique is a highly selective tool for investigating the structure and dynamics of hydrogenous materials. This situation contrasts with the problem of exploring the dynamics, say, of hydrogenous molecules with photon spectroscopy where the experimenter can be confronted with relatively low absorption intensities of vibrations of bonds between an element and hydrogen atoms. Consequently this leads to some of the most interesting applications of inelastic neutron scattering in surf-

Table 1

Neutron, X and γ ray cross sections for some elements of importance in catalysis

	Neutrons			X-rays	γ-rays
	σ_{coh}	σ_{inc}	σ_{abs} (λ = 1.08 Å)	σ_{abs} (λ = 1.54 Å)	σ_{abs} 14.4 KeV
H	1.76	79.7	0.19	0.7	
D	5.59	2.0	-	0.7	
Be	7.53	-	-	22	6.3
C	5.56	-	-	92	
O	4.23	-	-	305	58.9
Aℓ	1.54	-	0.13	2,180	4.3
Si	2.22	-	0.06	2,830	
S	0.99	0.2	0.28	4,740	
Fe	11.3	0.5	1.4	28,600	6.33
Ni	13.3	4.7	2.7	4,450	
Cu	7.26	1.2	2.2	5,580	
Mo	5.98	-	1.4	25,900	
Pd	4.52	0.3	4.0	36,400	
Pt	11.3	0.9	5.0	64,900	

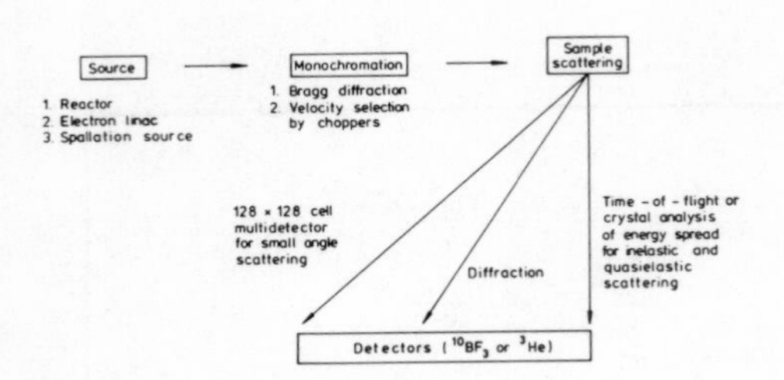

Fig. 1

Experimental neutron scattering; a mnemonic

ace vibrational spectroscopy. It can also be seen that protium and deuterium have different cross-sections, a fact that can be used by the experimenter to emphasise or de-emphasise certain aspects of the scattering from a molecule at will.

Finally, it is of interest to observe the relative sizes of the cross-sections of pairs of catalytically important elements such as silicon and aluminium or copper and nickel. In the first case the approximate doubling of the difference in the coherent cross-section relative to the X-ray cross-section difference makes it useful to apply the neutron method for obtaining further information that can be obtained with X-rays on the structure say, of zeolites, where there are many questions remaining to be answered on silicon, aluminium ordering.

The experimental aspects of neutron scattering are very adequately covered by existing text books[4] and very little is said about them in this particular article. Fig. 1 is included as a hopefully useful mnemonic.

2. INELASTIC SCATTERING

The inelastic scattering of neutrons allows an experimenter to determine the vibrational energy levels of a particular material, and as a consequence it is a technique that is complementary to infra-red, Raman, and inelastic electron scattering. Obviously it is at its most powerful when its advantages are fully exploited:

(1) The ability of neutrons to penetrate materials comprising elements of high atomic number.
(2) The absence of selection rules in inelastic neutron scattering.
(3) The ability of an experimenter to predict quantitatively the scattering intensities.
(4) The cross-section differences which enable one to examine selectively the presence of one nucleus amongst others.

The first advantage is well-illustrated by the ability of neutron scattering to observe and characterise hydrogen within and on the surface of metals, not just the simple ones like palladium, but the increasing number of metal chalcogenides which change their band structure on hydrogen sorption often undergoing semi-conductor - metal transitions. Tungsten trioxide and molybdenum trioxide for example, both absorb hydrogen and the resulting materials have been investigated spectroscopically and structurally with neutrons[5,6,7]. From their vibration spectra it is straightforward to identify the binding of the hydrogen in the different phases either in terms of hydroxyl groups or hydrates. Fig. 2 shows a spectrum recorded from $H_{0.4}WO_3$ which, with its single excitation at 1146 cm^{-1}, shows clearly that it is a metallic (in both senses) hydroxide.

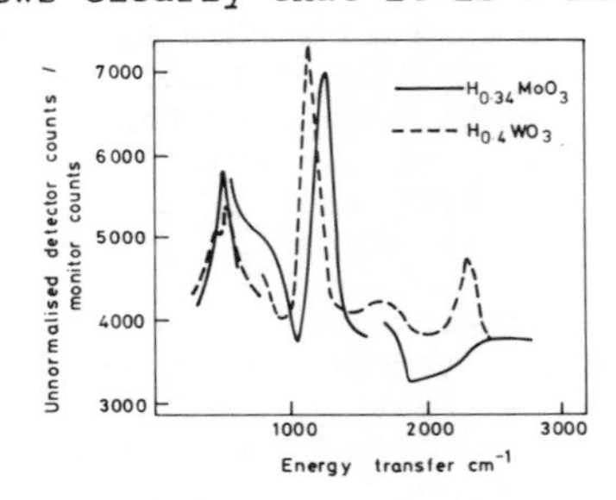

Fig. 2

Inelastic neutron scattering of $H_{0.4}WO_3$ and $H_{0.34}MoO_3$

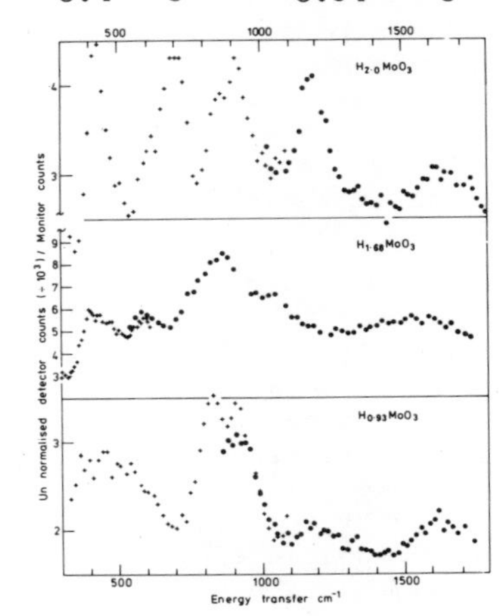

Fig. 3

Inelastic neutron scattering spectra of $H_{0.93}MoO_3$, $H_{1.68}MoO_3$

Fig. 2 also shows that $H_{0.34}MoO_3$ has the same structure, whereas Fig.3 shows that on increasing the hydrogen concentration $H_{1.0}MoO_3$, $H_{1.6}MoO_3$ and $H_{2.0}MoO_3$ reveal the spectra of hydrates. It is important to emphasise that when $H_{0.4}WO_3$ was investigated with Raman scattering no useful information could be obtained due to its high absorption[8]. Supported molybdenum trioxide is a very widespread catalyst, and it is tempting to suggest that investigations similar to those carried out on the bulk materials would show corresponding changes.

The absence of selection rules becomes important when an experimenter wishes to characterise all the hydrogen that may be present within a material, and such a case occurs with the sorption of hydrogen by molybdenum sulphide catalysts. It has been known for a considerable time that molybdenum and tungsten sulphide prepared in a form of small crystallite size and thus containing a large number of defects, would absorb hydrogen, but not until recently had there been any attempt to discover the structure of the hydrogen molybdenum sulphide system and how it might influence the properties of the catalyst.

The first investigations of degassed "catalytic" molybdenum sulphide showed it to have an inelastic scattering spectrum very different to that of MoS_2 prepared _in-vacuo_ by direct synthesis from its constituent elements[9] - Fig. 4. The differences can be explained only by assuming the presence of an appreciable quantity of hydrogen which still remained dissolved in the lattice of the "catalytic" material and which, by exchange measurements, was found later to amount to $H_{0.011}MoS_2$. Subsequent investigations show this hydrogen to be bound to the sulphur atoms of the sulphide rather than the metal atoms, and in this way the structure differed from that of H_xMS_2 (M = Ta, Nb).

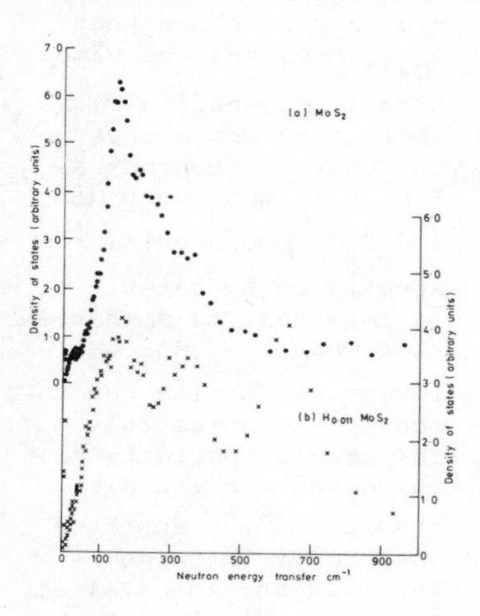

Fig. 4

Inelastic neutron scattering spectra of (a) MoS_2 and (b) $H_{0.011}MoS_2$.

This structural investigation is a highly important and necessary staging post on the way to obtaining a clear view of how the flow of hydrogen into, through and from the bulk of the catalyst influences catalytic activity. In particular this presence of hydrogen may be of importance in replenishing the surface of the catalyst with hydrogen at times of temporary depletion. Thus a parallel can be drawn with the oxygen storage in the catalysts referred to in the chapter by Acres[10].

Further important benefits that arise from the absence of selection rules are the ability to observe surface vibrations with amplitudes parallel to a metal surface and surface librations, both classes of excitation being difficult to observe in other ways. Extensive work has been carried out on adsorption by Raney nickel[11,12], initially the characterising of the hydrogen adsorbed by this material, but then later using this information to demonstrate surface reactions. One set of experiments showed that the adsorption of water upon the metal surface led at coverages, of 0.2 to 0.3 of a monolayer, to molecular dissociation, whereas at higher coverages $Ni\text{-}OH_2$ bonds were formed[13]. It is important to emphasise here the convincing nature of the neutron data which, being sensitive to all the hydrogenous species in the sample, meant that the experimenters have no need to consider other species which could be present on the surface but which, for some reason, were not giving rise to scattering.

Another set of experiments on Raney nickel made use of the final advantage mentioned above, which was that the intensity of inelastic neutron scattering can be predicted quantitatively from a knowledge of force constants, masses and cross-sections[14†], an important contrast with optical spectroscopy where the intensities are determined by the unknown response of electrons to nuclear displacements. The experimenters predicted the inelastic scattering spectrum of benzene adsorbed on Raney nickel using force constants that they had obtained previously from benzene-$Cr(CO)_3$. Fig. 5 shows that although they found good agreement at some energies between the experimental and predicted spectra, in order to reproduce exactly the experimental data it was necessary to include a component of scattering at around 1000 cm^{-1}. Since the form of this extra component was identical to that produced by scattering from hydrogen adsorbed upon Raney nickel, the implication was that 15% of the benzene molecules on the surface had undergone dissociative loss of hydrogen.

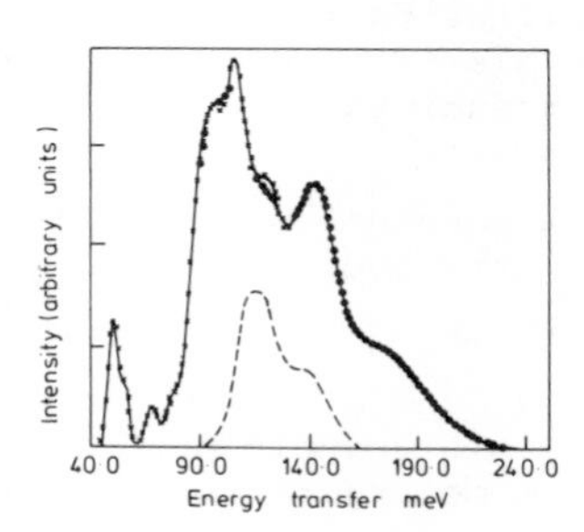

Fig. 5

Inelastic neutron scattering spectra from benzene absorbed upon Raney nickel (o and x experimental points, ----- the additional contribution required to equate the experimental with the predicted data).

3. QUASI-ELASTIC SCATTERING

Before the advent of quasi-elastic neutron scattering, the only insight into the diffusive behaviour of adsorbed gases, apart from field ion microscopy measurements under U.H.V. conditions, came from the measurement of mass flow. The experi-

---

† The neutron scattering cross-section from an incoherently scattering atom executing simple harmonic motion can be written:

$$\frac{d\sigma^2}{d\eta dE} = \sigma \frac{k}{k_o} \exp\left(\frac{-hQ^2}{2M\omega} \coth \frac{h\omega}{2k_BT}\right) I_n \left(\frac{hQ^2}{2M\omega} \operatorname{cosech} \frac{h\omega}{2k_BT}\right)$$

$$\times \exp\left(- h\omega/2k_BT\right)$$

where M is the mass of the scatterer and I_n is a modified Bessel Function of the first kind. For molecular vibrations the cross-section, although more complex, has the same components[15].

menter would compare the total diffusion rate of a gas through a sintered disc of the material in question with the diffusion rate of a "non-interacting" gas such as helium. If the approximation could be made that in the second case the only diffusion was Knudsen diffusion, then, after making appropriate corrections for mass differences, subtraction of the two rates lead to the surface diffusion coefficients. Doubts over the reliability of this method, because of the approximations involved, together with its macroscopic nature, has meant that little advance in catalytic understanding occurred as a consequence of the measurements. It is also apparent that the technique is limited to a study of unsupported surfaces. The neutron technique relies on the observation of quasi-elastic broadening of a neutron energy distribution by the diffusing atoms, and from the shape of this quasi-elastic broadening and its dependence upon the angle of scatter, then information both about the rate of the microscopic steps that make up the diffusion process and its geometric form can be deduced. For example, an atom diffusing in two dimensions in a jump-like manner will lead to a scattering law $S(Q,\omega)$ which is proportional to the scattered intensity[16]:

$$S(Q,\omega) = \frac{\frac{\ell^2}{4\tau} Q^2}{\omega^2 + \left(\frac{\ell^2}{4\tau} Q^2\right)^2} \quad \text{when } \frac{\ell^2}{4\tau} Q^2 < \text{resolution width of the spectrometer}$$

where τ is the time interval between jumps, and ℓ is the jump length involved. Most experiments with this technique have been on physisorbed gases[1], but recently some experiments relevant to catalysis have also been undertaken. For instance, an attempt has been made to enquire into whether the different behaviour of nickel and platinum with respect to hydrocracking and hydrogenolysis could be due to the different diffusion coefficients of hydrogen over the respective surfaces[17,18]. Raney nickel and a platinum Y zeolite have been investigated and residence times on the metal sites of $2.7 \pm 0.5 \times 10^{-9}$ seconds at 150°C and $5.5 \pm 2.5 \times 10^{-9}$ seconds at 100°C have been measured respectively. These initial experiments require some refining before a definitive answer to the original question is obtained. On the other hand, the experiments show what could be achieved if one wished to investigate quantitatively, frequently invoked surface phenomena such as spillover.

To take an example from another field; in recent years considerable information has come forward on the diffusion coefficients of hydrogen atoms within metals and how they are affected by the presence of impurity atoms within the metal which act as traps[19]. One can easily envisage how the same techniques could be applied to the surface diffusion of active species, and the experimenter could investigate how surface composition and surface charge influenced surface mobility.

Rotational diffusion of adsorbed molecules is another process which gives rise to quasi-elastic scattering and one also that is valuable for characterising the steric interactions experienced by molecules trapped, for example, within a zeolite cage. Measurements on ethylene adsorbed by a sodium 13X zeolite[20], Fig. 6, show the quasi-elastic scattering to consist of two components, one due to the translational diffusion, the narrow one, and the other due to rotational diffusion, the broad one. It can be shown from an analysis of these widths that the molecule was executing about 20 rotational jumps between each translational step within the cage. From an analysis of the angular dependence of the scattering it could be shown that the results were only compatible with the prediction of molecular reorientation about an axis such that the radius of gyration of the atoms was 1.5 Å. This would only be the case if the molecule was rotating about the line joining its mid-point to the sodium atom in the lattice. Further work of this type may be expected to reveal much of interest concerning the sensitivity of molecules within zeolite cages to the steric influences that surround them.

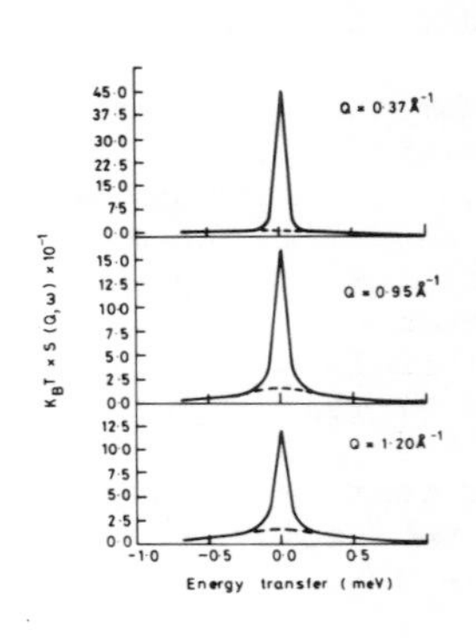

Fig. 6

Quasi-elastic scattering from ethylene absorbed by a sodium 13X zeolite.

4. TEXTURAL STUDIES AND SMALL ANGLE SCATTERING

In common with all other small angle techniques, the neutron method relies on diffraction by inhomogeneities such as pores or particles, the scattering length density of which differs from that of its surrounding matrix. The information obtained can then be used to determine the sizes or the surface areas of these inhomogeneities and, after more detailed analysis, their size distribution. Of interest in catalysis are the prospects the technique offers of characterising microporous materials, of following surface areas and pore sizes *in situ* during sintering and gassification processes, and of exploring the structure of catalyst support precursors.

Since the spectra arises from scattering length density differences, the technique is unable to differentiate between open and closed pores, when both of these are present, a factor that has to be borne in mind in any data interpretation. On the other hand, this difference between the information obtained from small angle scattering and gas adsorption can be used to advantage, for example, in exploring mechanisms of pore blockage. Contrasting the information available from the two techniques could lead to a distinction being possible between uniform pore and pore mouth blocking processes.

Neutron small angle scattering is complementary to the similar X-ray technique, and as elsewhere in this chapter, its relative advantages arise primarily from cross-section differences. Both *in situ* analysis and the observation of large and representative samples become possible with neutrons. To examine MoS_2 catalysts, for example, with X-rays, a sample of thickness of a few microns is necessary to avoid heavy absorption and multiple scattering by the sample, whereas thicknesses of the order of a few millimetres can be tolerated in a neutron experiment.

It is also of interest that in three component systems X-rays and neutrons may be sensitive to different regions within a material. For example, in a platinum-carbon catalyst containing a certain platinum concentration the X-ray small angle scattering will be dominated by the sizes of the platinum particles, whereas for the same sample the neutron scattering will be dominated by the sizes of the pores. This is a consequence of the similar neutron scattering length densities of platinum and carbon.

A small angle scattering curve, such as that shown in Fig.7, is capable of being fitted analytically in two regions[21] - that where QR_g is < 1 where R_g is the radius of the gyration of any inhomogeneity, and that where $QR_g > 4$. (These constants and the formalisms given below all make the assumption that the shape of the inhomogeneities is spherical and there are corresponding relationships for other geometries.)

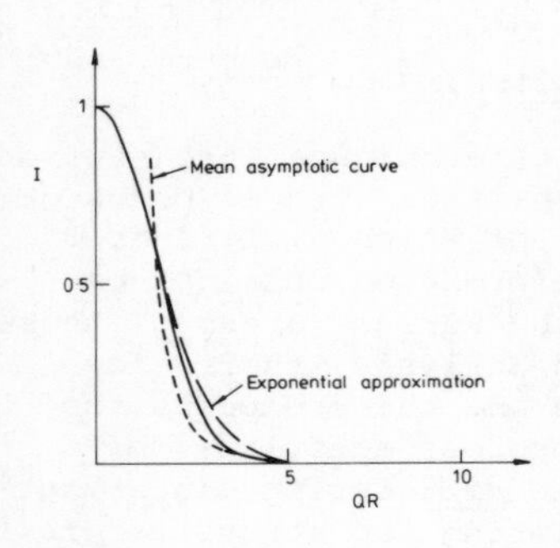

Fig. 7

A typical small angle scattering curve showing its relationship to analytical formulae. (The exponential approximation - Guinier's Law; and the mean asymptotic curve - Porod's Law.)

The first region is known as the Guinier region where the curve follows the law: $I \propto \exp \frac{\left(-Q^2R_g^{\,2}\right)}{3}$ and where plots of Log I v Q^2 enable R_g to be determined if the experimenter can ignore the problems of interparticle interference[21]. The second is the Porod region where the scattering decays as Q^{-4}. From the constant of proportionality in this region the total surface area of the inhomogeneities per unit mass of sample can be calculated :

$$S = \frac{1}{\rho} \pi c(1-c) \frac{\underset{Q \to \infty}{\text{Limit}} IQ^4}{\int_0^\infty Q^2 I(Q) dQ}$$

where ρ is the apparent density of the sample, and c is the volume fraction of the sample occupied by the pores or other inhomogeneity. More sophisticated analysis entails Fourier transforming the whole of the small angle scattering pattern so as to obtain the distribution of inhomogeneity lengths within the sample; but, to the author's knowledge, no neutron scattering data from porous materials has been treated in this way.

Some recent results have been concerned with two aspects[22]: the demonstration that for meso-porous materials, where there is little doubt about the validity of isotherm analysis, small angle scattering and isotherm techniques yield the same answer, and secondly, the determination of the surface areas of microporous materials. It is well known that to characterise materials with pores of diameters < 20 Å there are difficulties in using conventional isotherm analysis for determining surface areas because of the slow rate of gas uptake by such pores at low temperatures due to activated diffusion. On the other hand, these pores may be critically important in the higher temperature regimes in which catalysts are used. The importance of obtaining experimental data on micropores from an independent technique such as small angle scattering would be that it would enable the basis on which adsorption measurements in this area are interpreted to be examined further, for there is presently no good reason for assuming that the accepted molecular area of an adsorbate is of special significance in calculating the surface area of micropores.

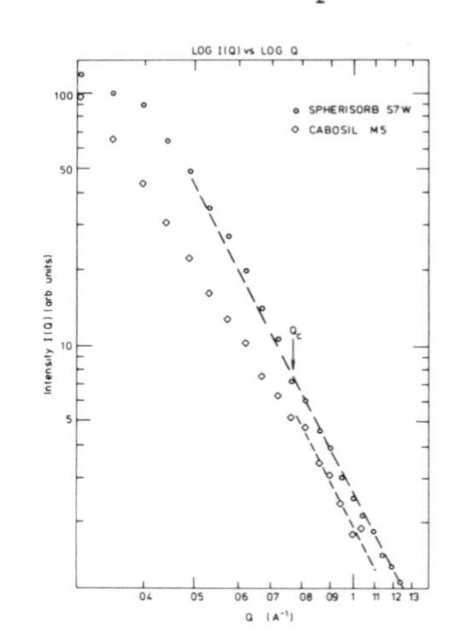

Fig. 8

The small angle scattering from Spherisorb S7W silica showing lines of slope -4.

One typical material examined in the meso-porous class is a silica, Spherisorb S7W, whose small angle scattering is represented in Fig. 8 as a log log plot. The dashed line has a slope of -4 as predicted by Porod. The surface area obtained from this curve is 246 ± 10 m^2g^{-1} which compares with the result obtained from a B.E.T. analysis of the nitrogen isotherm of 232 m^2g^{-1}. It is gratifying that these two results, both based on quite different assumptions and underlying physical principles, should yield such similar values for the surface areas.

The microporous materials that have been investigated include a number of coal chars, materials produced during coal gasification procedures, and a wood charcoal. In Table 2 the surface areas measured with nitrogen at 80 K, by a one point flow method, are contrasted with the results obtained from small angle scattering.

Table 2

Determinations of the surface areas of microporous materials by absorption and small angle scattering measurements

Surface Area m^2/gm \ Material	coal char (a)	coal char (b)	Alderwood Charcoal
N_2 at 80^oK (One point flow determination)	151	0.2	16
Small Angle Scattering	130	110	200

It can be seen that, because of activated diffusion, the nitrogen results bear no relation to the surface area that might be accessible at high temperatures, and the conventional way of overcoming this is to measure the gas uptake with CO_2 at a number of temperatures 195, 273, 293 and 313 K, using high pressures. The small angle scattering method offers a much simpler and more straightforward single determination approach. It is worth emphasising that recent developments in small angle scattering have enabled area detectors now to be generally available which collect the scattering from a sample over all small angles simultaneously so that surface areas can now be measured in minutes or at the most an hour or two. This is to be compared with the innumerable and lengthy equilibrations needed to characterise microporous materials by gas adsorption.

Lastly, but very importantly, the neutron scattering technique can be used to characterise the gel precursors of catalyst supports. Because there is a difference between the scattering length densities of water and alumina or silica, for example, the nascent structure in the colloid from which the support originates can be measured. By following the variation of structure with parameters such as time, concentration and pH, insight is obtained into ways of modifying the final support structure. Fig. 9 shows the radial distribution function of the interparticle separation in a concentrated silica sol[23], obtained by transforming the small angle scattering data.

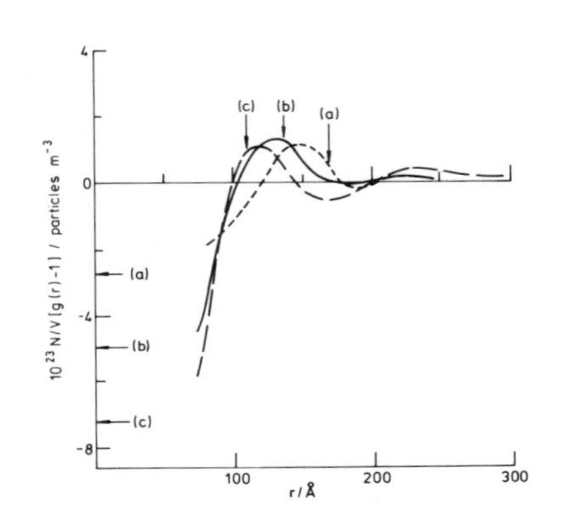

Fig. 9

Radial distribution functions of silica sols (R ≈ 4 nm) of different concentration (% W/W SiO_2): (a) 15, (b) 23, (c) 35. The volume fraction is 0.074, 0.12 and 0.19 respectively.

5. *IN SITU* ANALYSIS

In this chapter *in situ* analysis has been chosen for treatment as a separate topic so as to emphasise what the author contends will be a very important application of neutron scattering in the future. Catalysts are frequently used under conditions of high temperatures and gas pressures where the physical properties of the materials that make up these catalysts are generally insufficiently characterised. That is true of the bulk properties as well as the surface properties. A difficulty associated with this characterisation is that apart from neutrons most measuring probes suffer from high absorption by the materials that are necessary to contain samples under the appropriate conditions. Referring to Table 1 illustrates just how low neutron absorption cross-sections are in comparison with those of other probes, especially if one's interest centres on the types of material that are able to withstand chemical attack, pressure and high temperature.

Some equipment has recently been constructed at Harwell for *in situ* experiments, and its design considerations nicely illustrate some of the potential uses and the problems that the technique offers. The equipment has been constructed to explore the structures of hydrosulphurisation catalysts which are known to absorb hydrogen at atmospheric pressure and 300°C and which, from inspection of the isotherm, Fig. 10, probably absorb even more under the 100 atmospheres of pressure in which they can be used commercially. The catalysts have also been shown recently to have activities which are critically dependent upon the partial pressures of hydrogen and hydrogen sulphide in which they

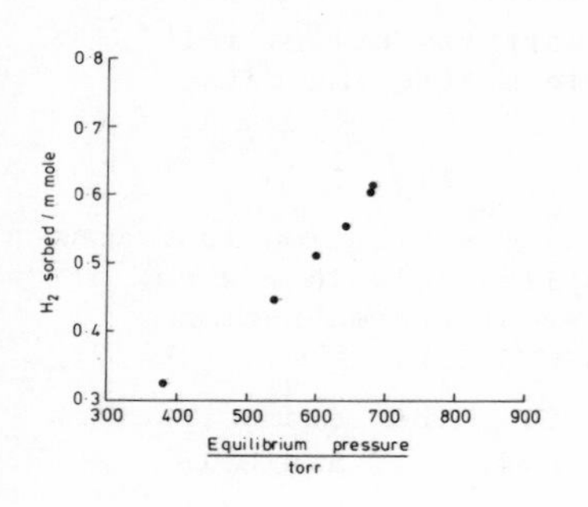

Fig. 10

The hydrogen sorption isotherm for "catalytic" molybdenum sulphide at 300°C.

are immersed, and it has been suggested that this is due to a possible sensitivity of the structure of these catalysts to partial pressure variations[24].

To determine the structure of these materials under such conditions, a container has to be selected which is inert to hydrogen sulphide, which will withstand the necessary pressure and temperature, and whose properties under these conditions are sufficiently well known that it can be designed to be safe without the necessity for lengthy test procedures. If neutrons are usable, then the problem becomes simple, for there are a number of stainless steels that are suitable as a containment material whose transmissions are high. On the other hand, if one should wish to use X-rays, then the experimenter is forced to use beryllium, a metal whose resistance to corrosion under the differing operating conditions that face a chemical engineer is largely unexplored.

The one-inch outside diameter micro-reactor that has been constructed has walls of 0.762 mm thick stainless steel through which hydrogen will diffuse at a rate of 6 cc's per hour during the experiment - an important design consideration, and which will have a neutron transmission of 84% for an incident neutron wavelength of 1.4 Å.

6. NEUTRON DIFFRACTION

In view of the general awareness of the advantages and disadvantages of neutron diffraction techniques and because many of its applications in catalysis are similar to those already referred to in this chapter, only a few general points will be made here.

The technique, like all neutron techniques, is one that measures bulk properties and so is successful in looking at surfaces if the material is all "surface" such as a zeolite, or if the experimenter is either able to highlight the surface with an atom of high cross-section or use a high surface area material. By these latter methods, many experiments have been done characterising the structures of physisorbed gases, but as yet similar work has not been completed with chemisorbed systems.

Neutrons will always have the advantage over X-rays for characterising hydrogenous phases or layers, and as such have been used to characterise $H_{0.4}WO_3$[4], $H_{0.36}MoO_3$[6] and $H_{0.01}MoS_2$[8].

Fig. 11 shows the diffraction pattern from "catalytic" MoS_2 before and after the sorption of deuterium (which has a higher ratio of coherent to incoherent scattering than hydrogen, see table 1), together with the difference pattern. Analysis of this difference pattern shows the hydrogen to be entering the lattice between the layers of the sulphur atoms. Another area where neutron diffraction will always be usable to advantage is in the in-situ examination, not only of gas/solid but also of liquid/solid interfaces. An illustration of the applicability of the technique to the study of electrodes in-situ has been described recently in a paper showing how the bulk structure of an electrode changed as a result of electrolysis in aqueous solution[25].

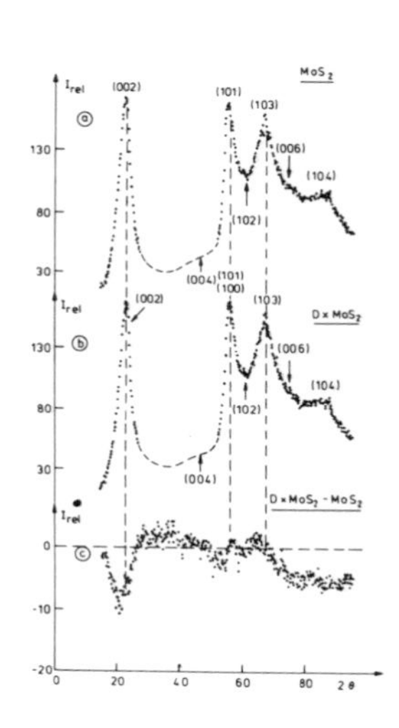

Fig. 11

The neutron diffraction patterns of "catalytic" molybdenum sulphide before and after deuterium sorption, and their difference.

REFERENCES

1. P.G. Hall and C.J. Wright in "Chemical Physics of Solids and their Surfaces", Vol. 7, Chem.Soc. (1978).
2. G.E. Bacon "Neutron Diffraction", Oxford University Press, London (1975).
3. Handbook of Chemistry and Physics 38th Edn. C.R.C. Press Cleveland (1956).
4. G.C. Stirling in "Chemical Applications of Thermal Neutron Scattering", Oxford University Press, Oxford (1973).
5. P.J. Wiseman and P.G. Dickens, J.Solid State Chem., 6, 374 (1973).
6. C.J. Wright, J. Solid State Chem., 20, 89 (1977).

7. P.G. Dickens, J.J. Birtill and C.J. Wright, J. Solid State Chem., 28, 185 (1979).
8. M.J. Sienko and H. Oesterreicher, J. Amer. Chem. Soc., 90, 6568 (1968).
9. C.J. Wright, D. Fraser, R.B. Moyes, C. Riekel, C.S. Sampson and P.B. Wells (In press).
10. G. Acres (this volume).
11. A.J. Renouprez, P. Fouilloux, G. Coudurier, D. Tochetti and R. Stockmeyer, J.C.S. Faraday I, 73, 1 (1977).
12. C.J. Wright, J.C.S. Faraday Trans. II, 73, 1497 (1977).
13. A.J. Renouprez, P. Fouilloux, J.P. Candy and J. Tomkinson, Surface Science, 83, 285 (1979).
14. A.J. Renouprez, H. Jobic and J. Tomkinson (to be published).
15. W. Marshall and S.W. Lovesey. Theory of Thermal Neutron Scattering. Oxford (1971).
16. R. Stockmeyer, Ber Bunsenges physik Chem., 80, 625 (1976).
17. A.J. Renouprez, P. Fouilloux, R. Stockmeyer, H.M. Conrad and G. Goeltz, Ber Bunsengersetschaft phys. Chem., 81, 429 (1977).
18. A.J. Renouprez, R. Stockmeyer and C.J. Wright, J.C.S. Faraday Trans., (In press).
19. D.H. Broderick, G.C.A. Schuit and B.C. Gates, J. Catal., 54, 94 (1978).
20. C.J. Wright and C. Riekel, Mol. Phys., 36, 695 (1978).
21. A. Guinier and G. Fournet. Small Angle Scattering of X-rays, John Wiley (1955).
22. P.G. Hall, A. Pidduck and C.J. Wright (to be published).
23. J.D. Ramsay, Discussions of the Faraday Soc., 65, 139 (1978).
24. D. Richter, J. Töpler and T. Springer, J. Phys. F., 6, L93 (1976).
25. C. Riekel, H.G. Reznik, R. Schöllhorn and C.J. Wright, J. Chem. Phys., 70, 5203 (1979).

PART II: THE APPLICATION OF ION BEAMS TO CATALYST RESEARCH

By

J.A. Cairns

1. INTRODUCTION

The impression that catalysis is a subject which lies somewhere between art and science arose in part because it was often difficult to reproduce catalyst performance, or to account for the success of a particular preparative route. However, this situation is changing rapidly, due in large measure to the influence of modern physical techniques, capable of elucidating the precise combination of parameters which ensure desired performance. Most of the current techniques now in vogue involve irradiating the catalyst with electrons (e.g. Auger electron spectroscopy) or photons (e.g. X-ray photoelectron spectroscopy), although at least one uses ion bombardment (viz. secondary ion mass spectrometry). It is intended here to demonstrate that ion bombardment offers a much wider prospect of obtaining analytical and structural information from catalyst materials. In order to place this in context, Table 1 lists the most commonly applied current techniques and indicates the complementary nature of the ion beam applications to be described here.

From these techniques little information will emerge concerning the chemical state of the atoms probed, because we will be inducing collisions directly with target nuclei, or at least, very close to them. On the other hand, it should become clear that some of the techniques offer prospects which could be of great potential interest to a catalyst chemist. For example, the use of prompt nuclear reactions allows the detection of light elements (including hydrogen) with high sensitivity and can map out a distribution of such elements with ca 2 μm spatial resolution across a catalyst; proton-induced X-ray analysis can detect trace elements, even in non-vacuum conditions; and Rutherford backscattering provides, non-destructively, a depth distribution profile of a heavy element in a light matrix with a depth resolution of as little as 10 nm.

Now there may be an understandable reluctance on the part of many catalyst chemists to become involved with ion accelerators. These machines were designed originally for research in nuclear physics, and, as such, tended to be rather temperamental, complicated devices. One of the most important aims of this article is to dispel this image. Most accelerators nowadays are designed for materials analysis, and as such are easy to use. An increasing number of industrial companies and university departments are installing their own accelerators, as they begin to appreciate their special attributes. For applications which require access to a large machine there will be usually a specialist operator available to provide the required beam at the appropriate energy on target so that the user is free to concen-

Table 1*

Summary of some current Analytical and Diagnostic techniques, including those based on Ion Beams

Bombardment of Target with:-

ELECTRONS	PHOTONS	IONS	Emission from Target of:-
LOW ENERGY ELECTRON DIFFRACTION AUGER ELECTRON SPECTROSCOPY	X-RAY PHOTO-ELECTRON SPECTROSCOPY		ELECTRONS
ELECTRON PROBE MICROANALYSIS	X-RAY DIFFRACTION X-RAY FLUORESCENCE	PROTON-INDUCED X-RAY EMISSION THIN LAYER ACTIVATION ION-INDUCED LIGHT EMISSION	PHOTONS
		PROMPT NUCLEAR INTERACTIONS RUTHERFORD BACKSCATTERING SPECTROMETRY LOW ENERGY ION SCATTERING RESONANT ION BEAM BACKSCATTERING (POROSITY) SECONDARY ION MASS SPECTROMETRY	IONS

* Personal communication from J.F. Ziegler, IBM Research, New York.

trate his attention on the target chamber. It is appropriate, therefore, at this point to consider the basic attributes of a typical accelerator and target chamber system.

2. THE ION ACCELERATOR AND TARGET CHAMBER

All accelerators have certain common features, which may be appreciated by reference to Fig. 1, which illustrates the Harwell Cockcroft-Walton machine, capable of producing a wide range of energetic ions at energies up to several hundred keV.

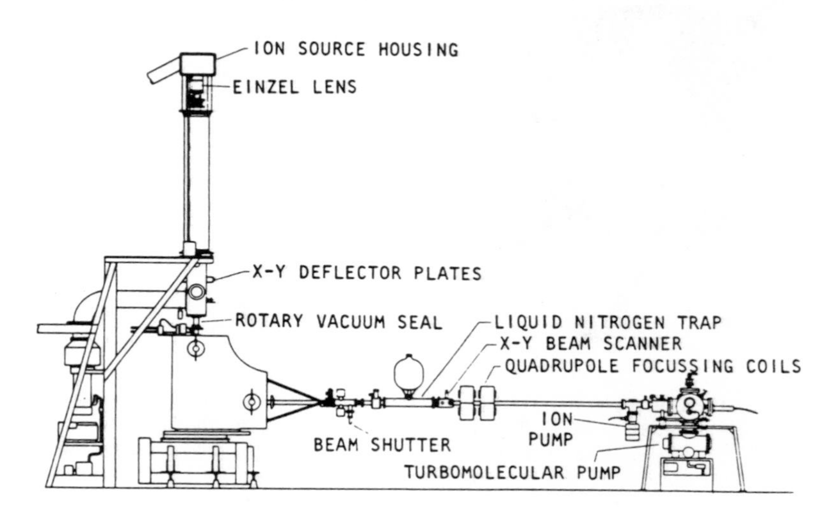

Fig. 1

A typical accelerator system. The ions emerging from the vertical tube are selected by the magnet and pass along the horizontal beam line, within which they can be steered and focussed, before entering the target chamber on the extreme right.

Fig. 1 shows, at top left, the source where the ions are created, and from which they are extracted at an appropriate potential. The particular ions of interest are selected by the large magnet, from which they pass along the horizontal beam tube, where they can be scanned, steered and focussed, before passing into the target chamber, which is shown here pumped by a turbomolecular pump. The target chamber is designed to observe the interactions between the ion beam and the specimen. For this reason it usually contains, or allows access to, one or more detectors, which measure the consequences of these interactions. A schematic illustration of a typical target chamber assembly is shown in Fig. 2. The detector here was installed to measure the characteristic x-rays which are produced when accelerated protons

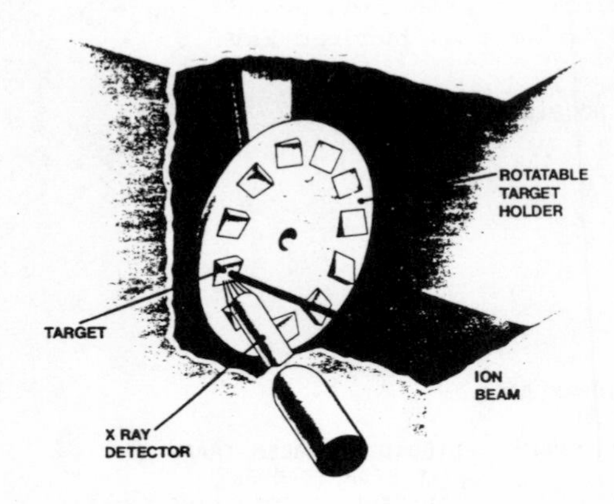

Fig. 2

A simple target chamber assembly. The targets, mounted on a disc, can be moved into the path of the beam sequentially by means of an externally-mounted vacuum-tight drive. The x-ray detector shown here measures the characteristic x-rays arising from interactions between the ion beam and the targets.

interact with solid targets, although other detectors (e.g. to measure energetic charged particles) could also be present. In order to cover the whole spectrum of x-rays produced, two types of x-ray detector may be employed: a gas flow proportional counter to cover the x-ray region from ~ 100 eV to several keV (such as the end-window variable geometry proportional counter as described by Cairns et al.[1]) and a Si(Li) solid state detector to cover the energy range from ~ 1 keV to several tens of keV. It may be of interest to note that the lower energy limit of the Si(Li) detector is determined by the beryllium window which is installed conventionally to protect the cooled detector from water condensation. If this window is replaced by a thin plastic window in a specially designed system, then the Si(Li) detector can be used for even ultra-soft x-rays, as shown in Fig. 3 (Cairns et al.[2]) which illustrates also the superior resolution of the Si(Li) detector as compared to a gas flow proportional counter. If even higher resolution is required, then x-ray spectrometers[3] may be used, as shown in Fig. 4.

This equipment would be suitable for the first ion beam technique to be described, namely proton-induced x-ray emission.

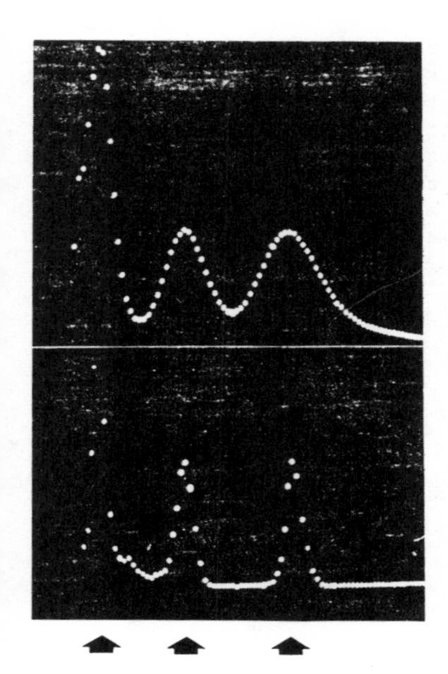

Fig. 3

A comparison of the resolving powers of a gas-flow proportional counter (top) and a Si(Li) detector for carbon K, copper L and silicon K x-rays. The Si(Li) detector in this case was mounted in a special assembly in which its conventional berrylium window was replaced by a 6 μm mylar window, in order to transmit the carbon K x-rays.
(Cairns et al.[2])

Fig. 4

A schematic illustration of the different detectors which may be used to resolve x-rays arising from ion bombardment of solid targets. The advantage of the Si(Li) or Si(Ge) detector is that it can provide an instantaneous spectrum of the x-rays, whereas the crystal spectrometer requires each x-ray to be examined sequentially. On the other hand, the spectrometer provides superior resolution and can be used also for the detection of very low energy (few hundred eV) x-rays.
(Folkmann[3])

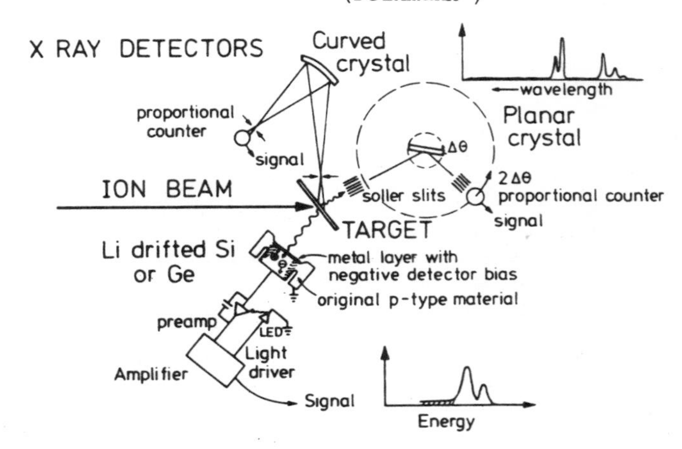

3. PROTON-INDUCED X-RAY EMISSION (PIXE)

The classical method of generating x-rays from solid targets, and the one still used in instruments such as the electron probe micro-analyser, is by electron irradiation. It is the purpose of this section to explain that there are sound reasons for substituting proton bombardment (at 1-2 MeV) for electron irradiation (which usually is performed at 10-30 keV). Incidentally, proton bombardment causes electron ejection from the target by direct Coulomb interaction, with subsequent x-ray emission, as illustrated in Fig. 5 (see Folkmann[3]) which shows also a proton-induced x-ray spectrum of the trace elements present in a dust sample.

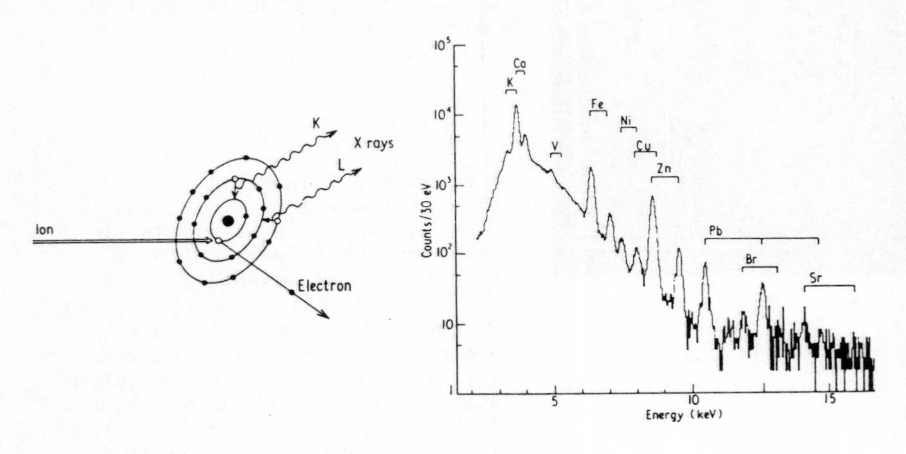

Fig. 5

Characteristic x-ray production by proton irradiation. This process arises as a consequence of direct electron ejection, as shown schematically on left. The spectrum on the right was obtained from a sample of air-borne particulates. (Folkmann[3]).

Basically there are two striking advantages: first, the x-rays arising from proton irradiation are essentially free from the continuous background of electromagnetic irradiation which accompanies electron-induced x-rays. This gives rise to a significant improvement in sensitivity, particularly for light elements in thin targets. (In thick targets, background radiation can arise from secondary processes.) The second attribute of protons, which is relevant here, is their much higher penetrating power: protons in the MeV energy range lose only a few percent of their energy in traversing thin (few μm) target material. This allows the elemental composition of such target material to be quantified, since the cross-sections for characteristic x-ray generation from all elements on bombardment with protons of specific energy are known.

These attributes have been turned to striking advantage by Cahill[4] in a large analytical programme which includes the detection of trace elements in air-borne particulates. He has shown that fast (several thousand samples per day) quantitative analysis of all elements above sodium in atomic number can be achieved at very competitive cost. This approach has been extended to the analysis of the elemental composition of heterogeneous catalysts by Cairns et al.[5]. The catalysts were finely ground, suspended in water, then applied as a thin film to a plastic support and irradiated with protons. A typical spectrum is shown in Fig. 6. Some of the trace elements detected

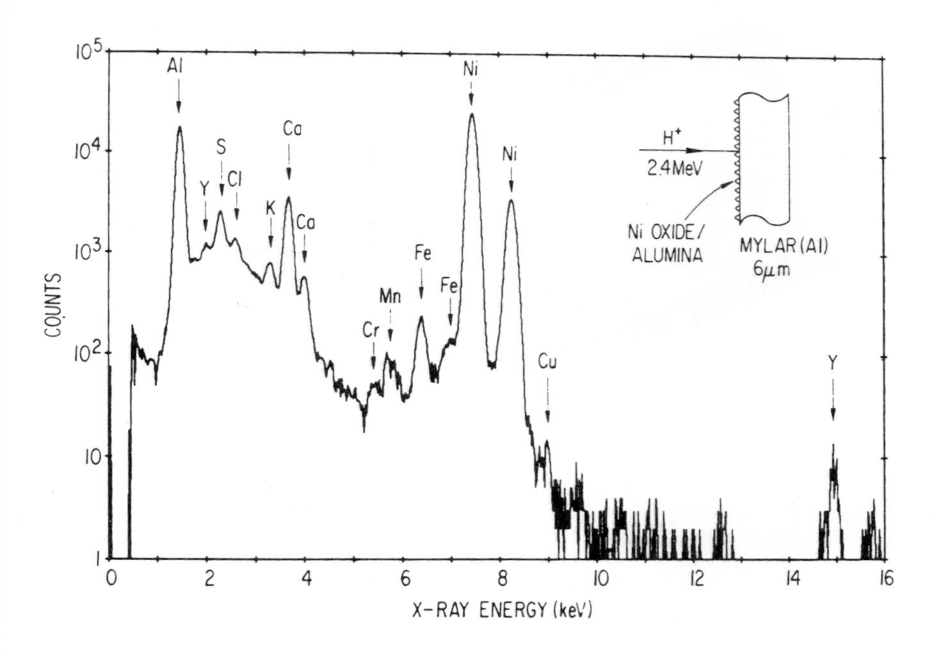

Fig. 6

Proton-induced x-ray spectrum of a nickel oxide/alumina catalyst mounted on a 6 μm mylar film. (Cairns et al.[5])

were found to have been introduced from the water used to make the suspension. This illustrates the potential sensitivity of the technique; it has been pointed out by Johansson and Johansson[6] in an interesting review article that, in this respect, PIXE is generally superior to x-ray fluorescence. For example, minimum amounts of matter as small as 10^{-12}g can be detected routinely. In addition, the variation in cross-section for elements above sodium is within a factor of 3 across the whole periodic table; and furthermore, spatial resolution down to *ca* 2 μm can be achieved by focussing the proton beam.

A further application of PIXE to catalysis has been described recently by Cairns and Cookson[7]. This concerned the analysis of catalysts developed for vehicle exhaust emission control. A convenient method of installing these catalysts in an exhaust system is to bond them to a ceramic monolith. However, there are certain advantages to be gained from using a metallic monolith, made from thin (0.005 cm) sheet (Pratt and Cairns[8]). One material which is suitable for this purpose is an aluminium-containing ferritic steel known as Fecralloy(R) steel, which has excellent high temperature oxidation-resistant properties and can be fashioned into a suitable monolith, as shown in Fig. 7. A Pt/Al_2O_3 catalyst

Fig. 7

A catalyst substrate constructed from Fecralloy(R) steel sheet, designed for installation in a vehicle exhaust system.

bonded to Fecralloy steel, having been installed in an engine test bed, was found to exhibit a progressive deterioration in catalyst performance, particularly for hydrocarbon oxidation. Its PIXE spectrum (Fig. 8) gave two valuable clues to account for the poor performance: the Pt loading was too low (note that the Pt signal is barely visible) and the Al_2O_3 surface area was also too low to accommodate the massive Pb contamination and leave the Pt sites relatively unaffected. The catalyst was redesigned accordingly and was now found to exhibit a much improved, sustained performance. Its PIXE spectrum (Fig. 9), recorded after a comparable engine test, showed that there was now a strong Pt signal; the catalyst had again been obliged to accumulate a heavy Pb contamination, but the larger number of available alumina sites had enabled it to do so without sacrificing performance. Note that the proton beam in these cases has penetrated the Pt/Al_2O_3 layer to give a convenient reference signal from the underlying steel, and although it would have been possible by a combination of other techniques to arrive at these conclusions, the use of PIXE offered an attractive combination of multielement sensitivity and speed. (It took only a few minutes to record each of the spectra shown in Figs 8 and 9.)

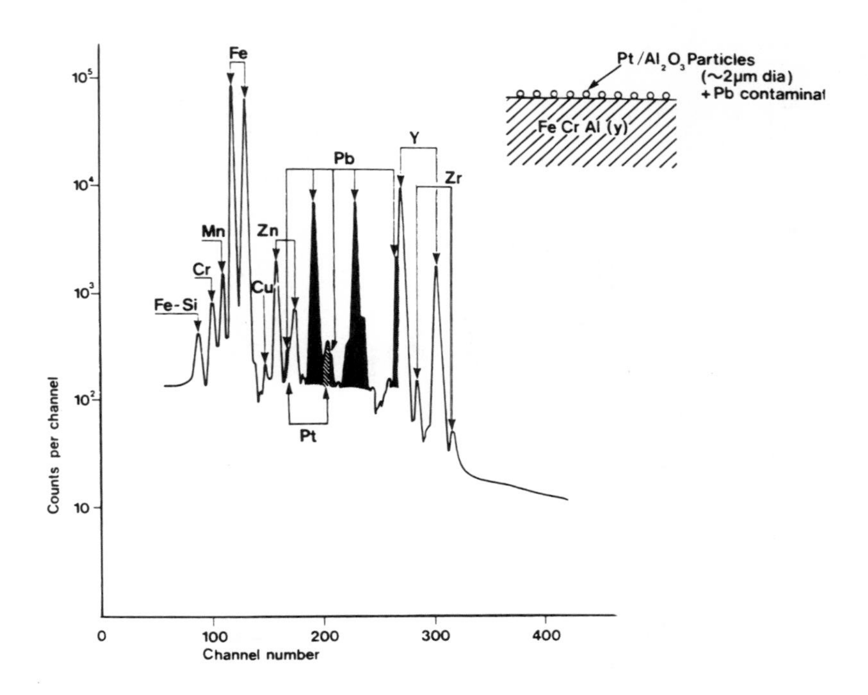

Fig. 8

A proton-induced x-ray spectrum of a $Pt/A\ell_2O_3$ catalyst bonded to a Fecralloy$^{(R)}$ steel substrate after use in controlling exhaust emissions from a petrol engine. This spectrum confirms that Pb contamination has completely dominated the Pt signal. (Cairns and Cookson[7])

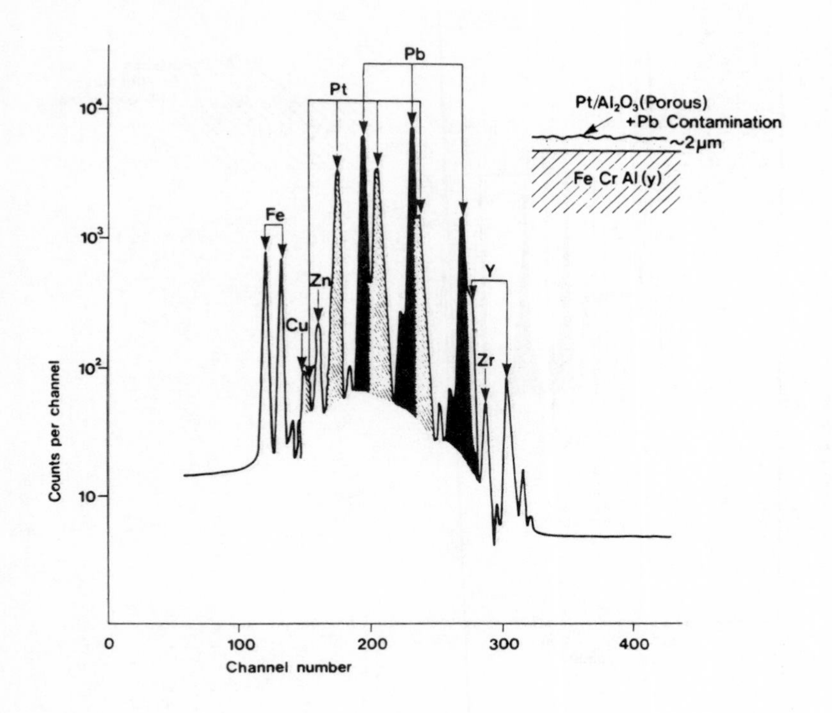

Fig. 9

A redesigned version of the catalyst described in Fig. 8. In this case the Pt signal is still strong, despite the heavy Pb contamination. (Cairns and Cookson[7])

A further application of the penetrating power of protons is in non-vacuum analysis. In this case a thin foil is placed over the end of the accelerator beam line and the emerging protons are used to generate characteristic x-rays from materials which would be vacuum-sensitive, including liquids. This approach has been used by Deconninck and Bodart[9]. Fig. 10 shows its application to the analysis of trace elements present in a liquid, which is presented to the beam as a thin stream from a pipette. It is clear that this technique could have extensive potential application to the catalyst industry in particular and to the chemical industry in general, as a means of fast monitoring of trace element impurities in feedstock.

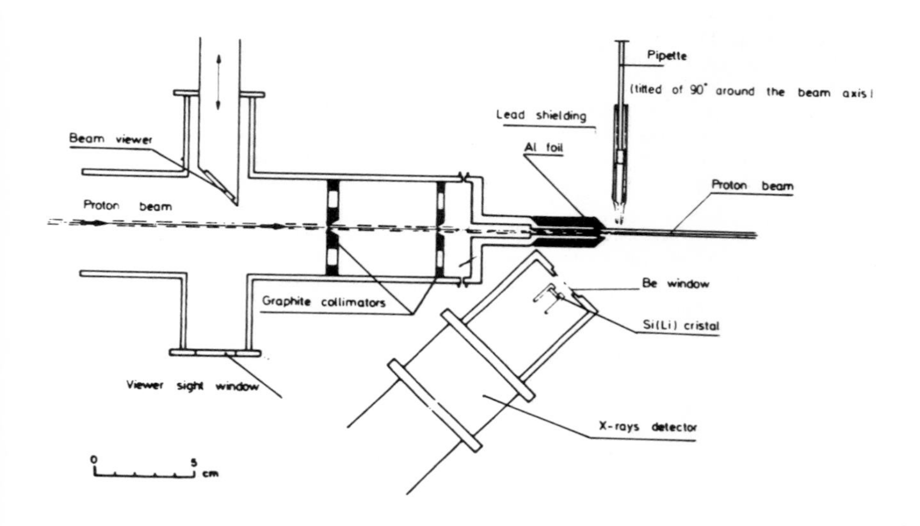

Fig. 10

Schematic illustration of the equipment used to obtain proton-induced x-ray spectra from samples in non-vacuum conditions. The collimated proton beam passes through the thin aluminium foil into the atmosphere, where it interacts with a liquid sample emerging from the pipette. The x-rays thereby generated are measured by the nearby Si(Li) detector. (Deconninck and Bodart[9])

4. THE DETECTION OF LIGHT ELEMENTS BY PROMPT NUCLEAR REACTIONS

It will have become apparent by now that the PIXE technique, for all its attractions, is unsuitable for the analysis of light elements. There are several reasons for this: the Si(Li) detector, which is used most widely in practice, has a poor detection efficiency for elements below sodium in atomic number because of the heavy absorption of soft x-rays (e.g. carbon K x-rays: 284 eV) by its beryllium window. In addition, quantification becomes very difficult because of absorption of such x-rays within the targets themselves; and hydrogen, one of the most important elements in catalysis research, cannot be detected.

This deficiency can be rectified by having recourse to an alternative ion beam technique, namely prompt nuclear reaction analysis, by which an incoming ion beam induces a nuclear transmutation within the target, thereby causing the emission of e.g. γ-rays or energetic particles, which can be measured by nearby detectors. The projectiles used for this purpose, which normally are in the few MeV energy range, are incapable of causing nuclear transmutations of heavy nuclei, because of the Coulombic repulsion barrier, and so the technique is genuinely applicable only to light (atomic number $\leq$ 20) elements. As stated above, we are concerned here with the measurement of emissions which occur essentially at the moment of impact between the ion beam and the target nuclei, rather than with emissions which may occur from the nucleus over a longer period of time, the latter being more properly referred to as activation analysis.

Let us now consider the main characteristics of the technique, with a view towards its potential application to catalyst systems. {For more general background information reference may be made to the review article by Cachard et al.[10].} Whenever an incoming ion beam causes a nuclear transmutation to occur in the target, the energy difference in the reaction is referred to as the Q value. When the reaction releases energy, the Q value is positive, whereas it is negative for endoenergetic reactions. In many cases, only one of the isotopes of the target may have a positive Q value, the other then being ignored, regardless of its concentration. For example, in the determination of oxygen using a proton beam, only ^{18}O is detectable, via the $^{18}O(p,\alpha)^{15}N$ reaction: the ^{16}O is ignored. Hence, by choosing the appropriate projectile, we may detect not simply light elements, but specific isotopes of light elements, even in the presence of heavy substrates.

The technique probes the first few μm of the target, is non-destructive and is independent of the chemical or physical state of the target, because the yield is only nuclear-dependent. In addition, the yields can be rendered quantitative because cross-section curves for interactions between specific projectiles and target nuclei are available in the nuclear physics literature. The sensitivity can be high: down to 10^{12} atoms cm^{-2}.

From the tabulated cross-section curves, one may choose a projectile energy range over which the cross-section varies only slightly, or for other applications, an energy near to a sharp resonance. In the latter case a concentration profile can be extracted from measurement of the reaction yield as a function of the projectile energy. In addition, spatial distributions of selected elements across a target may be obtained by focussing the projectile beam down to a fine spot (typically 2-3 μm). Table 2 lists various nuclear reactions which have been used for the identification of hydrogen, together with their depth resolutions and relative sensitivities. It includes also a selection of reactions which are appropriate for the detection of some other light element isotopes.

As an example of an application of the technique to catalysis, reference may be made to a recent determination by Wright et al.[11] of the spatial distribution of carbon across various

Table 2

Detection of Light Elements by Prompt Nuclear Reactions

ELEMENT	NUCLEAR REACTION	ION ENERGY (MeV)	DEPTH RESOLUTION (Å)	PROBING DEPTH (μm)	SENSITIVITY (at/cm^2)
H	$H(^{7}Li,\gamma)^{8}Be$	2.7-6	1,000(R)	5	10^{12}
H	$H(^{19}F,\alpha\gamma)^{16}O$	16-18	400(R)	0.5	10^{13-14}
H	$H(^{11}B,\alpha)^{8}Be$	1.5-2.5	500(R)	0.5	10^{13-14}
H	$H(^{15}N,\alpha\gamma)^{12}C$	6-8	100(R)	3	10^{14}
^{6}Li	(p,α)	1.9			
^{7}Li	(p,α)	1.3			
^{9}Be	(p,γ)	0.992(R)			
^{10}B	(d,p)	1.5			
^{12}C	(d,p)	1.0			
^{14}N	(d,p)	1.7			
^{16}O	(d,p)	1.49			
^{18}O	(p,α)	0.629(R)			

Note (R) indicates a resonance reaction.

used catalyst pellets. This was done by means of a focussed deuterium beam, using the reaction $^{12}C(d,p)^{13}C$. Fig. 11 illustrates the carbon distribution obtained from a catalyst pellet. This quantitative measurement was achieved in only a few minutes.

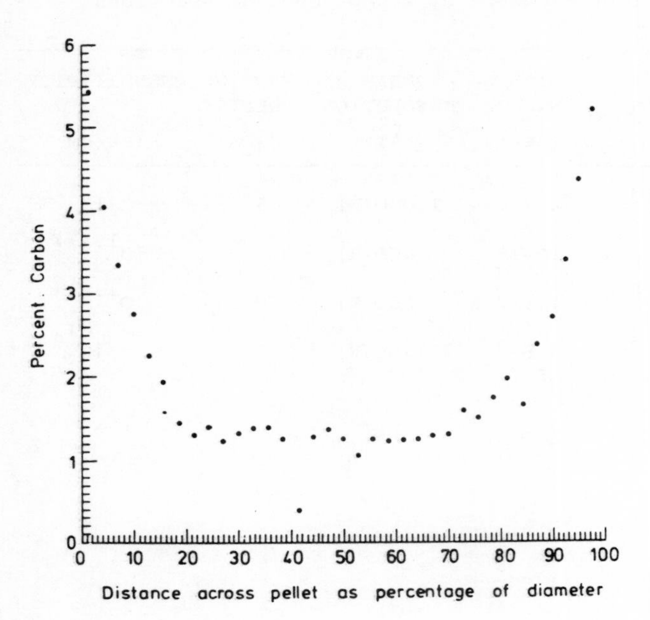

Fig. 11

The variation in carbon distribution across a used catalyst pellet, measured point by point by means of the $^{12}C(d,p)^{13}C$ prompt nuclear reaction. (Wright et al.[11])

5. RUTHERFORD BACKSCATTERING SPECTROMETRY

This technique derives directly from what is generally accepted to have been one of Rutherford's greatest discoveries. Many will be familiar with the sequence of events which began in 1909 when Marsden and Geiger, working in Rutherford's laboratory, investigated the deflection of α-particles from a radioactive source through a gold foil, using a zinc sulphide screen to detect the deflected α-particles. The unexpected result was that a small but significant portion of the projectiles ($\sim$ 1 in 20,000) were reflected back towards the source. Rutherford, realising the significance of this observation, deduced that the bulk of the mass of the target atoms was concentrated within a small volume, and thereby laid the foundation for the whole subject of nuclear physics.

The modern technique which has been derived from these original observations uses virtually the same experimental set-up, but since it is now clear that the scattering cross-sections can deviate somewhat from those given in the Rutherford formula, it is often referred to more generally as Nuclear Backscattering Spectrometry. It consists in bombarding a target with a beam of α-particles or protons (from an accelerator) and detecting, with a solid state silicon surface barrier detector, those α-particles backscattered from close nuclear encounters with the target nuclei. The ion beam energy must be kept below a value which is likely to cause nuclear transmutations. For this reason helium projectiles are employed at an energy of 1-3 MeV whereas protons would be used at a lower energy range (200-400 keV). The general principles of the technique may be appreciated by reference to the descriptive article by Ziegler[12]. An incoming projectile of low mass, in undergoing a close nuclear encounter with a heavy target atom, transfers only a small fraction of its initial energy. Thus the projectile will be reflected back to the detector with little loss in energy. On the other hand, if the projectile encounters a lighter nucleus, it will transfer a larger proportion of its initial energy. Therefore the backscattered spectrum would be expected to consist of a series of peaks which represent the residual energy of ions backscattered from nuclei of different masses. This is shown diagramatically in Fig. 12. From this we see that helium ions backscattered from heavy targets such as Pb would have a residual energy close to the initial value (E_o), whereas those backscattered from low atomic number atoms such as carbon would lose a significant fraction of their energy. This figure illustrates also that the

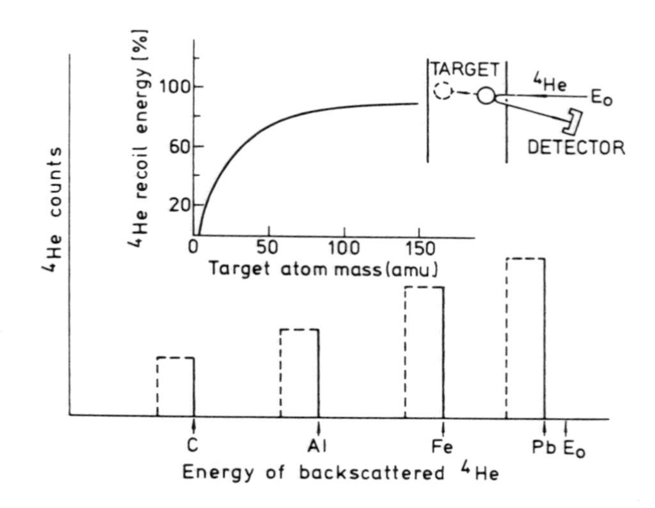

Fig. 12
Schematic illustration of the spectrum obtained from the backscattering of He^+ ions (of initial energy E_o) from a target containing Pb, Fe, Al and C. (Ziegler[12])

yield of backscattered projectiles decreases as a function of atomic number of the target; (cross-section is in fact proportional to Z_2^{-2}), and that the concentration of target atoms is a function of the area under the appropriate backscattered peak. A backscattered spectrum taken from the work of Mitchell and Eschbach[13] is shown in Fig. 13, from which we see two well-defined peaks from Cu and Bi. The target was silicon, with a layer of Cu on its surface and a Bi dopant, introduced previously by ion implantation. We may note that although the heavier elements are seen clearly, there is a broad continuum due to

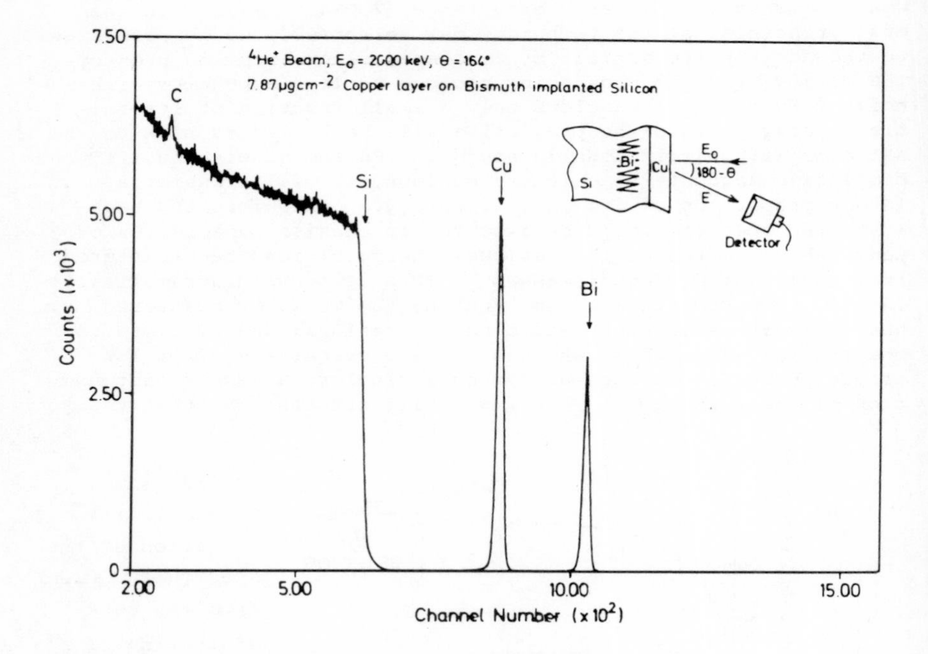

Fig. 13

A genuine example of nuclear backscattering spectrometry, from a target consisting of a thin copper deposit on bismuth-implanted silicon. (Mitchell and Eschbach[13].)

backscattering of the projectile from successively deeper layers of silicon atoms.

The main attractions of the technique in general are as follows :

1. It can be rendered quantitative, the area under the backscattered peak being proportional to the amount of an element of a particular mass;
2. It is non-destructive;
3. It is rapid, requiring in general only a few minutes per spectrum, and indeed represents one of the best ways of measuring diffusion profiles non-destructively with a resolution of the order of 10 nm over several μm.

However, it can be applied to full advantage only to relatively simple systems, i.e. to those containing only a few elements, because of limitations in the energy resolution of the backscattered ion detector; and it is best suited to heavy elements in a light matrix. With these limitations in mind it becomes clear why its special features have been most appreciated by the semiconductor industry where there is a need to study the diffusion of elements in pure low atomic number targets, of which silicon is the prime example. (For an in-depth treatment of the subject, the reader is referred to Chu et al.[14].)

So far as the relevance of the technique to catalysts is concerned, there have been two interesting recent applications: the study of surface structures, including relaxation effects in single crystals, and the evaluation of pore size distributions in catalyst pellets.

5.1 The application of Rutherford Backscattering Spectrometry to the measurement of surface relaxation effects in single crystals

This technique combines nuclear backscattering spectrometry with another fascinating attribute of ion beams, namely their ability to channel through single crystals. We begin then with a brief description of the channelling phenomenon (see review article by Gemmell[15]).

Whenever an energetic ion beam enters a crystal lattice in a direction which is close to a major row or plane of atoms, it experiences a series of Coulombic repulsion forces which tend to keep it confined to channels defined by the close-packed rows of atoms. The result is that such channelled ions are unable to approach closer to the target atoms than a distance which is dependent on their initial energy but may be typically of the order of 0.01 nm. Hence they cannot enter into close nuclear encounter phenomena, such as Rutherford backscattering. Thus, the backscattering yield will exhibit a dramatic reduction whenever the ion beam enters a channelling direction. This situation may be appreciated by reference to Fig. 14a (from the work of Davies[16]) from which we see three possible ion beams labelled A, B and C incident on a single crystal target. The random beam C exhibits no reduction in the Rutherford backscattering yield, but the well-channelled beam A shows a dramatic reduction (Fig. 14b). Incidentally, beam B is interesting because it has a higher probability of making close nuclear collisions with near-surface atoms and therefore exhibits a higher statistical backscattered yield. It tends to be responsible for the shoulder on either side of the channelling dip (Fig. 14b).

Let us now consider the situation in which a beam of ions is directed towards a single crystal, mounted on a goniometer and rotated gradually to a position at which the beam is incident along a well-defined crystal axis. If now we measure the yield of backscattered ions at this stage, we find that although it is much smaller than would be obtained if the beam were incident on the crystal in a random direction, as explained above, there is nevertheless a small but significant signal of relatively high energy backscattered ions. These come from close-nuclear collisions between the incident beam and atoms on the surface of the

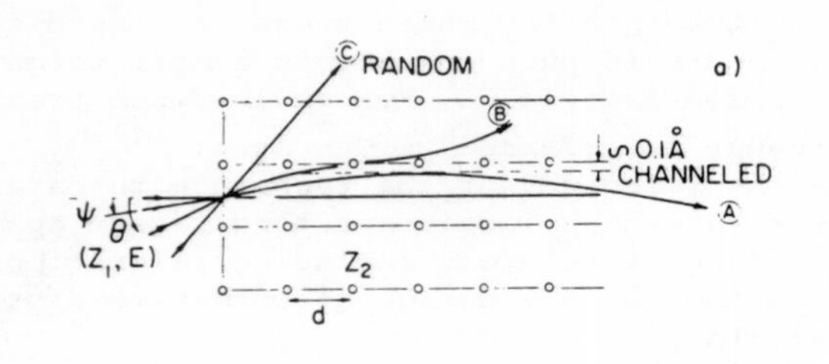

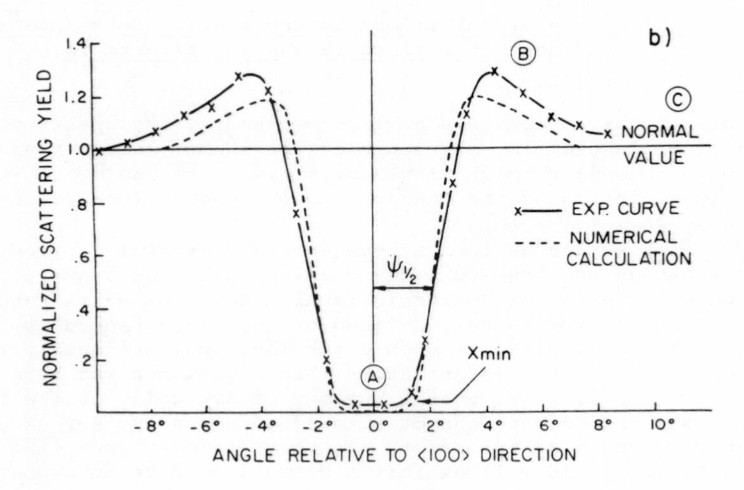

Fig. 14

(a) Three possible trajectories for a beam of ions incident on a single crystal. Only (A) becomes channelled.

(b) Experimental (x) and calculated (---) variation in back-scattered yield for 480 keV proton bombardment of tungsten, as a function of incident angle relative to the ⟨100⟩ direction. (Davies[16])

crystal - i.e. the area of this surface peak is related directly to the number of surface atoms which are in direct line-of-sight to the beam, and its magnitude is sensitive to crystal surface changes induced by phenomena such as reconstruction, adsorbed impurity atoms, and particularly surface relaxation effects.

The latter may be understood by reference to Fig. 15, which again is taken from the work of Davies[17] and illustrates the (111) planar spacing. We see from this that when the ion beam is incident perpendicular to the surface (⟨111⟩ in Fig. 15), the second atom in each row is perfectly shadowed, and so the peak

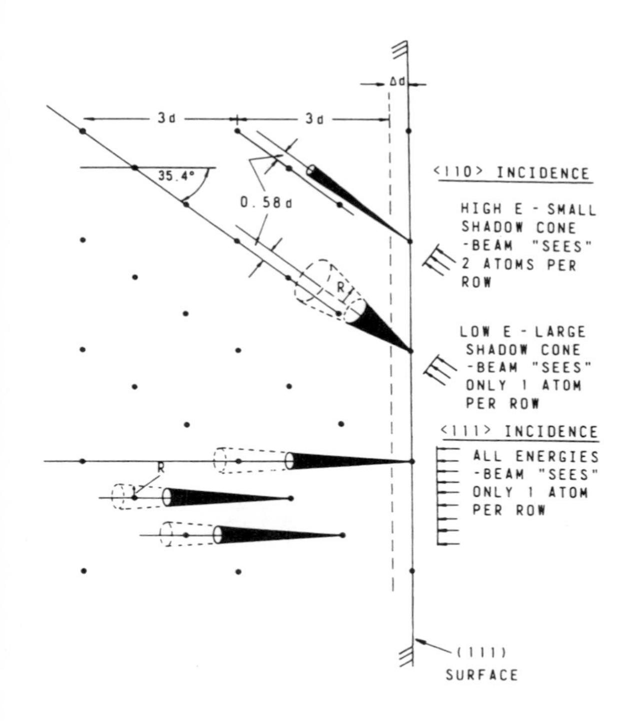

Fig. 15

Atomic configuration in the vicinity of the surface of a (111) Pt crystal, illustrating how the channelling phenomenon may be used to measure the surface relaxation Δd, where d is the bulk (111) planar spacing. (Davies[17])

area of the backscattered ions will be equivalent to 1 atom per row (i.e. 1.5×10^{15} Pt atoms cm^{-2}), provided the two-dimensional lattice vibrational amplitude is smaller than the collisional shadow cone radius R. Hence the surface relaxation Δd has no effect on this ⟨111⟩ shadowing. On the other hand, when the beam is directed along a non-perpendicular axis, such as the ⟨110⟩ in Fig. 15, the surface displacement Δd shifts the shadow cone relative to the underlying row of atoms, thereby causing the backscattered surface peak to increase towards a value of 2 atoms per row. Note also that since the cone radius R varies inversely as a function of incident beam energy E, the surface peak will increase from a value of 1 atom per row at low E to $\sim$ 2 atoms per row at high E. Therefore the magnitude of Δd can be obtained by measuring the value of the surface peak as a function of E for various low-index directions.

This technique, which naturally requires access to a high precision goniometer in a UHV target chamber connected to a suitable accelerator, (Fig. 16: Davies[16]), has been pursued by several Groups: (Turkenburg et al.[18]; Davies et al.[19]; Van der Veen et al.[20]; Zuhr et al.[21].

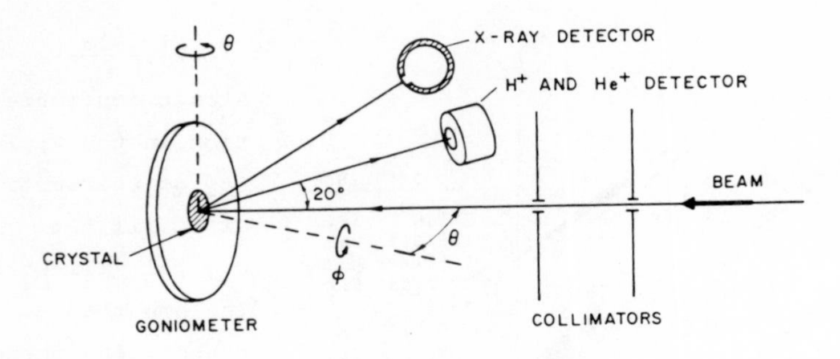

Fig. 16

Schematic diagram of the experimental assembly for a typical channelling experiment. (Davies[16]).

5.2 The use of resonant ion beam backscattering to measure pore size distributions in catalyst pellets

The most widely used techniques for measuring porosity are those based on gas adsorption and mercury penetration (see Sing's Chapter in this book). However, if the pore size dimensions are less than $\sim$ 2 nm then mercury penetration is not applicable, and gas adsorption is restricted because capillary condensation is no longer possible. An alternative approach is to use an ion beam technique. This has the advantages of being non-destructive, able to sample smaller pore sizes, and, by varying the energy of the incoming ion beam, capable of measuring pore size distributions as a function of depth within the pellet. It has the additional advantage of being capable of measuring porosity changes across the surface of the pellet with a spatial resolution of a few μm.

The technique consists in bombarding the sample with energetic ions, and measuring the energy spread of the backscattered ions. For example, Armitage et al.[22] bombarded silica samples of varying porosity with 3.3 MeV He^+ ions and observed the energy spread of those ions backscattered after undergoing a resonant elastic scattering with ^{16}O at 3.05 MeV. They made use of the fact that in porous materials individual backscattered ions pass through different numbers of pores before leaving the sample, and so exhibit a broadening in residual energy distribution. This is apparent from Fig. 17, which shows their comparison between the energy spectrum of He^+ ions backscattered from porous silica (full line) and non-porous silica (dotted line). This broadening of the energy spectrum can be used to fit a model of the pore size distribution. As a test of its validity, the

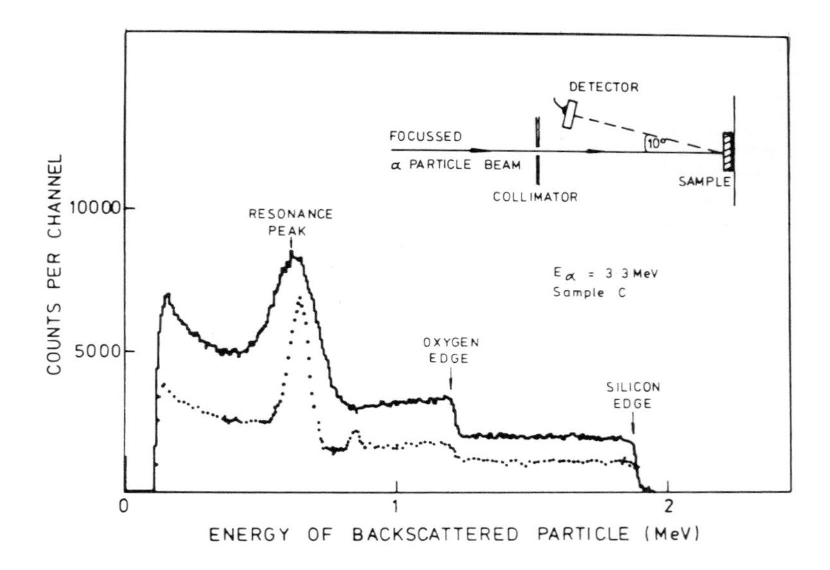

Fig. 17

Typical energy spectra of He^{+} ions backscattered from porous silica (full line) and non-porous silica (dotted line). The experimental arrangement is shown also. (Armitage et al.[22])

ion beam technique has been applied to porous materials which are amenable to conventional porosity measurement, and been found to yield good agreement.

6. LOW ENERGY ION SCATTERING

This technique follows in principle from the previous one, because it examines also the consequences of ion scattering. However, the ion energy is now much lower (usually less than 10 keV), and so we are concerned here with energy loss phenomena which take place as a result of electronic interactions between the ion beam and the target atoms, rather than with scattering from the target nuclei. The technique consists in measuring the residual energy of a well-defined ion beam after it has scattered from the surface of a solid. This energy spectrum is equivalent to a mass spectrum of the surface atoms, since it represents the energy transfers which take place in single collision events between the ions and the surface atoms. The first application of the method to surface analysis was described by

Smith[23]. This has inspired a considerable literature over the following years (Heiland et al.[24]; Brongersma[25]). The sensitivity of the technique to the outermost surface layers of the target is now well established: in fact, Harrington and Honig[26] analysed the (100) and (111) surfaces of a silicon crystal and demonstrated that the scattered ion intensities from these faces reflected the differences in packing densities of the top layers. This has been extended by others to measurements of the locations of surface atoms with respect to the underlying lattice. For example, Theeten and Brongersma[27] have examined the distribution of sulphur atoms on a Ni(100) surface and shown that it was consistent with the corresponding LEED pattern. Naturally the technique is suitable also for studying surface relaxation phenomena, and since it does not rely on being combined with channelling measurements, it may be more versatile.

All of this suggests possible applications to systems of relevance to catalysis, and indeed there are some recorded examples. For example, Brongersma and Buck[28] confirmed that Cu/Ni alloys exhibit a strong surface enrichment on annealing. In addition, Fig. 18 (Brongersma et al.[29]), shows the spectrum obtained from He^+ ion scattering from Bi_2MoO_6 (full line), as compared to two other specimens enriched in 4% Mo and 4% Bi respectively. It is clear from this that a small change in the bulk concentration of Bi or Mo has had a pronounced effect on their surface compositions.

Hence overall the technique of low energy ion scattering has the attraction of being genuinely surface-sensitive. However, it is a fairly sophisticated one to apply in practice, and, unlike the nuclear backscattering technique (section 5 above), the measured yields cannot be related readily in a quantitative manner to the specimen under investigation without having recourse to calibration against standards.

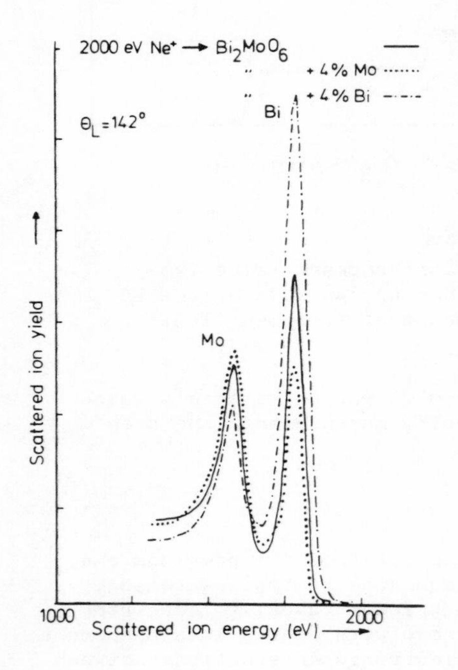

Fig. 18

Helium ion scattering spectra from bismuth molybdate catalyst (full line), compared to others having small differences in bulk composition.
(Brongersma et al.[29])

7. ION-INDUCED OPTICAL EMISSION

When ions in the low to medium energy range (30 eV to 100 keV) impinge on a solid target, they can induce the emission of optical radiation; in fact this can sometimes be seen by eye even at moderate beam current densities. This is the basis of the analytical technique to be discussed here. Examination of the optical spectra from a variety of targets has established that there are at least five different components. These may be summarised as follows :

(i) The interaction of the ion beam with the solid target results in the ejection of sputtered atoms from its surface. A fraction of these sputtered atoms escape from the surface in excited states, and decay optically when well clear of the surface, where the perturbation of atomic levels by the solid is negligible. This results in the emission of discrete atomic lines. However, the excited species may decay also by non-radiative de-excitation processes, and it is the variability of this probability from one target to another which makes it difficult to relate quantitatively the concentration of a surface element to its corresponding optical emission yield.

(ii) The backscattered ions may themselves emit discrete atomic or ionic emission lines.

(iii) There may be a broad band continuum from the bulk of the solid resulting from excitation of electrons in the solid or radiative recombination of electron-hole pairs.

(iv) There is another possible source of broad band emission, arising from molecular species ejected from the surface during irradiation.

(v) Optical radiation may arise also from excited molecules on the target surface. This radiation is believed to be associated with exciton recombination at the surface.

The first of these processes has been the one investigated most fully as a possible sensitive technique for surface analysis, since sputtered particles originate from the first few monolayers of the solid (Tolk et al.[30]).

Fig. 19 (White[31]) shows the optical spectra resulting from 4 keV Ar^+ bombardment of copper (upper section) and nickel. Radiation is observed from low-lying excited states of neutral Cu, Ni and from contaminants such as Na and CH. Note also the presence of H (perhaps from hydrocarbon surface contamination.)

However, there are problems: the intensity of the irradiation depends upon the chemical state of the target - it is generally lower from metals than from oxides. It has been suggested by Thomas[32] that the enhanced emission intensity of the latter is associated with sputtering from the surface of substrate atom-oxygen quasimolecules, for then the final atomic states would be populated by dissociative excitation at distances from the surface where the non-radiative de-excitation process would not operate.

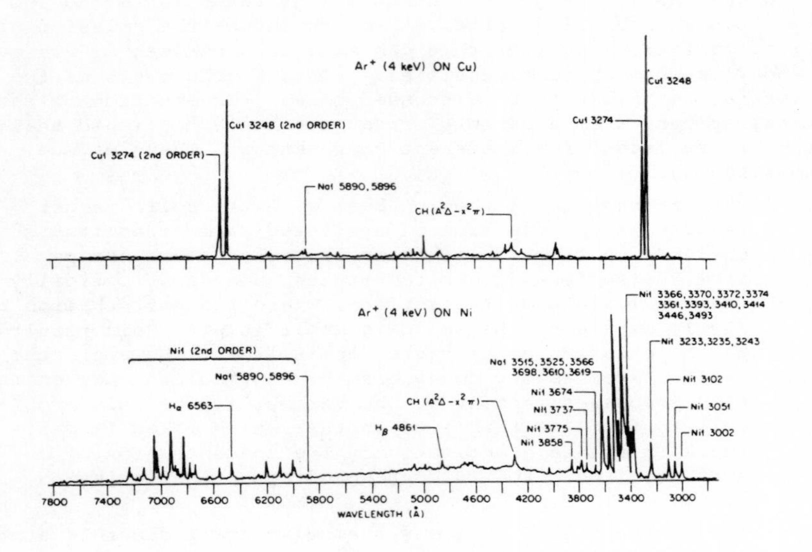

Fig. 19

Optical spectra produced by 4 keV Ar^+ bombardment of copper (upper section) and nickel surfaces. (White[31])

In summary, this is a technique which would seem to have some potential in catalysis research, although so far it does not appear to have been applied seriously to this field. It has the possibility of multielement (including hydrogen) detection capability with high sensitivity in certain cases. In fact its detection sensitivity is a very critical function of the chemical state of the target surface, and this feature, which is so frustrating to analysts, may prove to be its attraction to surface chemists.

8. THIN LAYER ACTIVATION

This technique is included here primarily for its potential interest to chemical engineers concerned with the operation of catalyst plant, because it offers a very convenient and accurate means of measuring low levels of wear and corrosion continuously without dismantling apparatus or machinery. However, it may also offer the prospect of measuring the erosion rate of fixed bed catalysts, such as Pt/Rh gauzes. The technique arose from the realisation (Conlon[33]) that the penetration into materials of ion beams in the few tens of MeV energy range is comparable to the range of wear tolerated in many situations before replacement of a component becomes essential. Conventional methods of measuring wear usually demand partial or total removal of the component from its installation. This, of course, may be economically unacceptable.

The principle of the method is straightforward, and has been described by Conlon[33]. The surface of interest is bombarded by an ion beam consisting of either protons, deuterons, ^{3}He ions or heavier particles, so that a tiny proportion of some particular element in the target is transmuted to a radioactive atom. (The fraction of atoms involved in this process is typically only about 1 in 10^{10}.) The radioactive isotope decays to a stable isotope within a period characterised by the half-life, and in so doing emits gamma rays. As the surface is worn away, its radioactivity diminishes, and thus a measure of the total activity remaining gives a sensitive indication of the amount of surface removed by wear. Correction can be made, of course, for natural decay of the radioisotope with time. The detection system consists of a gamma ray detector, usually a sodium iodide scintillation crystal, together with appropriate electronics. The emitted gamma rays can penetrate practical thicknesses of most material with only moderate attenuation, so that measurements can be made through reactor vessel walls.

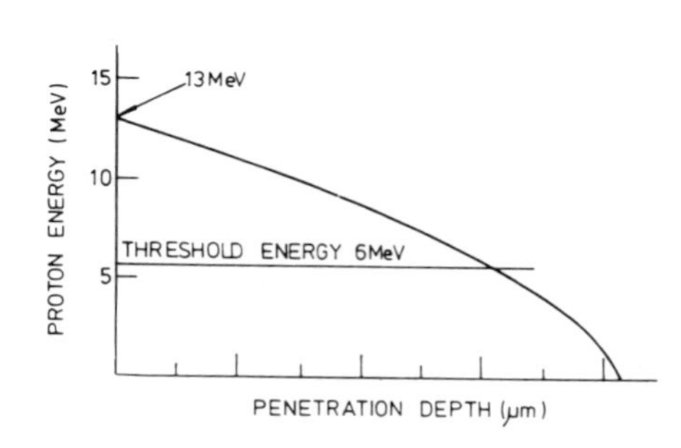

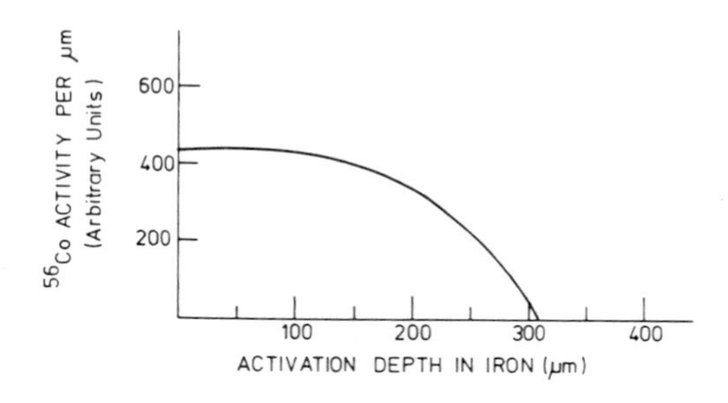

Fig. 20

Upper curve: proton energy versus penetration depth in iron target. Lower curve: induced ^{56}Co activity as a function of depth in iron target. (Conlon[34]).

Consider the situation in which an iron component has been irradiated to produce a ^{56}Co isotope under the influence of proton irradiation. The upper part of Fig. 20 (Conlon[34]) shows the penetration depth of the protons of initial energy 13 MeV. Protons lose energy mainly as a result of collisions with target electrons. However, a very small proportion, namely about 1 in 10^{6}, undergo direct interaction with the nuclei of the target to produce radioactive ^{56}Co atoms with a probability shown in the lower part of Fig. 20. The threshold energy for this transmutation is 6 MeV. The resulting total activity/surface loss curve can be obtained by integrating the lower curve, and is, in fact, close to linear over most of the activated region. From available nuclear data this curve can be calculated accurately for most reactions. The half-life of ^{56}Co is 77 days, and since the monitoring period can extend over several half-lives, the useful life of a component irradiated in this way is about a year. Other radioactive nuclei can be induced in iron, steel and other materials, thereby extending the period over which measurements can be made to several years or even to tens of years. In the process of decaying back to ^{56}Fe, the ^{56}Co emits gamma rays ranging in energy from 0.511 MeV to 3.5 MeV. The penetrating nature of these gamma rays allows irradiated objects to be observed <u>in situ</u> through layers which may be several inches (iron equivalent) in thickness. The ultimate sensitivity of the technique could be improved, of course, by using heavy ion irradiation to activate very thin layers of only a few μm or less. Such layers would offer a sensitivity of a few nm.

If a fluid circuit is available in the system, to carry the wear debris to a collection point, then its radioactivity may be measured continuously with high accuracy. For example, it is quite possible to measure thickness losses as small as 0.025 μm. In the past, neutron activation analysis for wear measurement has required the whole component to be placed in a nuclear reactor, thereby generating a significant level of radioactivity. The great attraction of the ion beam is that only a very small volume of the whole component is made radioactive. In addition, the total depth activated can be chosen to be comparable to the anticipated wear range and any preselected area can be treated. The result is that only tiny amounts of radioactive material are produced. (In fact, the amount of activity is typically no more than that found in old-fashioned luminous watches.)

Among the applications of the technique at the present time are <u>in situ</u> studies of wear in cylinder liners, valve seats, fuel injection equipment, and on the cutting edges of machine tools. The technique is being used also on a pilot scale to measure the rate at which plant and machinery erode.

So far as its potential to catalysis is concerned, there could be a number of interesting applications. First, as mentioned above, there is the prospect of continuous monitoring of chemical plant for erosion. Then, turning to catalysts themselves, there are several possibilities. For example, the preferential loss of certain elements could be observed <u>in situ</u>, provided the activation did not produce a chemical transmutation which rendered those atoms non-typical of the catalyst. (In fact Conlon[35] has shown recently that a variation of the technique can introduce chemically identical but radioactive atoms into the same sample; an alternative approach would be to implant the radiotracer.) In addition, gross physical erosion of appropriate fixed bed catalysts, such as Pt/Rh gauzes, can be monitored dynamically, so long as the catalyst is maintained in a fixed reference position with respect to the detector.

CONCLUSION

It will be noticed, by reference to Table 1, that there remains one ion beam technique which has not been discussed in this review: Secondary Ion Mass Spectrometry (SIMS). This has been a deliberate omission, because SIMS is by now well-established, with a large literature of its own. (For an in-depth treatment, see Blaise[36]; for surface applications, see Benninghoven[37].)

The object of the present work was rather to bring to the attention of the catalyst community the possibilities now emerging or still latent in the application of ion beams to catalyst characterisation. For this reason, each technique has been described in sufficient detail to elucidate only its essential relevant features. The references have been chosen to provide further background if necessary. The promise is there in abundance; with the increasing availability of suitable ion accelerators, the applications should follow.

ACKNOWLEDGEMENTS

Grateful thanks are expressed to R.S. Nelson and J.A. Davies for most helpful critical comments.

Footnote: (R) Fecralloy steel is the registered trade mark of the United Kingdom Atomic Energy Authority for a specific range of steels.

REFERENCES

1. J.A. Cairns, C.L. Desborough and D.F. Holloway, Nucl. Instr. Meth., 88, 239 (1970).
2. J.A. Cairns, A.D. Marwick and I.V. Mitchell, Thin Solid Films, 19, 91 (1973).
3. F. Folkmann, "Ion Induced X-rays: General Description", in Material Characterisation Using Ion Beams (Eds J.P. Thomas and A. Cachard), pp.239-281, Plenum Press, New York (1978).
4. T.A. Cahill, "Ion-excited X-ray Analysis of Environmental Samples", in New Uses of Ion Accelerators (Ed. J.F. Ziegler) p.1, Plenum Press, New York (1975).
5. J.A. Cairns, A. Lurio, J.F. Ziegler, D.F. Holloway and J.A. Cookson, J. Catalysis, 45, 6 (1976).
6. S.A.E. Johansson and T.B. Johansson, Nucl. Instr. Meth., 132, 473 (1976).
7. J.A. Cairns and J.A. Cookson, Nucl. Instr. Meth. (1979) to be published.
8. A.S. Pratt and J.A. Cairns, Platinum Metals Rev., 21, 74 (1977).
9. G. Deconninck and F. Bodart, Nucl. Instr. Meth., 149, 609 (1978).
10. A. Cachard, J.P. Thomas and E. Ligeon, "Microanalysis by Direct Observation of Nuclear Reactions" in Material Characterisation Using Ion Beams (Eds J.P. Thomas and A. Cachard) pp.367-402, Plenum Press, New York (1978).

11. C.J. Wright, J. McMillan and J.A. Cookson, J.C.S. Chem. Comm., (1979) to be published.
12. J.F. Ziegler, "Material Analysis by Nuclear Backscattering" in New Uses of Ion Accelerators (Ed. J.F. Ziegler), p.75, Plenum Press, New York (1975).
13. I.V. Mitchell and H.L. Eschbach, Nucl. Instr. Meth., 149, 727 (1978).
14. Wei-Kan Chu, J.W. Mayer and M.A. Nicolet, Backscattering Spectrometry, Academic Press, New York (1978).
15. D.S. Gemmell, Rev. Mod. Phys., 46, 129 (1974).
16. J.A. Davies, "Channelling: General Description" in Material Characterisation Using Ion Beams (Eds J.P. Thomas and A. Cachard), p.405, Plenum Press, New York (1978a).
17. J.A. Davies, "Application of MeV Ion Chanelling to Surface Studies" in Material Characterisation Using Ion Beams (Eds J.P. Thomas and A. Cachard) p.483, Plenum Press, New York (1978b).
18. W.C. Turkenburg, W. Soszka, F.W. Saris, H.H. Kersten and B.G. Colenbrander, Nucl. Instr. Meth., 132, 587 (1976).
19. J.A. Davies, D.P. Jackson, J.B. Mitchell, P.R. Norton and R.L. Tapping, Nucl. Instr. Meth., 132, 609 (1976).
20. J.F. Van der Veen, R.G. Smeenk, R.M. Tromp and F.W. Saris, Surface Science, 79, 219 (1979).
21. R.A. Zuhr, L.C. Feldmann, R.L. Kauffman and P.J. Silverman, Nucl. Instr. Meth., 149, 349 (1978).
22. B.H. Armitage, J.D.F. Ramsay and F.P. Brady, Nucl. Instr. Meth., 149, 329 (1978).
23. D.P. Smith, J. Appl. Phys., 38, 340 (1967).
24. W. Heiland and E. Taglauer, Nucl. Instr. Meth., 132, 535 (1976).
25. H.H. Brongersma and T.M. Buck, Nucl. Instr. Meth., 132, 559 (1976).
26. W.L. Harrington and R.E. Honig, 22nd Annual ASMS Conference, Philadelphia (1974).
27. J.B. Theeten and H.H. Brongersma, Revue Phys. Appl., 11, 57 (1976).
28. H.H. Brongersma and T.M. Buck, Surface Science, 53, 649 (1975).
29. H.H. Brongersma, L.C.M. Beirens and van der Ligt, G.C.J., "Applications of Low Energy Ion Scattering" in Material Characterisation Using Ion Beams (Eds J.P. Thomas and A. Cachard), p.65, Plenum Press, New York (1978).
30. N.H. Tolk, I.S.T. Tsong and C.W. White, Anal. Chem., 19, 16A (1977).
31. C.W. White, Nucl. Instr. Meth., 149, 497 (1978).
32. G.E. Thomas, Radiation Effects, 31, 185 (1977).
33. T.W. Conlon, Wear, 29, 69 (1974).
34. T.W. Conlon, Tribology International, April 1979, p.60 (1979a).
35. T.W. Conlon, "Indirect Recoil Implantation following Nuclear Reactions: Theory and Potential Application", AERE (Harwell) Report No. R9340 (1979b).

36. G. Blaise in Material Characterisation Using Ion Beams (Eds J.P. Thomas and A. Cachard) Plenum Press, New York (1978).
37. S. Benninghoven in Chemistry and Physics of Solid Surfaces, (Eds R. Vanselow and S.Y. Tong), C.R.C. Press (1977) Cleveland.

RADIOISOTOPE AND RELATED TECHNIQUES OF ASSESSING CATALYST PERFORMANCE

By

S.J. Thomson

1. INTRODUCTION

The application of the methods of radioactivity can contribute to the characterisation of catalysts before, during and after use, and to the understanding of their regeneration. Catalysts never seem to remain unchanged during catalysis and surface radioactive studies throw light on this situation: indeed it could be argued that during the use of a catalyst instantaneous pictures are required of a constantly changing situation.

Many examples of the use of radioactivity in surface science can be found in an earlier review[1]: in the present account applications are restricted to characterisation and observation of working catalysts. Since many molecular species are available in radioactive forms in the millicurie per millimole range, the sensitivity of their detection allows the catalyst chemist to examine surface populations corresponding to fractions of monolayer coverage. Arrival and departure of molecules can be followed and surface concentrations can be obtained, though with some difficulty because of the problems associated with absolute counting. Surfaces can be monitored directly, or the experimenter may work with flow systems coupled to gas-chromatographs and radiation detectors. By this latter means it is easy to follow reactions, adsorptions, desorptions and regeneration of catalysts.

The use of radioactivity has many merits: the possibility of working in a long pressure range, 0 to 760 torr; the use of mixtures of reactants where only one is labelled; the absence of disturbance to the system during observations and the possibility of studying adsorption and catalysis independently of one another.

The application of radioactivity has its limitations and one should be explicit about these. Times of observation must be long enough to overcome difficulties associated with the random nature of radioactive decay. Direct observation of surface processes can only be made within certain temperature limits set by the detector, e.g. -10 to 100°C for a G.-M. counter. Most important of all is the reminder that was tersely stated by a recent referee of one of our papers[2]:- 'Since one can count only ^{14}C-containing molecules, and there can be any one of a number of species present......., the detailed interpretation of the data should be viewed cautiously'.

Radioactivity also has a role in intercalibration among modern surface-physics methods. Thus Peralta et al.[3] used a radioactive method for AES calibration of sulphur on Mo(110) and Heegeman et al. used it for sulphur on platinum.

2. ADSORPTION IN STATIC SITUATIONS

In developing the theme of three periods in the life of a catalyst, we shall begin with studies of adsorption of molecules in static systems. The same molecules will reappear when an account is given of their behaviour during the second period in the life of a catalyst, the working phase.

2.1 Adsorption of Acetylene, Ethylene and Carbon Monoxide on Catalysts

The work to be described has been concerned with the adsorption of ethylene, acetylene and carbon monoxide on supported metal catalysts. The work began in the early 1960's and the first paper[5] was concerned with the relationship between chemisorption and catalysis. At that time the velocity of a catalysed process was thought of in terms of concentrations of all adsorbed molecules. Yet when ^{14}C-labelled molecules were adsorbed on nickel films it was found that only a fraction of the adsorbed species participated in subsequent catalytic hydrogenation reactions, Fig. 1.

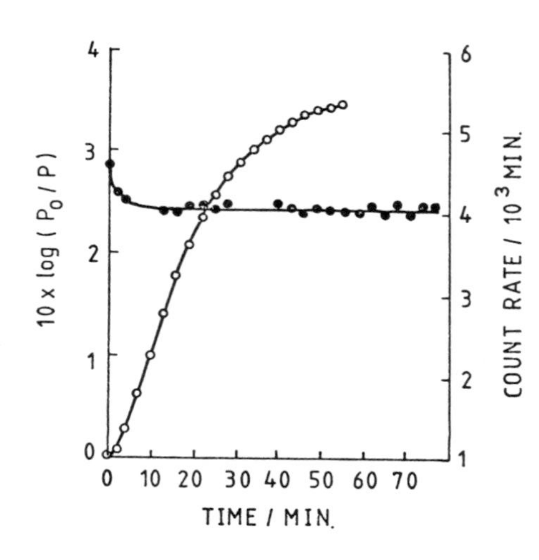

Fig. 1.

^{14}C-ethylene was adsorbed on a nickel film to give a surface count rate ●. Non-radioactive ethylene and hydrogen were introduced and hydrogenation occurred, o. The surface count rate fell only by a small amount.
S.J. Thomson and J.L. Wishlade, Trans. Faraday Soc., 58, 1170, 1962.
(reproduced with permission)

This phenomenon of retention of adsorbed species on the surface of a catalyst was found[6] to be true also of other metals in supported forms which gave varying fractions of the surface inactive in catalysis viz. palladium, 63.5%: nickel, 24%: rhodium, 22.5%: iridium, 16%: platinum, 6.5%.

The implication of this for characterisation of catalysts by adsorption methods is clear. It is that the meaning of adsorption results as far as catalysis is concerned is indeed unclear. What is now obvious is that we should characterise catalytic systems in terms of metal with its associated adsorbed species.

In order to do this we shall have to examine the behaviour of adsorbed species in more detail. We do this by looking at the adsorption of labelled molecules over a range of pressure. Perhaps we should avoid the use of the term 'adsorption isotherm' for the adsorptions, for this suggests studies of systems in equilibrium; and it suggests also that the molecules under study retain their integrity. In practice neither condition is met: molecules undergo reaction during adsorption and only some may retain their identity.

Adsorption behaviour is best seen in terms of the shape of a generalised plot taken from a number of papers[7] of amount adsorbed versus pressure for ethylene, acetylene or carbon monoxide, Fig. 2.

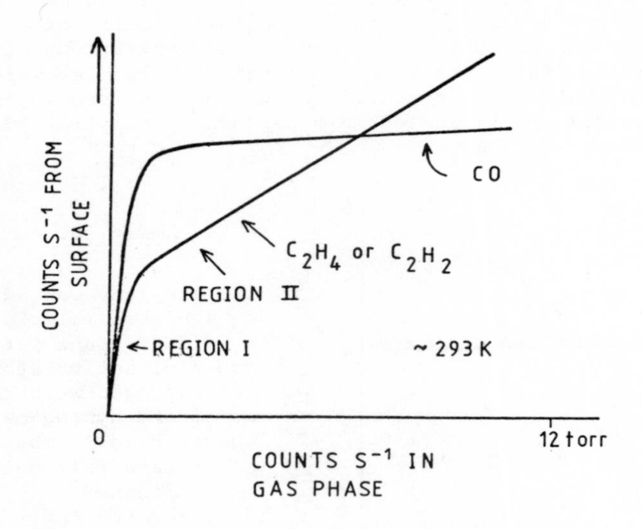

Fig. 2

The general shape of plot of surface count rate versus gas phase count rate (pressure) for adsorption of ^{14}C-CO, ^{14}C-acetylene or ^{14}C-ethylene on a supported noble metal. The turning point occurs at under 1 torr.

Olefin adsorption rises rapidly at low pressures in region I to a point at which coverage is approximately twice that for carbon monoxide. It is likely that the turning point between regions I and II corresponds to coverage of the metal. In region II there is a linear increase in adsorption, this time with a gentler slope. In the case of carbon monoxide there seems to be only a very slight increase in adsorption in region II.

What is now immediately available to us is the possibility of investigating the reactivity of species in regions I and II.

This can be done by introducing labelled molecules which give adsorption in region I followed by non-radioactive molecules in region II or vice versa. What emerges is interesting and surprising, and is beautifully shown in recent papers by Al-Ammar and Webb[8]. In region I adsorption takes place with the accompaniment of self-hydrogenation.

The shape of the adsorption graph at 294-298K for ethylene or for acetylene is shown in Fig.3 for rhodium on silica and iridium on silica. The pressure range was 0 to 12.5 torr. The graphs show the steep non-linear region, I, and a linear secondary region II. For both acetylene and ethylene the gas phase present in region I consisted entirely of ethane, i.e. self-hydrogenation had occurred. In region II ethane and the adsorbate hydrocarbon were present. No ethylene was ever observed when acetylene was the adsorbate. The catalysts used in these experiments had been reduced in hydrogen at temperatures up to 623K and then evacuated. Catalysts cooled in hydrogen had diminished regions I.

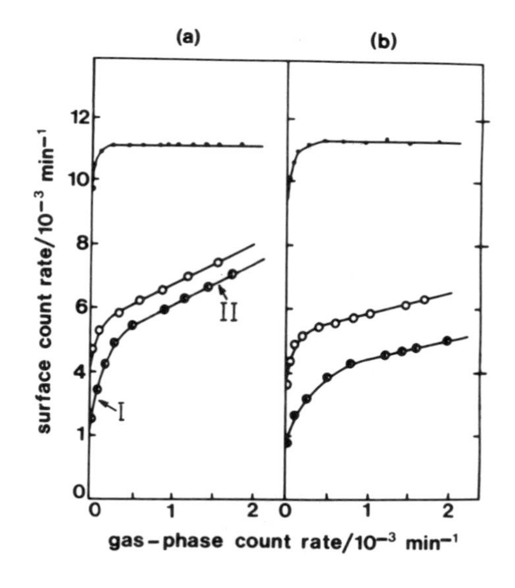

Fig. 3

Surface count rate vs gas-phase count rate for adsorption of ^{14}C-carbon monoxide ●, ^{14}C-acetylene o, ^{14}C-ethylene ◐ on

(a) freshly reduced Rh/SiO_2

(b) freshly reduced Ir/SiO_2.

A.S. Al-Ammar and G. Webb J.C.S. Faraday Trans.I, 74, 195 (1978) (reproduced with permission).

Once catalysts had been used, the adsorption behaviour was quite different. When successive hydrogenation runs had produced steady working states, e.g. Fig. 4a, adsorptions on iridium had the appearance shown in Fig. 4b. The primary region I disappeared and the capacity to adsorb carbon monoxide had been much diminished. When catalysts were examined which had been exposed to ethylene or acetylene to produce regions I and II, evacuation or molecular exchange did not change the surface count rates. Acetylene did not show the ability to displace species produced from ethylene on rhodium: a few percent could

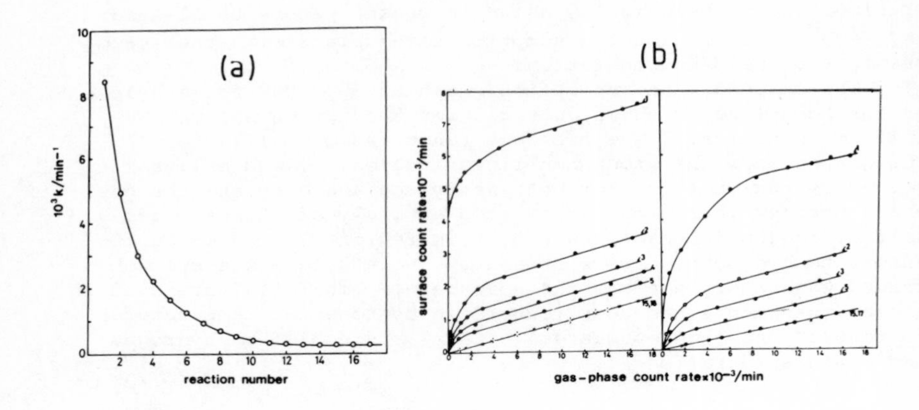

Fig. 4

(a) Fall in first order rate constant for successive hydrogenations of acetylene over Ir/SiO_2 at 298K.

(b) Adsorption behaviour of ^{14}C-ethylene (on left), of ^{14}C-acetylene (on right) on Ir/SiO_2 at 298K after varying numbers of reactions have been carried out.

A.S. Al-Ammar and G. Webb, J.C.S. Faraday Trans. I, 74, 652 (1979). (reproduced with permission.)

be displaced on iridium. The effect of introducing a mixture of hydrogen plus ethylene to the ^{14}C-ethylene-precovered surfaces was, approximately, to halve the surface count rates on rhodium or iridium.

One of the most striking of Al-Ammar and Webb's results was on co-adsorption. Admission of ^{14}C-ethylene to a steady state catalyst in the presence of acetylene in the gas phase gave an adsorption behaviour for ^{14}C-ethylene which was identical to that for a steady-state catalyst without the presence of acetylene. This is clearly shown for iridium in Fig. 5.

The idea that one adsorbed species does not sense the presence of another coadsorbed species has also been stated[9] as a result of LEED and UPS studies of hydrogen plus carbon monoxide on Ni(111) at temperatures under 450K: above this temperature carbon monoxide breaks down. UPS investigation of electronic levels for the coadsorbed molecules showed them to be the same as for the single components.

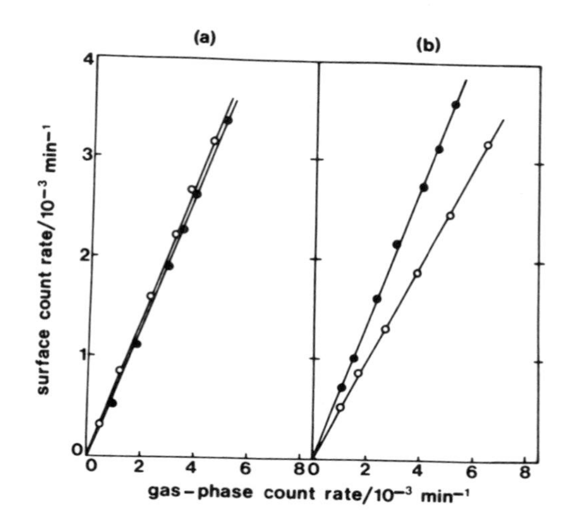

Fig. 5

Adsorption of ^{14}C-ethylene on (a) Ir/SiO_2 and (b) Rh/SiO_2 for a steady state catalyst in the absence ● and in the presence o of 12.5 torr acetylene.

A.S. Al-Ammar and G.Webb, J.C.S. Faraday Trans. I, 74, 195 (1979). (reproduced with permission.)

2.2 Metal and Associated Species

One can now see that there is a serious problem associated with characterisation of catalysts, in the sense that it would appear from what has been discussed in this section that the catalyst is not simply a metal but metal plus associated species. It should be recognised that much work in infrared spectroscopy of adsorbed species and in sophisticated electron spectroscopy of surfaces may not be examining species of interest to catalyst chemists.

There are, however, reports by a number of observers using these techniques of species adsorbed in a reactive or in a secondary mode which may be of catalytic significance. Thus Prentice et al.[10] have observed a weakly bound reactive π-species in their infrared study of ethylene adsorptions on platinum and palladium. The π-species co-existed with σ-bonded $M\text{-}CH_2\text{-}CH_2\text{-}M$. Conrad et al.[11], working in the field of coadsorption of C_2N_2/H_2 on Pt(100) surfaces, found a new feature in their UPS spectrum which might have arisen from a surface intermediate in the production of HCN.

In a molecular beam, AES study of the adsorption of ethylene on Ni(110), Zuhr and Hudson[12] found that the initial interaction which produced adsorbed C_2H_2 species was followed by the appearance of a secondary, non-dissociated, phase which they say may be the active phase in olefin hydrogenation. Duś, from a surface potential study of hydrogenation of ethylene, has reported[13] on two forms of ethylene on evaporated palladium. The first layer was adsorbed with decomposition: the second layer was that which could be hydrogenated.

On α-Fe(100), Brucker and Rhodin also discuss secondary adsorption[14]. On clean iron, acetylene and ethylene break down to give CH_n species with n 1 or 2 respectively, with $C=CH_2$ also a possibility in the case of ethylene. But adsorption of these olefins on a carbided surface resulted in formation of adsorbed forms of these molecules in which a degree of C-C stretching had occurred. Perhaps these are the catalytically active forms. The formation of a new species in studies of adsorption on an intentionally contaminated surface has also been reported by Demuth[15]. Acetylene was used to modify a Ni(111) surface to the point at which a (2x2) overlayer was present. Further adsorption of acetylene at room temperature produced a new species, identified as probably -CH, which would be of significance in Fischer-Tropsch synthesis, methanation or oxidation reactions.

It was the consideration of the role of adsorbed carbonaceous species in catalysis which led Thomson and Webb[16] to propose surface MC_xH_y species as being active as catalytic centres.

2.3 The Behaviour of Hydrogen

Whilst the previous sections have shown how the use of carbon-14 labelled species has revealed unexpected behaviour on surfaces, it should be noted that radioactive methods have also been used to show how hydrogen behaves on surfaces. Taylor et al. and Altham and Webb[17] saw hydrogen spillover from metal to the support when they used tritium as tracer. Paál and Thomson[18] used the same nuclide to show the heterogeneity of behaviour of hydrogen on platinum both in desorption and in its catalytic reaction with ethylene.

It appears that unexpected forms of hydrogen can occur on hydrocarbon-covered surfaces. The example[19] which shows this clearly is the temperature-programmed release of hydrogen from polycrystalline rhenium and Re(0001) surfaces covered by residues from acetylene or ethylene, $C_2H_{1.2}$ and $C_2H_{1.4}$ respectively. Thermal desorption produced only hydrogen, where the curve shapes in TPD (see Chapter by McNicol) differed significantly from those for hydrogen on the clean metal.

2.4 Sub-Surface Species: Facets

The conclusion which emerges from all the foregoing examples is that characterisation of metal catalysts may be evolving into characterisation of metal with its associated adsorbed species. In addition, the profound effect of sub-surface species is now well established. As an example, Palmer and Vroom[20] used AES to monitor surface cleanliness of cobalt and nickel foils in a study of methanation. Sub-surface oxygen produced very active catalysts.

In addition to changes which occur in catalysts because of

adsorption or the generation of sub-surface species there are changes which occur in the morphology of catalysts during reaction.

The most elegant recent demonstration of the development of different crystal facets was made by Flytzani-Stephanopoulos et al.[21]. They prepared small 0.06 cm diameter spheres of Pt single crystals and observed the changes which occurred during oxidation of ammonia, propane and carbon monoxide or the decomposition of ammonia.

3. ADSORPTION DURING CATALYSIS

We now turn to the question of whether the behaviour of a catalyst as evinced by the methods of adsorption applies to the catalyst in its working situation. In the case of adsorption of olefins on metals, we shall see that quite different adsorption situations arise when the catalysts are functioning in reaction mixtures. There is already a paper in the literature which serves as a warning on this subject. Laider and Townsend[22] investigated the kinetics of the hydrogenation of ethylene under various regimes of order of admission of the reacting gases. Admission of hydrogen first gave a rate three times that for prior admission of ethylene in the case of iron and nickel films. Simultaneous admission of reactants gave an intermediate rate. Thus it is clear that behaviour in catalytic systems should be studied during catalysis.

3.1 Adsorption During Catalysis

The central problem to which we address ourselves is that in a catalysed process we are usually ignorant about how a population of molecules is behaving on a surface and unaware of whether conversions are occurring rapidly on a small number of sites or slowly on a large number of sites. It could be argued that this is the central problem in characterising catalysts. Furthermore, a fundamental description of catalytic conversions requires knowledge of the extent of the various kinds of adsorption which may be occurring.

As long ago as 1925 Taylor[23] recognised that catalytic reactions would be found where the fraction of the surface which was active would depend on a small number of active centres up to the involvement of every surface atom. In the majority of quantitative studies of catalysis, specific rates, which could be quite erroneous, are calculated on the basis of surface areas. Brunauer, Emmett and Teller's 1938 physisorption method[24] is used widely for area determination (see Chapter by Sing). For supported metal catalysts selective adsorption on the metal can be measured by using carbon monoxide, hydrogen or oxygen, although none of these is ideal.

From these measurements, and from X-ray or electron microscopy, information on surface areas of metals can be estimated and used in calculations of specific rates. Some guide as to whether metal surfaces would be employed in whole or in part during catalysis can be obtained from the division of reactions into two main categories. Thus Boudart[25] divided catalysed reactions into structure-sensitive and structure-insensitive types, the latter perhaps occurring with equal facility on faces, edges and corners. Ethylene hydrogenation is of this type, as was shown by Schuit and van Reijen[26] and Schlatter and Boudart[27]: other facile reactions are reported on by Dautzenberg and Platteeuw[28] and Aben, Platteeuw and Stouthamer[29].

Structure-sensitive reactions are less frequently reported. Balandin[30] suggests that the active sites involved may be metal atoms with particular numbers of near neighbours. Hydrogenolysis of neopentane over platinum is structure-sensitive[31].

Whereas in kinetic studies attention is focussed on molecules which undergo rapid conversions, chemisorption studies revealed that some fraction of a population of molecules may be irreversibly adsorbed on a surface. Since it has always been assumed that chemisorption is a requirement for catalysis, many attempts have been made to correlate the two. Confusion must arise unless it is realised that molecules involved in strong chemisorption may have little to do with catalysis.

Tamaru, whose work will be referred to later, was one of the earliest investigators to state clearly the need for simultaneous measurements of adsorption and catalysis. He emphasised the view that catalysis was something occurring on a covered surface, not a bare metal.

3.2 Examples of Concomitant Studies of Adsorption and Catalysis

This is a somewhat neglected field of study in catalysis but has been reviewed by Maatman[32]. A few examples of other methods will be given before we examine radioactive methods.

There have been gravimetric studies of ammonia synthesis[33], oxidative dehydrogenation of butene to butadiene over a Zn/Cr/Fe oxide[34], and the dehydration of ethyl alcohol[35]. Volumetric methods have been used by Tamaru and his colleagues[36] in ammonia decomposition over tungsten and the decomposition of formic acid over silver. A rapid pressure response method[37] has been used in study of the dehydration of t-butyl alcohol over alumina.

Gas chromatographic methods show great promise because of the wide range of applications. They have been used in adsorption work[38] and extended into simultaneous studies of adsorption and catalysis by making the chromatographic column act also as a catalytic column. Retention times on the catalyst/column packing can be measured during reaction to give information on the adsorption behaviour of the reactant. Thus the decomposition of formic acid on palladium[39] and the water-gas shift on iron[40] have been investigated. Isomerisation[41] and dehydrosulphurisation[42] have also been studied. Phillips and his colleagues[43] have made a very significant advance in the chromatography/ catalysis method by introducing a stop-flow variation to obtain reaction kinetics.

Infrared methods have also been used in concomitant studies, e.g. of formic acid decomposition[44] and in hydrogenation of ethylene[45]. The kinetic and tracer approach has been successfully used in finding densities of active sites by Tétényi et al.[46].

Another approach to finding site densities for active sites on catalysts has been through the use of absolute rate theory. Using the symbolism of Glasstone, Laidler and Eyring[47],

$$\text{rate} = c_a(kT/h)\exp(\Delta S_a/R)\exp(-\Delta E_a/RT).$$

For a zero-order rate-determining step where only one species is involved, if the activation energy is known and ΔS_a is set equal to zero, then values of the site density c_a may be calculated. It is assumed that the entropy change between the adsorbed state and the transition state is small or zero.

For reactions on metals the results tend to give low site densities compared with the total possible sites if all exposed metal atoms were active. An example can be found in Maatman et al.[48] who dehydrogenated cyclohexane over platinum on alumina and found densities of 10^9 sites per cm^2 for platinum compared with a probable exposure value of 10^{14} sites per cm^2. Low site densities may indeed arise if kinks and terrace edges are the sites of catalytic activity[49].

Concomitant studies of chemisorption and catalysis were made by Lawson[50] for formic acid decomposition on silver using a radioactive tracer technique. His interest was in the activity of epitaxially grown silver films. He found, in an exceptionally fine study, adsorption on all exposed sites but that more detailed tracer analysis revealed differences of activity on different crystal faces.

3.3 Current Radioactivity Methods for Concomitant Studies

Radioactive methods may be employed in more complex systems in this field of concomitant study of adsorption and catalysis. In principle it would seem to be easy to label a molecular species and then observe, by direct monitoring of the surface of the catalyst, how the molecules behave under reaction conditions. This can be done and further checked and extended by the use of the Occupancy Principle[51].

It is important to remember that the radioactive method is one which does not determine the nature of the molecular species on the surface, but it does reveal something about the magnitude of the transient population and about the extent to which it is retained on the surface of a working catalyst.

We shall describe two examples of the adsorption of molecules during catalytic conversions. The first[52] concerns the hydrogenation of ^{14}C-ethylene over an iridium on silica catalyst: in this example there is a striking difference between adsorption of ethylene in the static and reaction modes.

3.4 Ethylene Hydrogenation over Iridium on Silica

To accomplish our quantitative measurements[52] we used direct monitoring of surface radioactivity and a modified version of the Occupancy Principle[51]. This principle has been enunciated from theoretical analysis of tracer dynamics in biological systems[53]. The object of these biological studies had been to elucidate otherwise inaccessible information on the total amounts and total flow of particular elements or molecules in body organs: the parallel problem in catalysis was to determine the size of the pool of active molecules on a catalyst and to determine the time history of the behaviour of these molecules.

In the application of the Occupancy Principle a pulse of radioactive tracer molecules was injected into a stream of molecules passing through the system. The rise and fall of radioactivity in the catalyst chamber was compared with the rise and fall of radioactivity generated by the same stream of molecules in a space of known capacity. The Occupancy Principle states that the ratio of occupancy to capacity is the same for all parts of the system and equals the reciprocal of the entry flow. From this, which we shall elaborate on later, the population of molecules on the catalyst can be determined.

The time history of a radioactive tracer passing over a catalyst could be determined at the same time. For this a novel fast observation technique had to be employed[54].

In the present study, pulses of ^{14}C-ethylene were passed through the iridium on silica catalyst in a stream of hydrogen. β^--emission was monitored for a direct estimate of adsorbed amounts and for application of the Occupancy Principle. For this latter, the pulses also passed through a monitored, known volume. Thus the size of the adsorbed pool of molecules could be determined by two methods.

There seemed to be three possible situations for the coverage of the catalyst during ethylene hydrogenation.

1. The metal provides the active sites but these represent only a small fraction of the metal area.
2. The whole of the metal surface may be active: this would be the situation for a facile reaction according to Boudart's classification[25].
3. The metal and the support might both participate in the catalysis.

The results for ^{14}C-ethylene passing over a 4.77% iridium on silica catalyst in the temperature range 267 to 310K were as follows. A typical count-rate versus time plot is shown for an experiment at 267K, Fig. 6(a). Maximum reversible adsorption corresponded to 2.5×10^{16} molecules mg^{-1} catalyst. The initial retention after the pulse had passed was 3.9×10^{15} molecules mg^{-1} and this fell slowly over 50 minutes to 7.0×10^{14} molecules. Conversion of the pulse to ethane was 99.9%. Changes in partial pressure of ethylene between 64 and 99 torr in hydrogen flow, where the partial pressure varied from 716 to 681 torr, gave results around 2.2×10^{16} molecules mg^{-1} adsorbed.

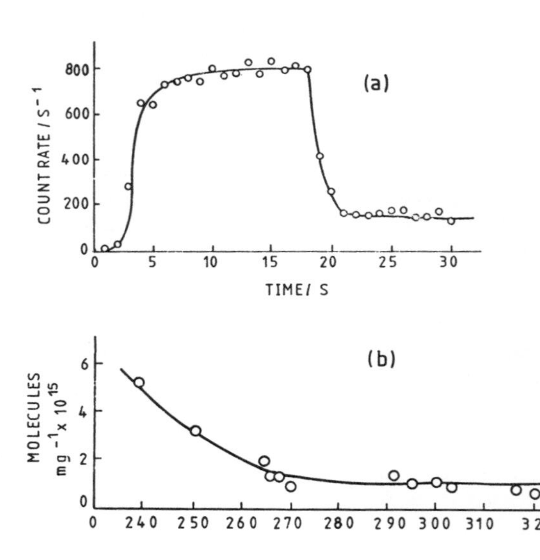

Fig. 6

(a) Count rate versus time for a pulse of ^{14}C-ethylene in hydrogen flowing through a catalyst chamber.

(b) Variation with temperature in population of labelled molecules on the surface of an Ir/SiO_2 catalyst when pulses of ^{14}C-ethylene in hydrogen passed over the catalyst. S.V. Norval, S.J. Thomson and G. Webb, Applications Surface Sci., to be published 1979. (reproduced with permission.)

When adsorption was studied as a function of temperature it appeared that the reversible adsorption tended towards a limiting value of $\sim 1.5 \times 10^{16}$ molecules mg^{-1}, for temperatures above 270K. In a non-competitive situation ethane alone gave an adsorption of about half that for ethylene at 294K. The extents of adsorption of ethylene measured by direct counting, direct observation, desorption studies and the Occupancy Principle all gave $\sim 1.5 \times 10^{16}$ molecules adsorbed per mg at 270K. Typical results are shown in Fig. 6(b).

We must now examine the extent to which this represents coverage of the metal. The area of metal atoms was measured by adsorption of carbon monoxide and by electron microscopy as 5.1 m^2 g^{-1} and 6.4 m^2 g^{-1} respectively. The area of metal found by gas adsorption was probably low and this was corrected for by taking account of the work of Sinfelt and Yates[55] who studied H/Ir and CO/Ir areas. We concluded that the metal exposure was probably 6.8 m^2 g^{-1}, and, if each cm^2 of Ir exposes 1.2×10^{15} atoms, then the atomic exposure was 8.1×10^{16} atoms Ir per mg of catalyst. This number can now be compared with the adsorption figures for ethylene, where 1.5×10^{16} molecules were adsorbed during reaction above $\sim$ 270K. Thus it seems that only one-fifth of the exposed surface was covered under our reaction conditions. This is in contrast to the behaviour in adsorption of ethylene over other Group 8 metals where, in a static system, full monolayer coverage was observed.

We have now to relate this result for pulses of ethylene and hydrogen passed over a fresh catalyst to the earlier static adsorption graphs for ethylene. In the pulsed experiment in a hydrogen-rich system, growth of permanent residues is inhibited but it does take place. About 10% of the radioactivity in a pulse was left behind and this behaviour corresponds to the situation reported by Al-Ammar and Webb[8]. They did a series of successive hydrogenations over fresh iridium and rhodium catalysts, and produced, after 8-10 runs, a steady-state catalyst in which the primary adsorption region had disappeared, Fig. 4(a).

The message is clear: each run changes the nature of the surface of a catalyst. A further good example is shown in the rate of allene hydrogenation: here the rate versus run number over rhodium on alumina goes through a reproducible maximum[56], Fig. 7.

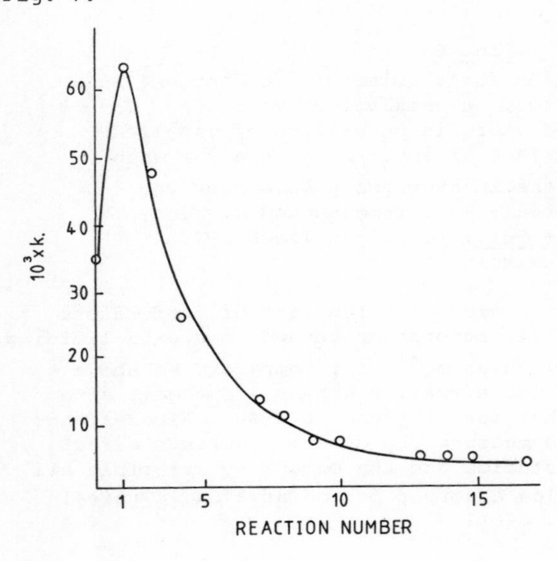

Fig. 7

Variation in the rate constant for the successive hydrogenation of allene over a Rh/Al_2O_3 catalyst at 298K. Pressures, allene 12 torr, hydrogen 36 torr.

N. Kuhnen, unpublished work.

3.5 Adsorption of Chlorobenzene During Hydrogenolysis

Quite different behaviour from that of gaseous ethylene is observed for the hydrogenolysis of chlorobenzene vapour to benzene and hydrogen chloride[57]. At about 295K pulses of ^{14}C-labelled chlorobenzene were passed over a palladium on silica catalyst and the surface observed by a G.-M. counter. A typical result is shown in Fig. 8. Conversions were between 15 and 33% and coverage of the catalyst was between 5 and 67% of the total area, i.e. in excess of the metal area. The support evidently acted as a reservoir for the low-boiling chlorobenzene with reaction presumably occurring on the metal. Desorption studies gave unexpected results. From pure silica desorption of chlorobenzene was a slow process, whereas from the catalyst system desorption of labelled molecules was fast.

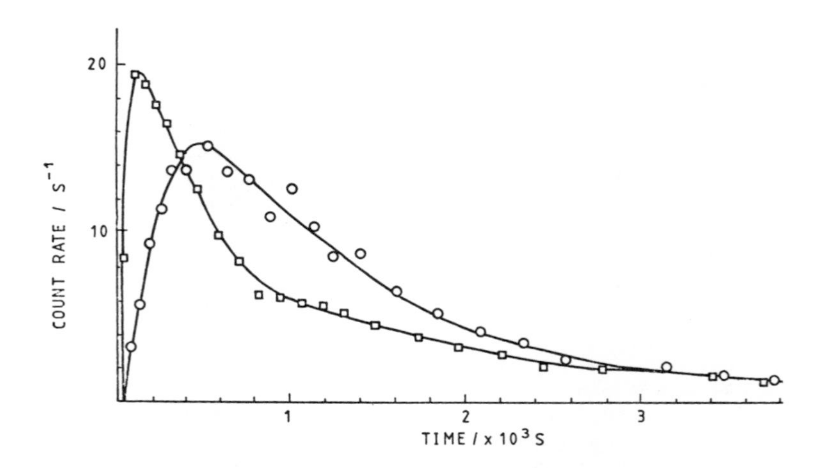

Fig. 8

Count rates from the surface of a Pd/SiO_2 catalyst, o , and from a standard volume, □ , when pulses of ^{14}C-chlorobenzene in hydrogen passed over the catalyst. S.V. Norval and S.J. Thomson, J.C.S. Faraday Trans. I, 75, 1798 (1979). (reproduced with permission.)

3.6 Silica

Silica as a support has been used in the last two examples. Once again radioactivity has contributed to the elucidation of its nature as a support. As an example, Altham and Webb[58] used tritium exchange and flow proportional counting to develop a method of detecting exchangeable hydrogen atoms on the support. By their method Webb[59] established that the 5% Ir/SiO_2 catalyst used in this work had 4.7×10^{17} exchangeable atoms per mg at $130^{o}C$: alternatively this may be expressed as 2.7×10^{14} exchangeable atoms per cm^2. This agrees well with the value of 3×10^{14} established by Hall et al.[60]: they employed a temperature-programmed deuterium exchange method: exchange was not significant until the temperature reached about $450^{o}C$.

3.7 Steam Reforming

The next working catalyst to be described behaves in a quite surprising fashion. Although extensively covered by carbon, it seems to continue almost undiminished in activity. It is the 75% nickel on silica steam-reforming catalyst. In experiments on steam reforming it is possible to choose such an amount of catalyst that it is just insufficient to bring about total hydrocarbon conversion. Thus alterations in the catalyst might be seen as a function of yield of products.

In the work of Jackson et al.[61] pulses of an aromatic or aliphatic hydrocarbon were passed in steam over nickel on silica at $475^{o}C$. ^{14}C-labelled materials were used and the products examined by chromatography and proportional or scintillation counting. Typically the yields (based on carbon input) were 35% CO_2, 35% CO + CH_4 with 0-10% unreacted hydrocarbon and 10 to 30% carbon residue. This carbon had the form of filaments, polymeric material, and yet further carbon which could be divided into reactive and unreactive forms. The polymeric material was about C_{30} in chain length and it contained $-CH_2$, -CHO and -CH=CH-CHO groups.

The results can be displayed in a schematic form, Table 1, which shows run number, relative units of all forms of carbon deposited, the extent of the free nickel surface as found by adsorption of ^{14}C-carbon monoxide, and the change in unreacted hydrocarbon passing through the system as the catalyst was repeatedly used.

Build-up of carbon with successive experiments was more rapid with aromatic feedstocks than with the aliphatics, and the free surface fell more rapidly. Yet the amount of hydrocarbons unreacted in both cases only changed by a few percent. This result suggests that the active surface is either a minute fraction of the metal surface or that the surface plus carbonaceous residues is the active surface. The Fig. 9 shows an electron micrograph of filaments on a working catalyst.

Table 1

Results of Study of Steam Reforming Catalysts (see text).

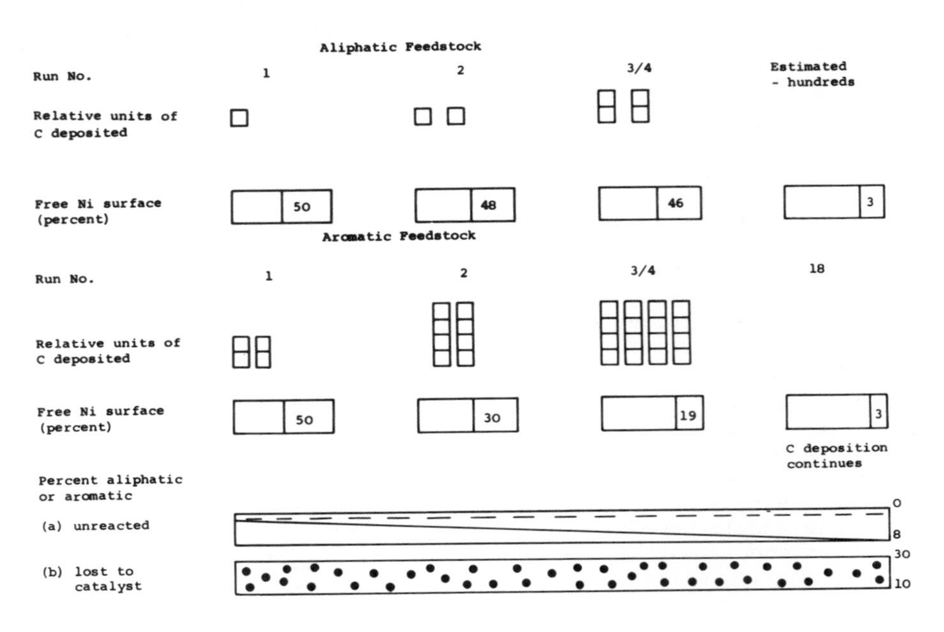

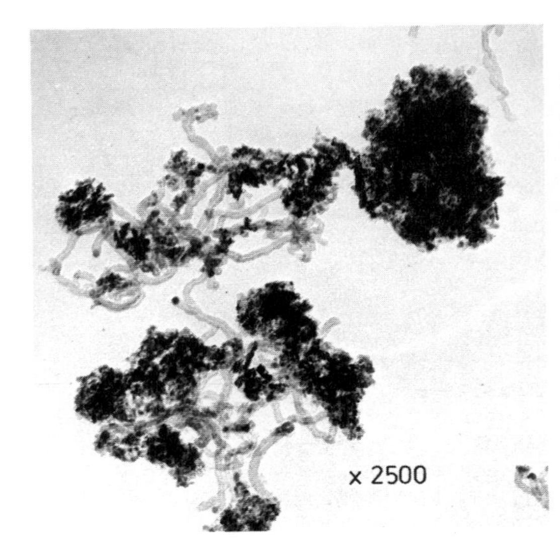

Fig. 9

Electron micrograph of a steam-reforming catalyst extracted from the system in its working state.
S.D. Jackson, S.J.Thomson and G. Webb, to be published.
(reproduced with permission.)

There are other examples in the literature of active carbon-covered surfaces. For example, Nieuwenhuys and Somorjai[62] found that iridium single crystal surfaces were always covered by a carbonaceous deposit during the dehydrogenation of cyclohexane. Thus, when they intentionally prepared Ir(111) surfaces covered by carbon through exposure to cyclohexane at $900^{o}C$ at 10^{-7} torr for thirty minutes, the activity for dehydrogenation was not affected.

A carbonaceous species which did not seem to interfere with catalysis was found by Lang et al.[63] in examining n-heptane reactions on platinum. On stepped surfaces isomerisation and hydrogenation reactions occurred which were evidently not affected by the presence of a carbonaceous deposit. Dehydrocyclisation to toluene occurred only if the stepped surface had an ordered layer of partially dehydrogenated carbonaceous material upon it. The ordered deposit occurred on stepped platinum and dehydrocyclisation did not occur on smooth Pt(111) faces.

Carbon overlayers were identified by Blakely and Somorjai[49] as of critical importance in cyclohexane dehydrogenation. Benzene was not produced on stepped platinum(111) surfaces until after an induction period and the growth of deposits. The ordered overlayer produced mainly benzene: the disordered overlayer more cyclohexene.

One should keep in mind, however, the notion that surface re-arrangements may always be giving rise to the exposure of catalytically active metal sites. In cleaner and simpler systems than steam reforming this has already been observed. It was, for example, proposed by Demuth and Eastman[64] in their study of ethylene adsorption on the (111) face of nickel. Acetylenic species were observed but bare sites capable of adsorption of hydrogen were created.

4. EXOELECTRONS

Having established the case that surfaces with carbon on them are working catalysts, we should now ask if there is a physical characteristic of catalysts which might be used in their characterisation. It should be coupled with catalytic activity and it should vary with the state of the surface, e.g. with carbon coverage.

One of the few properties of a surface which falls into this class is the ability of the surface to emit low energy electrons. This phenomenon of exo-electron emission has been known for many years, yet it figures hardly at all in the literature of catalysis. It has been reviewed recently by Ramsey[65].

A surprising variety of surface processes result in the emission of electrons of low energy, believed to be in the range 0 to 10 eV. These are abrasion or fatigue[66], exposure to radioactivity[67] or cathode rays[68], plastic deformation of metals[69], change of phase[70] and chemical reaction such as the hydrogen-oxygen reaction on gold[71], various gases on the alloy of sodium and potassium, NaK_2[72], oxygen on zinc[73] or copper[74].

In catalysis exo-electron emission accompanies the reaction of ethylene and oxygen to produce ethylene oxide over silver catalysts[75]. Catalyst samples were monitored inside a G.-M. counter while a mixture of argon, ethylene and oxygen passed through the counter. Electron emission occurred during reaction, and emission rate was proportional to rate of formation of ethylene oxide. The gases when introduced alone did not produce emission. The possibility of emission during adsorption is well known, e.g. Krylova[76] and Rakhmatullina and Krylova[77]. Sujak and his co-workers[78] have also demonstrated that optimum working temperature for a catalyst may be found by studying variation of emission with temperature. Hoenig and Utter[79] have also examined exo-electron emission during studies of oxidation of methane or carbon monoxide over palladium.

Cooper and Thomson[80] designed and evaluated a form of proportional counter for exo-electron studies suitable for operation with a wide variety of gases over a long temperature range. They inserted a platinum filament, which could be heated electrically, into their proportional counter, Fig. 10. They then heated this

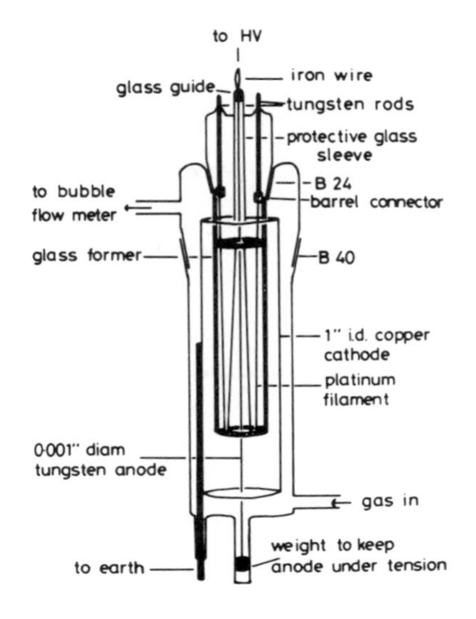

Fig. 10

Modified gas-proportional counter containing a platinum filament which could be electrically heated.

M. Cooper and S.J. Thomson, to be published. (reproduced with permission.)

wire, often to red heat, in a variety of gases and studied the decay of electron emission as the wire cooled to room temperature in the same gas. These cooling curves had a remarkable variety of structural features on them which the authors have interpreted as being due to adsorption or surface rearrangement: this latter was confirmed by electron microscopy. The figures show some examples: one of platinum cooling in hydrogen, Fig. 11, the other for cooling in a mixture of 90% argon 10% methane, Fig. 12.

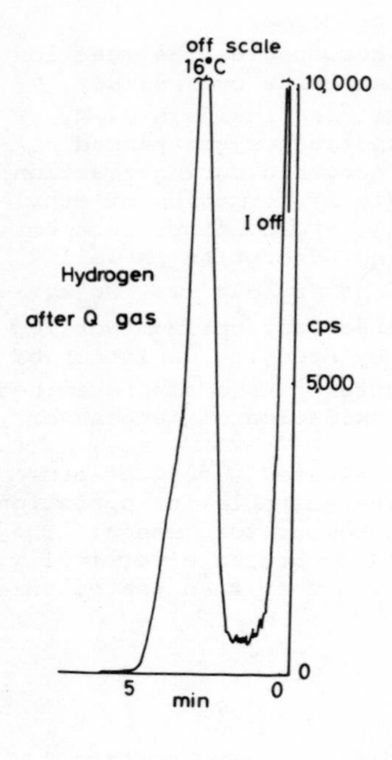

Fig. 11

Exoelectron emission from a platinum wire which had been heated to red heat and then allowed to cool in hydrogen. The wire had previously been heated in a methane 10%, argon 90% mixture (= Q gas).
M. Cooper and S.J. Thomson, to be published.
(reproduced with permission.)

Fig. 12

Exoelectron emission from a platinum wire cooling from red heat in an atmosphere of methane 10%, argon 90% (= Q gas).
M. Cooper and S.J. Thomson, to be published.
(reproduced with permission.)

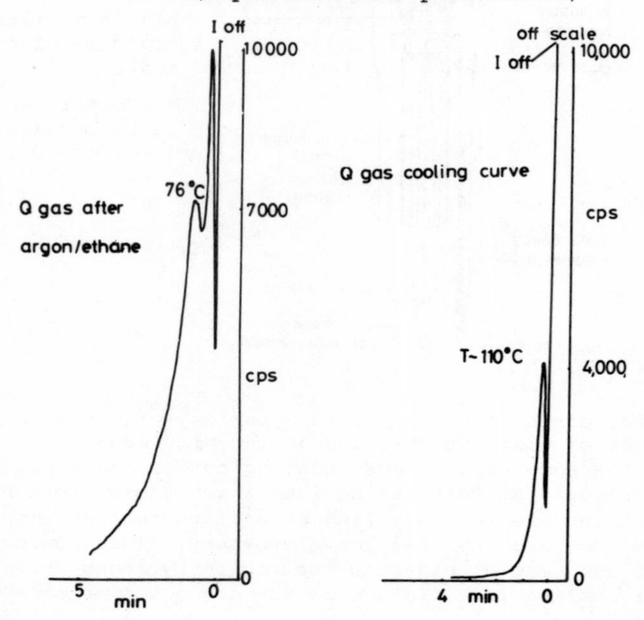

Indeed, the authors found that exo-emission only occurred after their platinum wires had been gently carbided in argon-
The presence of carbon seemed to have a profound effect on the ability of the platinum to release electrons. The dependence of emission on the probable presence of carbon is shown in Fig. 13, where a succession of heating and cooling cycles in ethane plus argon produced enhanced emission.

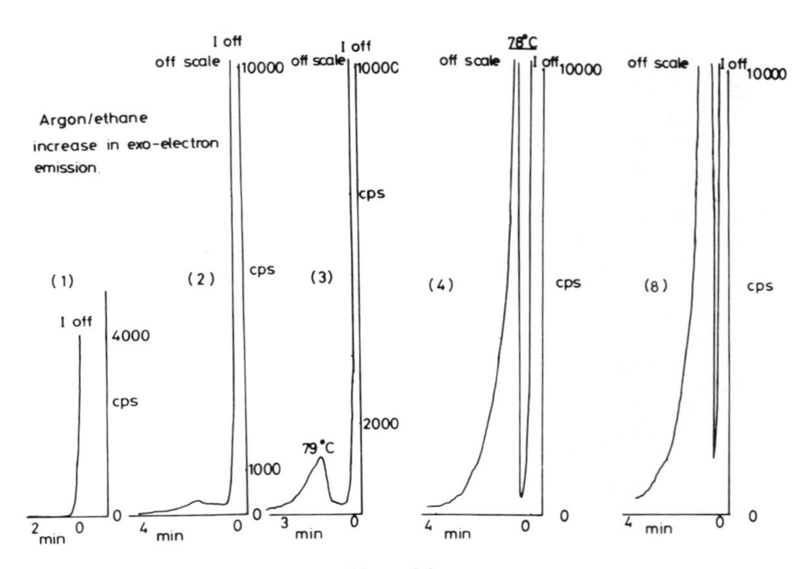

Fig. 13

A succession of exoelectron emission curves from a platinum wire heated to red heat in ethane 10%, argon 90%. Emission rises with increasing exposure to the ethane.
M. Cooper and S.J. Thomson, to be published.
(reproduced with permission.)

The authors are of the opinion that either the making and breaking of bonds at the surface during adsorption upon cooling, or surface rearrangements, similar to those observed by Lambert and his co-workers[81], are responsible for electron emission. The work function of their platinum wire is likely to be reduced by hydrocarbon adsorption[82].

5. CONCLUSION

The theme of this contribution has been that adsorption studies in static systems may not bear any relationship to catalysis, except in so far as adsorption creates the working surface of the catalyst. Characterisation of catalysts requires, therefore, two stages: first, an examination of the pristine catalyst:

second, detailed examination of the changes which then occur to the surface during catalysis.

The author acknowledges many stimulating discussions with Dr. K.C. Campbell and Dr. G. Webb.

REFERENCES

1. K.C. Campbell and S.J. Thomson, Progr. Surface and Membrane Sci., 9, 163 (1975).
2. Anon.
3. L. Peralta, Y. Berthier and J. Oudar, Surface Sci., 55, 199 (1976).
4. W. Heegeman, K.N. Meister, E. Bechtold and K. Hayek, Surface Sci., 49, 161 (1975).
5. S.J. Thomson and J.L. Wishlade, Trans. Faraday Soc., 58, 1170 (1962).
6. D. Cormack, S.J. Thomson and G. Webb, J. Catal. 5, 224 (1966).
7. J.U. Reid, S.J. Thomson and G. Webb, J. Catal. 29, 421, 433 (1973): 30, 372, 378 (1973).
   A.S. Al-Ammar, S.J. Thomson and G. Webb, Chem.Comm., 323 (1977).
8. (a) A.S. Al-Ammar and G. Webb, J.C.S. Faraday Trans. I, 74, 195, 657 (1978): (b) to be published (1979).
9. H. Conrad, G. Ertl, J. Küppers and E.E. Latta, 6th Int. Congress on Catalysis, London, 1976, Paper A33.
10. J.D. Prentice, A. Lesiunas and N. Sheppard, J.C.S. Chem. Comm., 76 (1976).
11. H. Conrad, J. Küppers, F. Nitschke and F.P. Netzer, J. Catal., 52, 186 (1978).
12. R.A. Zhur and J.B. Hudson, Surface Sci., 66, 405 (1977).
13. R. Duś, Surface Sci., 50, 241 (1975).
14. C. Brucker and T. Rhodin, J. Catal.,47, 214 (1977).
15. J.E. Demuth, Surface Sci., 69, 365 (1977).
16. S.J. Thomson and G. Webb, Chem. Comm., 526 (1976).
17. G.F. Taylor, S.J. Thomson and G. Webb, J. Catal., 12, 133 (1970): J.A. Altham and G. Webb, J. Catal., 18, 133 (1970).
18. Z. Paál and S.J. Thomson, J. Catal., 30, 96 (1973).
19. R. Ducros, M. Housley, M. Alnot and A. Cassuto, Surface Sci., 71, 433 (1978).
20. R.L. Palmer and D.A. Vroom, J. Catal., 50, 244 (1977).
21. M. Flytzani-Stephanopoulos, S. Wong and L.D. Schmidt, J. Catal., 49, 51 (1977).
22. K.J. Laidler and R.E. Townsend, Trans. Faraday Soc., 57, 1590 (1961).
23. H.S. Taylor, Proc. Royal Soc., A108, 105 (1925).
24. S. Brunauer, P.H. Emmett and E. Teller, J. Amer. Chem. Soc., 60, 309 (1938).
25. M. Boudart, Adv. Catal., 20, 153 (1969).
26. G.C.A. Schuit and L.L. van Reijen, Adv. Catal., 10, 242 (1958).

27. J.C. Schlatter and M. Boudart, J. Catal., 24, 482 (1972).
28. F.H. Dautzenberg and J.C. Platteeuw, J. Catal., 19, 41 (1970).
29. P.C. Aben, J.C. Platteeuw and B. Stouthammer, Rec. Trav. Chem. Pays-Bas, 89, 449 (1970).
30. A.A. Balandin, Adv. Catal., 19, 1 (1969).
31. M. Boudart, A.W. Aldag, L.D. Ptak and J.E. Benson, J.Catal., 11, 35 (1968). J.R. Anderson and N.R. Avery, J. Catal., 5, 446 (1966).
32. R.W. Maatman, Catal. Rev., 8, 1 (1974).
33. P. Mars, J.J.F. Scholten and P. Zwietering in 'The Mechanism of Heterogeneous Catalysis', Ed. J.H. de Boer, Elsevier, Amsterdam, 1960, p.66.
34. F.E. Massoth and D.A. Scarpiello, J. Catal., 21, 294 (1971).
35. S.M. Hsu and R.L. Kapel, J. Catal., 33, 74 (1974).
36. K. Tamaru, Bull. Chem. Soc. Jap., 31, 666 (1958). K. Tamaru, Trans. Faraday Soc., 55, 824 (1959): and 57, 1410 (1961). K. Fukuda, T. Onishi and K. Tamaru, Bull. Chem. Soc. Jap., 42, 1192 (1969).
37. E.R. Haering and A. Syverson, J. Catal., 32, 396 (1974).
38. G. Schay and G. Szekely, Acta Chim. Acad. Sci. Hung., 5, 167 (1954).
39. K. Tamaru, Nature, 183, 319 (1959).
40. J. Nakanishi and K. Tamaru, Trans. Faraday Soc., 59, 1470 (1963).
41. D.W. Bassett and H.W. Habgood, J. Phys. Chem., 64, 769 (1960).
42. P.J. Owens and C.H. Amberg, Canad. J. Chem., 40, 941 (1962).
43. C.S.G. Phillips, A.J. Hart-Davis, R.G.L. Saul and J.Wormald, J.Gas Chromatrography, 5, 424 (1967). R.M. Lane, B.C. Lane and C.S.G. Phillips, J. Catal., 18, 281 (1970). C.S.G. Phillips and C.R. McIlwrick, Anal. Chem., 45, 782 (1973).
44. J. Fahrenfort, L.L. van Reijen and W.M.H. Sachtler in 'The Mechanism of Heterogeneous Catalysis', Ed. J.H. de Boer, Elsevier, Amsterdam, 1960, p.23.
45. N. Sheppard, N.R. Avery, B.A. Morrow and R.P. Young in 'Chemisorption and Catalysis', Ed. P. Hepple, Inst. of Petroleum, 1970, p.135.
46. P. Tétényi, L. Babernics, L. Guczi and K. Schachter, Proc. 3rd Int. Congr. Catal., 1964, Vol 1, p.547 (1965). P. Tétényi and L. Babernics, J. Catal., 8, 215 (1967). L. Babernics and P. Tétényi, J. Catal., 17, 35 (1970).
47. S. Glasstone, K. Laidler and H. Eyring, 'The Theory of Rate Processes', McGraw-Hill, New York, 1941, Ch.7.
48. R.W. Maatman, P. Mahaffy, P. Hockstra and C. Addink, J. Catal., 23, 105 (1971).
49. D.W. Blakely and G.A. Somorjai, J. Catal., 42, 181 (1976).
50. A. Lawson, J. Catal., 11, 283, 295 (1968).
51. J.S. Orr and F.C. Gillespie, Science, 162, 138 (1968).
52. S.V. Norval, S.J. Thomson and G. Webb, Applications Surface Sci., to be published.

53. P.-E.E. Bergner, J. Theoret. Biol., 1, 120, 359 (1961): 6, 137 (1964) Acta Radiol., Suppl. 210, 1 (1962).
54. J.A. Hardy, private communication.
55. J.H. Sinfelt and D.J.C. Yates, J. Catal., 8, 82 (1967).
56. N. Kuhnen, unpublished work.
57. S.V. Norval and S.J. Thomson, J.C.S. Faraday Trans. I, 75, 1798 (1979).
58. J.A. Altham and G. Webb, J. Catal., 18, 133 (1970).
59. G. Webb, private communication.
60. W.K. Hall, H.P. Leftin, T.J. Cheselske and D.E. O'Reilly, J. Catal., 2, 503 (1963).
61. S.D. Jackson, S.J. Thomson and G. Webb, to be published.
62. B.E. Nieuwenhuys and G.A. Somorjai, J. Catal., 46, 259 (1977).
63. B. Lang, R. Joyner and G.A. Somorjai, J. Catal., 27, 405 (1972).
64. J.E. Demuth and D.E. Eastman, Phys. Rev. Lett., 32, 1123 (1974).
65. J.A. Ramsey, Progr. in Surface and Membrane Sci., 11, 117 (1976).
66. E. Rabinowicz, Sci. American, 236(1), 74 (1977).
67. P. Curie, Comptes Rendus, 129, 714 (1899).
68. J.C. McLennan, Phil. Mag., 3, 195 (1902).
69. A.H. Meleka and W. Barr, Nature (London), 187, 232 (1960).
70. F. Futschik, Thesis Viena 1955: see O. Bruna, K. Lintner and E. Schmid, Z. Phys., 136, 605 (1954).
71. H. Hartley, Proc. Royal Soc. Ser.A, 90, 61 (1914).
72. A. Denisoff and O. Richardson, Proc. Royal Soc. Ser. A, 148, 533 (1935).
73. J. Lohff, Z. Phys., 146, 436 (1956).
74. R. Reidl, Act. Phys. Aust., 10, 402 (1957).
75. H. Ohashi, H. Kano, N. Sato and G. Okamoto, J. Chem. Soc. Japan Ind. Chem. Section, 69, 997 (1966):
N. Sato and M. Seo, Nature (London), 216, 316 (1967):
N. Sato and M. Seo, J. Catal., 24, 224 (1972).
76. I.V. Krylova, Phys. Stat. Solid (a), 7, 359 (1971).
77. I.A. Rakhmatullina and I.V. Krylova, Russ. J. Phys. Chem., 42, 1407 (1968).
78. B. Sujak, T. Gorecki and L. Biernacki, Acta Phys. Pol. A39, 147 (1971).
79. S.A. Hoenig and M.G. Utter, J. Catal., 47, 210 (1977).
80. M. Cooper and S.J. Thomson, to be published.
81. P.D. Reed and R.M. Lambert, Surface Sci., 57, 485 (1976):
M.E. Bridge and R.M. Lambert, Surface Sci., 63, 315 (1977):
C.M. Comrie, W.H. Weinberg and R.M. Lambert, Surface Sci., 57, 619 (1976).
82. G.A. Somorjai, Adv. Catal., 26, 1 (1977).

EXTENDED X-RAY ABSORPTION FINE STRUCTURE STUDIES OF SUPPORTED CATALYSTS

By

Richard W. Joyner

1. INTRODUCTION

Two structural questions are of importance in catalysis: "What is the structure of the catalyst of interest?" and "What influence does this structure have on the catalytic reactivity?". Both of these questions are difficult to probe experimentally, especially for highly dispersed catalyst materials. For particle sizes < 100 Å, X-ray diffraction is useful only for determining the particle's size, and provides no information on particle structure[1].

Important progress has been made, however, in elucidating the relationship between surface structure and catalytic reactivity, through the use of Low Energy Electron Diffraction (LEED). The studies of Somorjai et al.[2] and others (see Roberts and McKee[3]) are well-known, as are the results of several industrially oriented laboratories such as the British Petroleum group[4]. The LEED technique, however, has obvious limitations, requiring single crystals as model catalysts and operating only in ultra-high vacuum.

It is now clear, and this chapter as well as the 'case history' of Cox in this book, describe how useful information on the structure of real catalysts and on structure-reactivity relationships for these materials may be obtained from the technique of extended X-ray absorption fine structure, EXAFS. By studying oscillatory variations in absorption cross-sections near absorption edges, this technique provides three types of information on the elements present in the catalyst. These are :

(1) Interatomic Distances, R_j

These can be obtained for first and subsequent coordination shells, with an accuracy currently estimated to be between ± 0.01 and ± 0.05 Å.

(2) Coordination Number, N_j

The accuracy quoted for the first coordination shell is usually ± 20 percent, although this may be optimistic (see A.D. Cox, elsewhere in this book).

(3) Debye-Waller Factor, which reflects the mean square fluctuation of the atom in question. Not enough work has yet been reported to assess the accuracy with which this, possibly the least relevant parameter to catalysis, can be determined.

The technique probes the local environment of the particular element or variety of elements under study. It is therefore applicable to amorphous materials as well as crystalline samples and can be used to study the active element (e.g. the metal) in the catalyst. Moreover, the catalyst may be studied in a controlled atmosphere or in vacuum.

The remainder of this chapter considers some theoretical aspects of EXAFS (Section 2.1), some experimental factors (Section 2.2) and describes the results of its application to catalysis (Section 3).

2. THE EXAFS TECHNIQUE

2.1 Theoretical Aspects

When the photon energy becomes just sufficient to cause emission of photoelectrons from any electron shell, a sharp increase occurs in the cross-section for photon absorption. This is the well-known phenomenon of the absorption edge. Above the edge an oscillatory variation in the absorption cross-section is often observed, (Fig. 1), which is sometimes designated Kronig structure. Kronig[5] himself suggested a plausible explanation. The origin of this was not fully understood until Sayers, Stern and Lytle[6] showed how it reflected the local structure surrounding the atom under study. The greater ease of interpretation of data in the range > 100 eV above the absorption edge has given rise to the name extended X-ray absorption fine structure.

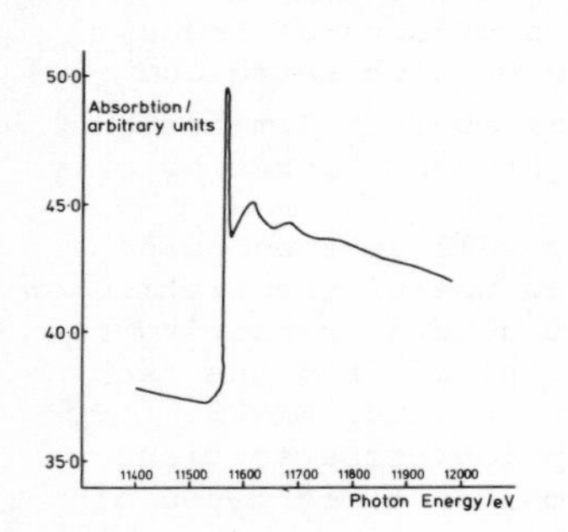

Fig. 1

L_3 absorption spectrum of 6% Pt/SiO_2 catalyst, EUROPT-1. (From Joyner (1979).)

The structural information is contained in the oscillations of absorption cross-section which are superimposed on a slowly varying background. The oscillations result from constructive or destructive interference between the outgoing photoelectron wave and its reflection from neighbouring atoms. The EXAFS function $\chi(\underset{\sim}{K})$ is defined as

$$\chi(\underset{\sim}{K}) = \frac{\mu - \mu_o}{\mu_o} \qquad (1)$$

μ and μ_o are atomic absorption coefficients, μ is characteristic of the atom in the material of interest, and μ_o of the atom in the free state.

$\underset{\sim}{K}$ is the photoelectron wave vector

$$\underset{\sim}{K} = \sqrt{2E} \qquad (2)$$

where K and E are in atomic units, reciprocal Bohr radii and Hartrees respectively.
(1 Bohr Radius = 0.529 Å, 1 Hartree = 27.2 eV)
μ is given by :-

$$\mu = \ln(I_o/I) \tag{3}$$

where I_o and I are respectively the incident and transmitted photon flux.
To yield $\chi(\underset{\sim}{K})$ the non-oscillatory part must first be removed from the experimental data, using a polynominal or spline fitting procedure.

The theory of EXAFS has been considered in detail by Ashley and Doniach[7] and by Lee and Pendry[8]. Most treatments are, however, based on the following equation due to Sayers et al.[6]:-

$$\chi(\underset{\sim}{K}) = \Sigma_j \frac{N_j}{\underset{\sim}{K}.R_j^2} \sin(2\underset{\sim}{K}.R_j+2\delta+\psi_j).|f_j(\pi)|\exp(-\bar{U}_j^2 \underset{\sim}{K}^2).\exp(-2V_iR_j/\underset{\sim}{K}) \tag{4a}$$

R_j and N_j are respectively the interatomic distance of the jth shell from the central atom and the coordination number of the jth shell. δ and ψ_j are phase shifts of the emitting atom and the backscattering atom in the jth shell respectively. $|f_j(\pi)|$ is the amplitude of the backscattering factor. $\bar{U}_j^2$ is the Debye-Waller factor $= 2\sigma_j^2$, where σ is the root mean square displacement of the emitter from the scattering atom. V_i is the imaginary part of the self-energy, as in LEED theory[9], and takes account of inelastic scattering of the photoelectron wave.

The terms in $|f_j(\pi)|$, $\bar{U}_j^2$ and V_i thus affect only the amplitude of $\chi(\underset{\sim}{K})$ which, for a single shell, usually decreases with increasing photoelectron energy. Equation (4a) thus may be expressed as :-

$$\chi(\underset{\sim}{K}) = \sum_j \frac{N_j}{\underset{\sim}{R}_j^2} \sin(2\underset{\sim}{K}.R + 2\delta + \psi_j)\ F(\underset{\sim}{K}) \tag{4b}$$

where $F(\underset{\sim}{K})$ is a slowly varying function of $\underset{\sim}{K}$.
For an atom surrounded only by a single shell we have

$$\chi(\underset{\sim}{K}) = \frac{N_j\ F(\underset{\sim}{K})}{R_j^2} \sin(2\underset{\sim}{K}.R + 2\delta + \psi_j) \tag{4c}$$

or if we (unjustifiably) neglect δ and ψ_j, then

$$\chi(\underset{\sim}{K}) = \frac{N_j F(\underset{\sim}{K})}{R_j^2} \sin (2\underset{\sim}{K}.R) \qquad (4d)$$

Thus we can see that the EXAFS oscillations, $\chi(K)$ are simply a sine function, of period $2\underset{\sim}{K}.R_j$. One way in which the value of R_j may be obtained is by Fourier transformation of the data :-

$$f(R) = N_j \int_{-\infty}^{\infty} \frac{F(\underset{\sim}{K})}{R_j^2} \sin (2\underset{\sim}{K}.R) \exp^{iCR} d\underset{\sim}{K} \qquad (5)$$

For an infinite, noise-free data set, the result would be a delta function at $R = R_j$. For a real data set, however, the result of Fourier transformation is functions such as those shown in Fig. 5 or Fig. 6.

We must now consider the influence of the phase shifts $(2\delta + \psi_j)$, which we have hitherto ignored. Effectively they shift the maximum in the Fourier transform by $\sim$ 0.1 - 0.3 Å to some $R < R_j$. Fig. 2 illustrates this, showing results for metallic copper[10]. Arrows show the interatomic distances for the first and subsequent coordination shells in the metal; the lower curve is the Fourier transform of the raw EXAFS data. The maxima are shifted uniformly by $\sim$ 0.2 Å to lower distance. If the effect of the phase shift is included, the upper curve results, and the maxima become aligned, with the correct interatomic distance.

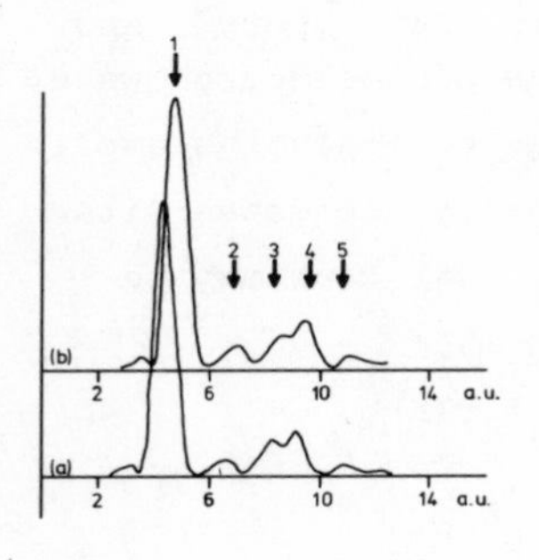

Fig. 2

EXAFS, Fourier transform for metallic copper. Lower curve: neglecting phase shifts. Upper curve: including the effect of phase shifts $(2\delta + \psi_j)$ in the transform. The arrows show the nearest-neighbour distances. (From Gurman and Pendry (1976).)

The "phase shift problem" must therefore be solved if accurate interatomic distances are to be obtained, and three main approaches to its solution have been developed. The simplest is to assume that the phase shift correction is the same for any given A-B bond, (A = central atom). This correction can be obtained from an EXAFS study of a compound whose structure is known from X-ray diffraction: it is then applied to measure any unknown A-B bond length. In support of this approach Fig. 2 shows that the correction is similar for the first four coordination shells of metallic copper. The method is, however, empirical, leading to interatomic distances of accuracy between ± 0.02 and 0.05 Å.

Alternatively the EXAFS of a compound of known structure can be used to extract ($2\delta + \psi_j$) as a function of photoelectron wave vector. Such an approach has been advocated by Citrin et al.[11], who argue that, provided the photoelectron energy exceeds ~ 100 eV, scattering will be predominantly from core electrons. The phase shifts should thus not be influenced by chemical environment and should therefore be transferable. Accuracies of ± 0.01 Å have been claimed from this approach.

The third solution to the problem involves the use of calculated phase shifts. Theoretical studies relating to LEED and photoemission have meant that such phase shifts are readily available[9], and a comprehensive listing for use in EXAFS calculations has recently been published[12]. Recent results using this approach are encouraging. As an example Fig. 3 shows experimental and calculated results for $CuSO_4.5H_2O$, which has been used as a model compound in a study of transition metal-exchanged

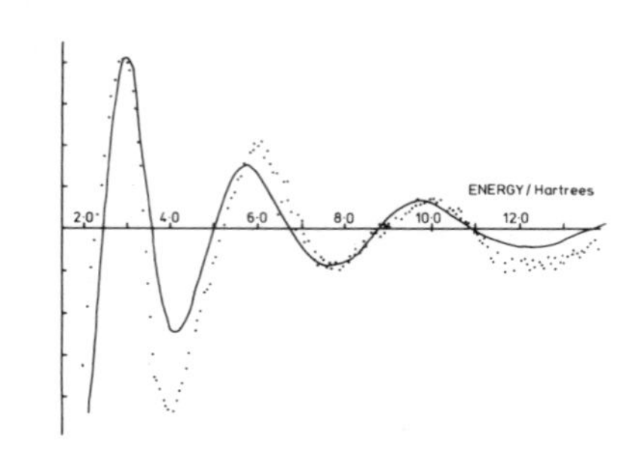

Fig. 3

Experimental points and calculated EXAFS (solid curve) for copper sulphate pentahydrate. The calculated Cu-O distance is 1.945 Å. (This author, unpublished results.)

zeolites, (Joyner, in course of publication). The Cu-O distance was treated as a parameter in the calculation and the best fit with the experimental results was obtained with a Cu-O nearest-neighbour distance of 1.945 Å. This compares with the X-ray and neutron diffraction results, which show an average Cu-O distance of 1.96 Å, suggesting that the EXAFS accuracy is ± 0.02 Å.

By comparison with the accuracy to which interatomic distances may be obtained, the accuracy for coordination numbers is rather poor. Coordination number is only one of the number of factors, including $|f_j(\pi)|$, $\bar{U}_j^2$ and V_i, which determine the amplitude of the EXAFS oscillations, $\chi(\underset{\sim}{K})$. A further problem is that the amplitude calculated from Eqn.(4) exceeds the experimentally observed amplitude, often by 30 percent. This is due to the occurrence of 'shake-up and shake-off', a phenomenon which frequently bedevils photoelectron phenomena[13]. In disordered systems, correlation between coordination number and Debye-Waller factor makes accurate determination of either parameter rather difficult (see the Chapter by Cox in this book).

Coordination numbers are often calculated by reference to the magnitude of the Fourier transform peak in a known reference compound K. For the unknown compound U, the coordination number of the jth shell is calculated from :-

$$N_{j(U)} = \frac{Mag_{j(U)} \; R^2_{j(U)}}{Mag_{j(K)} \; R^2_{j(K)}} \qquad (6)$$

This assumes that the Debye-Waller factor is the same in known and unknown compounds. Coordination numbers obtained in this way are often stated to be accurate to 20 percent, although even this may well be optimistic.

2.2 Experimental Aspects

The basic experiment involves the measurement of absorption coefficient, μ, as the photon energy is varied across and beyond the absorption edge. A typical arrangement is shown diagrammatically in Fig. 4. The photon source is either a synchrotron or storage ring, or the Bremsstrahlung from a conventional or rotating anode X-ray source. With conventional X-ray sources, data collection times are impracticably long, typically 1-2 weeks. Del Cueto and Shevchik[14] have reduced this considerably, using a rotating anode and a curved focussing crystal. There is little doubt, however, that a synchrotron (or storage ring), with its intense, polarised source over a very large range of photon energies, represents the best and most versatile photon source. In the energy range of interest, channel-cut silicon crystals are excellent monochromators. Photon counting is by inert-gas filled ionization chambers, and the whole is computer-controlled. With a synchrotron source, spectrum acquisition time is typically

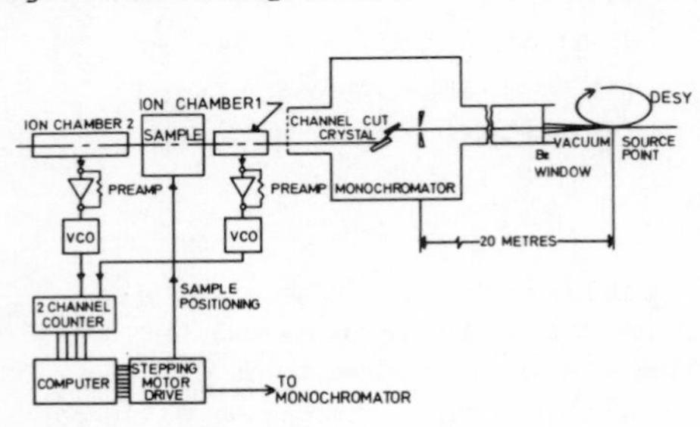

Fig. 4

Experimental arrangement for EXAFS. (After Kincaid (1975).)

20 minutes.

The sample may be a gas, solid, liquid, or a solution. For best signal/noise ratio in the spectrum the sample should absorb about two-thirds of the incident photons. The sample thickness, (τ), may be calculated from

$$\mu.\tau \cong 1 \quad (7)$$

values of μ being obtained from Tables[15].

For pure metals this criterion leads to a sample thickness of several microns, whereas for dilute catalysts, where the support is usually of very low X-ray absorption, the thickness is about 1 mm. The illuminated area is typically 15 x 3 mm and the sample may be studied in air, in vacuum or in a suitably controlled environment. Any temperature may be used, lower temperatures, e.g. 4K or 77K, however, improve signal intensity by suppression of the Debye-Waller factor.

The range of elements which can be studied depends on the operating energy of the synchrotron, which determines the maximum energies available, and also the presence of any "windows", which result in a low energy cut-off. For a 2 GeV synchrotron with conventional beryllium windows the photon energy range is typically 5 - 15 keV, ($2.5 > \lambda/\text{Å} > 0.8$). K shell edges for elements $21 < Z < 35$, (Ti - Br), are thus accessible as are the L shells, $54 < Z < 80$, (Cs - Au). With a higher energy synchrotron, or through the use of a "wiggler" magnet, elements in the intermediate range, $35 < Z < 54$, become amenable to study. The use of beryllium windows is not obligatory, so that with proper design the available range may also be extended to lower Z. For transmission EXAFS, however, the required sample thickness may become impracticably small.

2.3 Fluorescence and Surface EXAFS

Two modifications which extend the applicability of the EXAFS technique may be mentioned briefly. Jaklevic et al.[16] have shown that the sensitivity may be enhanced by detecting X-ray fluorescence due to core-hole filling, rather than simply measuring the absorption cross-section. Several orders of magnitude enhancement in sensitivity are achieved, making possible the study of much more dilute systems. It has, for example, recently been reported that ppm impurities of copper may be detected in a matrix of iron. This augurs well for the power of the X-ray fluorescence method of recording EXAFS signals[17].

Alternatively the technique may be rendered surface-sensitive by measuring Auger electrons generated by core-hole filling. The application of this technique to the study of, *inter alia*, iodine on silver is discussed by Citrin et al.[18,19].

Extended fine structure of similar origin to that in EXAFS has also been noted in appearance potential spectroscopy[20], and in electron energy loss spectroscopy[21,22].

3. APPLICATIONS OF EXAFS TO THE STUDY OF CATALYSTS

3.1 Noble Metal Catalysts

Supported noble metal catalysts are of considerable commercial significance and are routinely prepared with high dispersion, > 50 percent. EXAFS study of the metal thus provides information predominantly on the active surface of the catalyst.

Via, Sinfelt and Lytle[23] have examined osmium, platinum, and iridium catalysts after *in situ* reduction, and their results are summarised in Table 1. Some Fourier transforms are shown in Figs 5 and 6.

Table 1

Properties of Dispersed Metal Catalysts. Coordination number, N_1; Interatomic Distance, R_1, and Interatomic Distance in the Bulk Metal.

(From Via et al. (1979).)

METAL	SUPPORT	N_1	R_1/Å	R_1/Å (Bulk)
Os	SiO_2	8·3	2·702	2·705
Ir	SiO_2	9·9	2·712	2·714
Ir	Al_2O_3	9·9	2·704	2·714
Pt	SiO_2	8·0	2·774	2·775
Pt	Al_2O_3	7·2	2·758	2·775

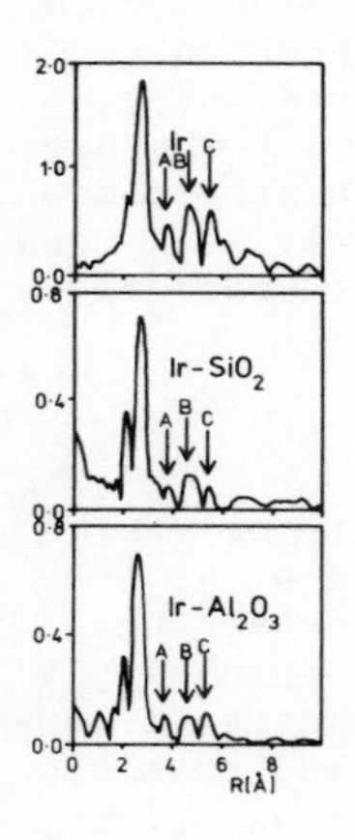

Fig. 5

Fourier transforms. Upper curve, iridium metal; middle curve, 1% iridium on silica; lower curve, 1% iridium on alumina. (From Via et al. (1979).)

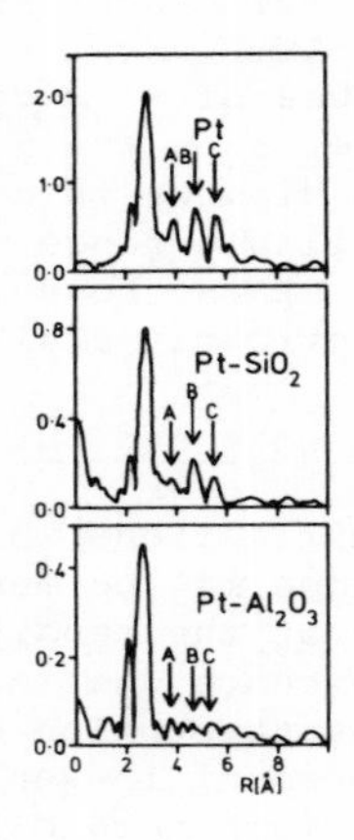

Fig. 6

Fourier transforms. Upper curve, platinum metal; middle curve, 1% platinum on silica; lower curve, 1% platinum on alumina. (From Via et al. (1979).)

Spectra were obtained with the Stanford synchrotron, SPEAR, with the sample held in a boron nitride vacuum cell. Interatomic distances were obtained using energy-dependent phase shifts calculated from the pure metal spectra, followed by a least-squares fitting procedure.

Their results show a basic resemblance of the supported catalysts to the pure metals. This is itself of considerable importance, since there is a corpus of theoretical work which suggests that icosahedral structures may be energetically preferred to the face-centred cubic structure of the bulk metals[24]. Calculations using the rather imprecise Lennard-Jones pairwise interaction potential suggest that the icosahedral structure will be preferred for particles of up to about 150 atoms ($\sim$ 15 Å diameter). We would expect EXAFS to reveal icosahedral structure if it should predominate in the supported catalysts. Table 2 shows nearest-neighbour distances and coordination numbers for 55 atom clusters, calculated from simple ball models for both icosahedral and f.c.c. structures. The icosahedral particle possesses a much greater number of nearest-neighbour distances than the f.c.c. entity. The catalysts studied by Via et al.[23], however, show only those nearest-neighbour distances, (a, $\sqrt{2}$a, $\sqrt{3}$a, 2a.) consistent with the f.c.c. crystal habit. The presence of icosahedral packing may therefore be discounted.

55 ATOM CLUSTER			
ICOSAHEDRON		F.C.C.	
R_j	N_j	R_j	N_j
1	6·18	1	9·3
1·051	0·59		
1·303	1·45		
1·376	0·73		
		1·414	3·93
1·451	0·36		
1·486	0·73		
1·539	0·36		
1·618	1·45		
1·70	0·36		
		1·732	9·45
1·834	1·45		
75% Dispersion			

Table 2

Calculated interatomic distances, R_j, and coordination numbers for 55 atom icosahedral and face-centred cubic structures.

The results of Figures 5 and 6 have a number of other interesting features. As would be expected, the coordination numbers (Table 1) are lower than in the bulk materials. The interatomic distances for the catalysts agree with bulk distances within the quoted experimental error of ± 0.01 Å. The one exception is the Pt/$A\ell_2O_3$ catalyst, where the distance, at 2.758 Å, is 0.017 Å less than the bulk value. Via et al.[23] suggest that this may be evidence of catalyst/support interactions, which are known to be particularly marked between platinum and alumina. The Pt/$A\ell_2O_3$ material also shows a lower coordination number, 7.2, than the rest of the catalysts, although selective chemisorption suggested a similar dispersion.

It is worth drawing attention to an interesting feature in the data of Via et al.[23], which is not discussed by them. It may be relevant to the question of "rafts" in supported catalysts considered earlier (see Chapter by Howie). The non-nearest-neighbour peaks in the Fourier transforms (Figs 5, 6) are indicated by the arrows A, B and C, corresponding respectively to $\sqrt{2}a$, $\sqrt{3}a$ and 2a distances in the catalyst. The respective coordination numbers of these shells in the bulk are 6, 12 and 12. Figures 5 and 6 show that the ratio of these peaks are very similar in the bulk and in the silica-supported materials. For Ir/Al_2O_3, however, the ratio $\sqrt{2}a/\sqrt{3}a$ is much higher than in the bulk, and for platinum supported on alumina the $\sqrt{2}a$ peak is the only non-nearest feature observed above the noise.

These results provide prima facie evidence for the existence of rafts of (100) crystallography on the alumina support. The distribution of non-nearest neighbours varies considerably with crystal plane. A single layer of (111) structure would show $\sqrt{3}a$ distances, but not $\sqrt{2}a$ lengths, whereas a raft of (100) structure would show $\sqrt{2}a$ distances predominantly compared to $\sqrt{3}a$ distances, as observed here. (Bulk $\sqrt{2}a/\sqrt{3}a$ ratios, as observed on SiO_2 supports, suggests the formation of 3-dimensional rather than raft-like particles.) Further EXAFS calculations, including a detailed consideration of the Debye-Waller factor in the non-nearest neighbour shells, would be required to support this conclusion.

The possibility of icosahedral structure has also been examined by Moraweck et al.[25]. These authors have used a zeolite-Y matrix to prepare small platinum particles which small angle scattering shows have rather uniform particle size, 13 ± 2 Å. Fourier transforms are shown in Fig. 7.

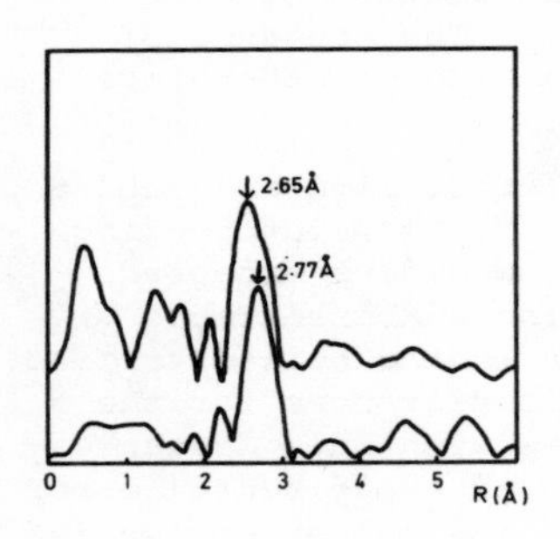

Fig. 7

Fourier transforms of results from 13 Å platinum particles in a zeolite-Y matrix. Upper curve, clean metal; lower curve, after exposure to hydrogen.

(From Moraweck et al. (1979).)

The authors conclude that the particles, which contain, on average, ~ 55 atoms, show no evidence of icosahedral structure. In accord with calculation for both the icosahedral and f.c.c. structures[26], the Pt-Pt distance is shorter than that in the bulk metal, by 0.12 Å. Intriguingly, the Pt-Pt interatomic distance relaxes to that of the clean metal on exposure to hydrogen.

Bassi, Lytle and Parravano[27] have studied supported gold and platinum catalysts. These catalysts had been subject to reduction, but not under in situ conditions. The Fourier transforms showed peaks assigned to M-O and M-M bonding (Fig. 8) and, in some cases, M-Cl bonding. A conventional X-ray set was used as a photon source and each spectrum was the average of 10 sets of data. Phase-shift corrections were determined from reference compounds. Based on the observation of M-O and M-M peaks, the authors concluded that two phases were present, one being metallic, with metal particles up to 100 Å in diameter.

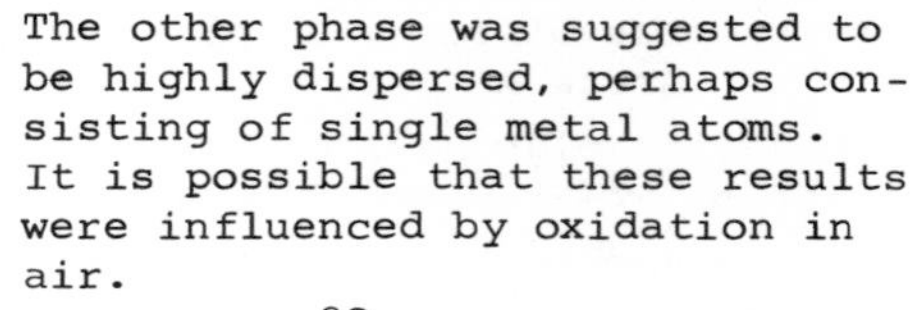

The other phase was suggested to be highly dispersed, perhaps consisting of single metal atoms. It is possible that these results were influenced by oxidation in air.

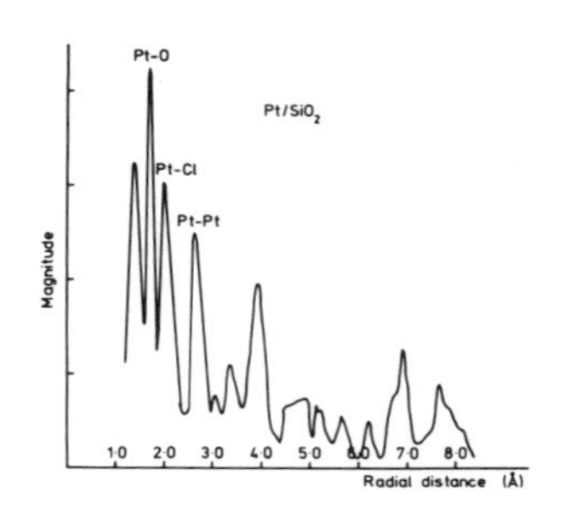

Fig. 8

Fourier transformed results from a 6 percent Pt/SiO_2 catalyst of 20 percent dispersion.
(From Bassi et al.(1976).)

Joyner[28] has studied two platinum catalysts, 6 percent on silica with 60 percent dispersion (the Council of Europe reference catalyst, EUROPT-1) and 0.47 percent on alumina, with ≃ 100 percent dispersion. The materials were reduced, but exposed to air before and during examination. The EXAFS spectrum of EUROPT-1 is shown in Fig. 1, and Fourier transforms in Figs 9 & 10. These are dominated by a single peak (arrow A) which is due to a platinum oxygen distance.

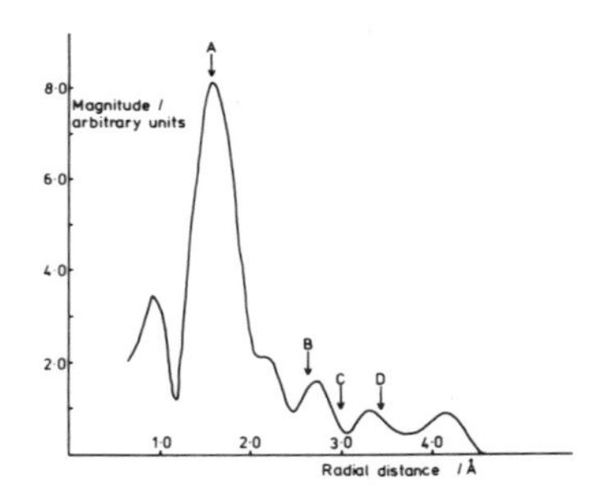

Fig. 9

Results from a 6 percent Pt/SiO_2 catalyst (EUROPT-1) of 60 percent dispersion. Peak marked A is due to a Pt-O distance. Arrows B, C and D show the positions where peaks would be expected from Pt-Pt bonds in Pt metal, (arrow b) and βPtO_2, (arrows C,D).
(From Joyner, 1979[28].)

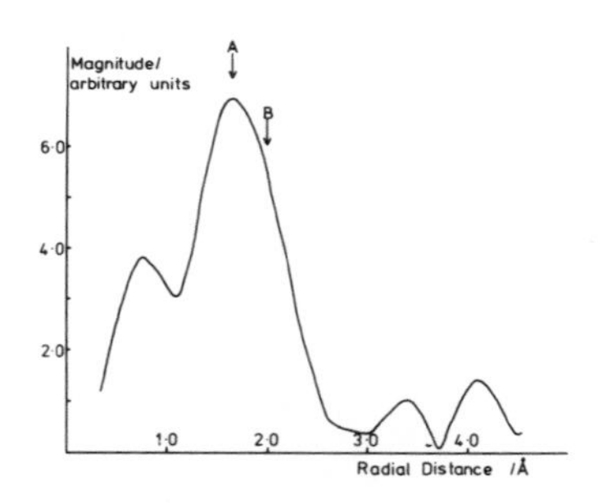

Fig. 10

Results from a 0.47 percent $Pt/A\ell_2O_3$ catalyst. Peak marked A is due to a Pt-O distance. Arrow B indicates where a peak due to Pt-Cℓ bonding would occur.
(From Joyner, 1979[28].)

After applying the phase-shift correction determined by Bassi et al.[27], this is shown to be 1.94 ± 0.05 Å, in agreement with the Pt-O distance in βPtO_2, 2.00 ± 0.02 Å.

The most interesting feature of these results is that there is no evidence of platinum-platinum bonding, either in Pt metal or PtO_2. Arrow B (Fig. 9) shows where a peak would be expected from metallic platinum, and arrows C, D indicate the likely position of Pt-Pt peaks from βPtO_2. Joyner concluded that the platinum was present in an amorphous structure similar to a glass, with the platinum bound in PtO_x units, ($x \leq 6$). This result can be contrasted with that of Bassi et al.[27] (Fig. 8), where a silica supported catalyst of similar metal loading but lower dispersion (20 percent) showed evidence of platinum metal as well as PtO_2. This suggests that very small particles are more active in oxygen chemisorption, as has also been suggested by Burwell et al.[29] and Boudart et al.[30]. By contrast, bulk platinum is very resistant to oxidation. Structural changes occasioned by exposing reduced catalysts to oxygen may have implications for the determination of metal surface areas by oxygen/hydrogen titration.

The Fourier transform of the 0.47 percent $Pt/A\ell_2O_3$ catalyst is similar to that for the silica supported material. The main peak occurs at the same interatomic distance, but is much broader. This reflects increased noise in the data from a more dilute sample, but is probably also due to the presence of some Pt-Cℓ bonding, which would give a peak in the position of arrow B (Fig. 10). After reduction the crystal had been exposed to hydrogen chloride (773 K, 3 hrs) to increase the dispersion, and analysis showed 0.99 percent chlorine by weight.

3.2 Alumina Supported Transition Metal Catalysts

Friedman, Freeman and Lytle[31] have described a comprehensive characterisation of a series of copper-alumina catalysts, using X-ray diffraction, ESR, X-ray photoelectron spectroscopy and diffuse reflectance spectroscopy as well as EXAFS. Some Fourier transforms are shown in Fig. 11. Depending on metal loading and calcination temperature, four environments of the Cu^{2+} ion were distinguished :-

- (a,b) supported on the γ alumina, in tetrahedral or octahedral sites. The latter are tetragonally distorted through John-Teller effects and are probably only 5-coordinate at the surface;
- (c) in a copper oxide phase;
- (d) in a copper aluminate phase.

Below ~ 4 percent copper, no copper-containing phase was detectable by X-ray diffraction. Reflectance spectroscopy shows the presence of Cu^{2+} in both octahedral and tetrahedral sites. The small peak in the EXAFS Fourier transform (Fig. 11c, d) at ~ 2.5 Å was suggested to result from octahedral coordination in the γ-alumina, which has the defect spinal structure. Above 4 percent metal loading, X-ray diffraction shows the presence of copper (II) oxide, which is also observable by EXAFS in the Fourier transform range 3-7 Å (cf Figures 11a,e). If the material is annealed above 770 K, however, these peaks are suppressed and the spectrum more closely resembles that of copper aluminate (cf Figures 11b,f). Friedman et al.[31] use coordination numbers extracted from the EXAFS result, through the procedure implied by Eqn. (6), to determine the ratio of octahedral to tetrahedral sites in the catalysts. Their paper also gives a critical review of other, quite extensive studies on the characterisation of these catalyst materials.

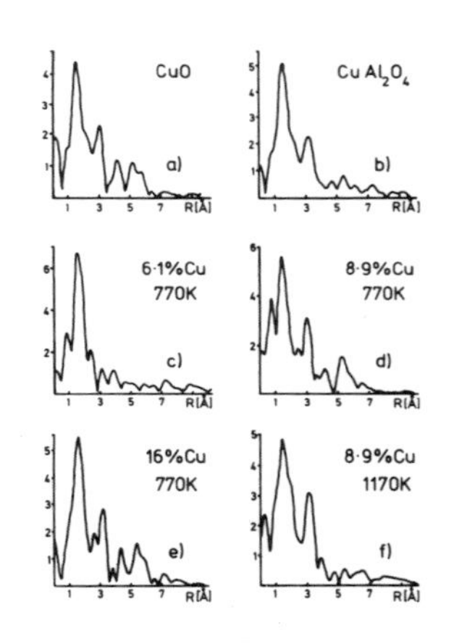

Fig. 11

Fourier transforms from Friedman, Freeman and Lytle (1978) for copper on alumina catalysts. Metal loadings and calcination temperatures are shown.

The author has recently made a preliminary study[32] of some rather similar, $Ni/A\ell_2O_3$ catalysts (prepared by his colleague, Dr. J.R.H. Ross). We are interested in the extent to which an aluminate phase is formed as a function of metal loading and calcination temperature. Fig. 12 shows some raw absorption spectra for energies just above the nickel K edge.

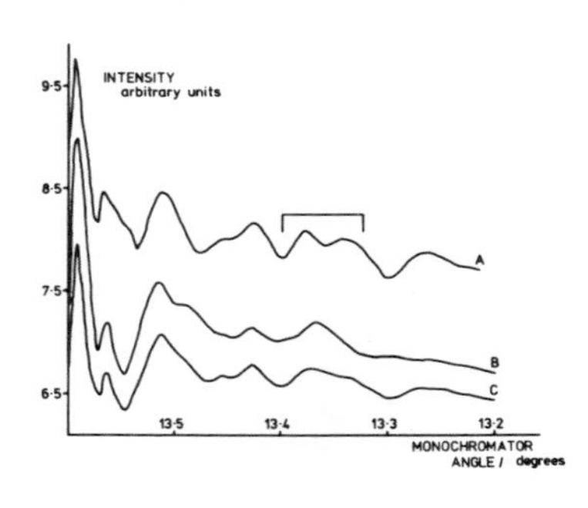

Fig. 12

Nickel K edge absorption spectra. The absorption edge is at the left and photon energy increases from left to right. Curve A, nickel oxide; curve B, nickel aluminate ($NiA\ell_2O_4$); curve C, $Ni/A\ell_2O_3$ catalyst, ~ 10 percent Ni by weight. Note that curve C is a sum of features from curves A and B.
(From Joyner, 1979[32].)

Comparison of curves A (NiO) and B ($NiA\ell_2O_4$) shows that EXAFS distinguishes quite readily between the two compounds. Curve C is that of a catalyst of low nickel content (~ 10 percent by weight) after calcination at 1270 K. The presence of both oxide and aluminate phases is clear, especially in the bracketed region. At much higher nickel contents only an oxide phase is detected.

Lytle, Sayers and Moore[33] have reported a study of an 8 percent copper, 7 percent chromium on alumina material designed as an auto-exhaust catalyst. Their results for this material, obtained before and after the U.S. Federal Test Cycle procedure, are shown in Fig. 13. Marked changes were noted in the chromium spectra, which were interpreted as a conversion of Cr^{5+} to Cr^{6+} in an octahedral environment on exposure to exhaust gases. No marked changes occur in the copper environment, which is that of Cu^{2+} bound to oxygen.

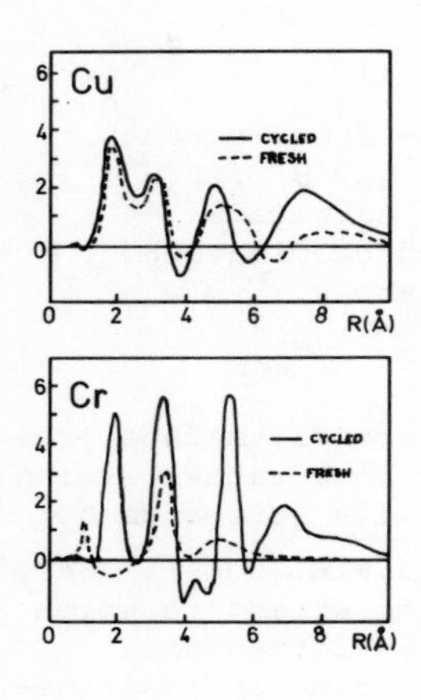

Fig. 13

Fourier transformed results from 8 percent Cu/7 percent Cr on alumina auto-exhaust catalyst before and after use in the Federal Test Cycle procedure.
(From Lytle, Sayers and Moore (1974).)

3.3 Other Studies

In this section we consider a number of diverse examples, where transmission EXAFS has been applied to studying catalysis or adsorption.

Lytle et al.[34] have studied a 1 percent ruthenium/silica catalyst which had been exposed to a 1 percent oxygen in helium mixture. Fourier transforms are shown in Fig. 14. Exposure to oxygen at 673 K lead to the formation of ruthenium (IV) oxide (cf Fig. 14c,e). Of more interest is the result of oxidation at 298 K, (Fig. 14b). The ratio of oxygen atoms to surface ruthenium atoms was calculated to be 3:1 and it was proposed that oxygen molecules were adsorbed "end on" in positions of three-fold symmetry on the ruthenium surface. This conclusion is difficult to accept in view of other studies of the interaction of ruthenium and oxygen[35,36].

It has been argued, from electron microscopic evidence[37], that the ruthenium clusters are in the form of rafts only a single metal atom thick. It is therefore of interest that the

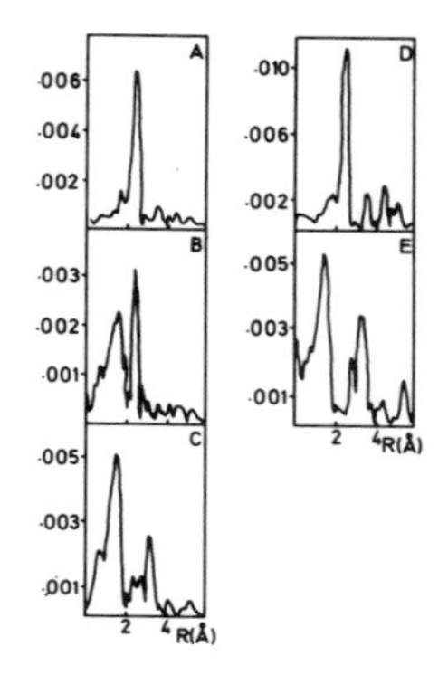

Fig. 14

Results from a 1 percent ruthenium/ silica catalyst. A) clean catalyst; B) after exposure to 1 percent oxygen in helium at 298 K; C) after exposure to oxygen at 673 K. Curves D (ruthenium metal) and E (RuO_2) are reference compounds.
(From Lytle, Via and Sinfelt (1977).)

$\sqrt{2}a$ peak in the reduced catalyst, which occurs at 3.5 Å, (Fig. 14a) is much larger than the $\sqrt{3}a$ peak at 4.35 Å. This contrasts with the result for the clean metal (Fig. 14d) but is similar to that noted above for alumina supported platinum and iridium.

Reed, Eisenberger and Hastings[38] have determined the Ti-Cℓ distance in γ-$TiC\ell_3$, which is the solid component of the Ziegler-Nalta stereospecific polymerization catalyst. The EXAFS spectrum is dominated by a single interatomic distance, due to Ti-Cℓ bonding, estimated to be 2.21 ± 0.01 Å.

A foretaste of the information which EXAFS can be expected to supply on bimetallic catalysts is given by the results of Lytle, Via and Sinfelt[39]. Their preliminary data shows that the EXAFS spectrum of Cu in Cu-Ru catalysts is markedly affected by the presence of ruthenium. Reduced catalysts after exposure to air showed no evidence of Cu-Cu bonding.

Lastly it is appropriate to mention the studies by Stern and co-workers of bromine adsorbed on graphite[40,41]. Use has been made of the polarised nature of the synchrotron source and the highly oriented nature of pyrographite to determine the orientation of molecular bromine on well-ordered 'grafoil' surfaces. At a coverage of 0.2 monolayers the Br-Br axis is thought to be perpendicular to the surface plane, with one bromine atom fixed in the hexagonal site on the basal plane[40]. A very different situation pertains in the coverage range 0.6 - 0.9 monolayers. Here the Br-Br bond is parallel to the surface, with the bromine atoms in adjacent hexagonal sites. The Br-Br bond is stretched by 0.03 Å to accommodate the lattice mismatch. The results provide an interesting insight into a system that LEED[42] and electron microscopy (see Eeles and Turnbull[43]) have already shown to have considerable structural complexity.

ACKNOWLEDGEMENTS

It is a pleasure to acknowledge stimulating discussions with Drs J. Bordas, A.D. Cox and R.F. Pettifer. Financial support from the Science Research Council and I.C.I. (Petrochemicals Division) is also acknowledged.

REFERENCES

1. J.R. Anderson, "Structure of Metallic Catalysts", Academic Press, London (1975).
2. G.A. Somorjai, Proc. 2nd European Conference on Surface Science (ECOSS2), in Surface Science (1979).
3. M.W. Roberts and C.S. McKee,"Chemistry of the Metal Gas-Interface", Clarendon Press (Oxford) (1978).
4. T. Edmonds and J.J. McCarroll, in "Topics in Surface Chemistry" (Ed. E. Kay and P.S. Bagus), Plenum Press, New York (1978).
5. R.L. Kronig, "The Optical Basis of the Theory of Valency" Cambridge University Press (1935).
6. D.E. Sayers, E.A. Stern and F.W. Lytle, Phys. Rev. Lett., 27, 1204 (1971).
7. C.A. Ashley and S. Doniach, Phys. Rev. B11, 1279 (1975).
8. P.A. Lee and J.B. Pendry, Phys. Rev. B11, 2795 (1975).
9. J.B. Pendry, "Low Energy Electron Diffraction", Academic Press, London (1974).
10. S. Gurman and J.B. Pendry, Solid State Comms., 20, 287 (1976).
11. P.H. Citrin, P. Eisengerger and B.M. Kincaid, Phys. Rev. Lett., 36, 1346 (1976).
12. B.-K. Teo and P.A. Lee, J. Amer. Chem. Soc., 101, 2815 (1979).
13. G. Beni, P.A. Lee and P.M. Platzman, Phys. Rev. B13, 5170 (1976).
14. J.A. Del Cueto and N.J. Shevchik, J. Phys. E. Sci. Instrum., 11, 616 (1978).
15. C.H. Macgillavry and G.D. Rieck, Editors, "International Tables for X-Ray Crystallography" Vol.3, Kynoch Press, Birmingham (1968).
16. J. Jaklevic, J.A. Kirby, M.P. Klein and A.S. Robertson, Solid State Comms., 23, 679 (1977).
17. J.B. Hastings, P. Eisenberger, B. Langeler and M.L. Perlman, Phys. Rev. Letts, 43, No.24, 1807 (1979).
18. P.H. Citrin, P. Eisenberger and R.C. Hewitt, Phys. Rev. Letts, 41, 309 (1978).
19. P.H. Citrin, Proc. 2nd European Conference on Surface Science (ECOSS2), in Surface Science (1979).
20. R.L. Park, P.I. Cohen, T.L. Einstein and W.T. Elam, J. Crystal Growth, 45, 435 (1978).
21. B.M. Kincaid, A.E. Meixner and P.M. Platzman, Phys. Rev. Letts, 40, 1296 (1978).

22. R.D. Leapman and V.E. Cosslett, in "Developments in Electron Microscopy and Analysis" p.133 (Ed. J.A. Venables) Academic Press (1976) (Proc. of EMAG 75, University of Bristol, September 1975).
23. G.H. Via, J.H. Sinfelt and F.W. Lytle, J. Chem. Phys., 71, 690 (1979).
24. J.J. Burton, Catal. Revs, 9, 209 (1974).
25. B. Moraweck, G. Clugnet and A.J. Renouprez, Surface Sci., 81, L631 (1979).
26. M.B. Gordon, Thesis, University of Grenoble (1978).
27. I.W. Bassi, F.W. Lytle and G. Parravano, J. Catalysis, 42, 139 (1976).
28. R.W. Joyner, to appear in J. Chem. Soc. Faraday Trans.I (1979).
29. R.L. Burwell, T. Uchijima, J.M. Herrmann, Y. Inoue, J.B. Butt and J.B. Cohen, J. Catalysis, 50, 464 (1977).
30. M. Boudart, A. Aldag, J.E. Benson, N.A. Dougharty and C.G. Harkins, J. Catalysis, 6, 92 (1966).
31. R.M. Friedman, J.J. Freeman and F.W. Lytle, J. Catalysis, 55, 10 (1978).
32. R.W. Joyner, in "Applications of Synchrotron Radiation to the Study of Large Molecules of Chemical and Biological Interest, pp.114 (Ed. R.B. Cundall and I.H. Munro), S.R.C. Daresbury Laboratory (1979).
33. F.W. Lytle, D.E. Sayers and E.B. Moore, Appl. Phys. Lett., 24, 45 (1974).
34. F.W. Lytle, G.H. Via and J.H. Sinfelt, J. Chem. Phys., 67, 3831 (1977).
35. J.C. Fuggle and D. Menzel, Surface Science, 52, 521 (1975).
36. J.C. Fuggle and D. Menzel, Surface Science, 53, 21 (1975).
37. E.B. Prestridge, D.J.C. Yates and L.L. Murrell, J. Catalysis 57, 41 (1979).
38. J. Reed, P. Eisenberger and J. Hastings, Inorganic Chemistry 17, 481 (1978).
39. F.W. Lytle, G.H. Via and J.H. Sinfelt, Petroleum Preprints, pp.366 (1976).
40. E.A. Stern, D.E. Sayers, J.G. Dash, H. Shechter and B. Bunker, Phys. Rev. Letts, 38, 767 (1977).
41. S.M. Heald and E.A. Stern, Phys. Rev. B17, 4069 (1978).
42. J.J. Lander and S. Morrison, Surface Science, 16, 1 (1967).
43. W.T. Eeles and J.A. Turnbull, Proc. Roy. Soc. Lond. A283, 179 (1965).

AN EXAFS STUDY OF RUTHENIUM CATALYSTS

By

A.D. Cox

1. INTRODUCTION

Extended X-ray Absorption Fine Structure (EXAFS) has become an important tool in the characterisation of catalyst systems. With the new generation of high energy electron storage rings providing greater access to the high X-ray fluxes required, and a growing community of scientists with experience in the field, this trend will continue.

In this paper we present in outline a method of analysis based upon an ab-initio calculation of the theoretical EXAFS expression and then go on to extract structural information from a series of ruthenium catalysts supported on silica. The theoretical calculation is based upon the unapproximated 'curved wave' formalism (Lee and Pendry[1]). This has seldom been used as a method of analysis before, because the complexity of the calculations has required extensive use of computer time. However, by rearrangement of the theoretical expression, it has been possible[2] to reduce the time involved by several orders of magnitude and provide a workable algorithm. Another benefit of this approach is that it is possible to place confidence limits on the results using statistical techniques, and to produce a table of correlation coefficients between the various structural parameters. The results for the particular example of ruthenium show that even the refined theory does not describe the EXAFS phenomena exactly; nevertheless it is still possible to obtain useful results that provide a new insight into the structure of these small particles. The results provide a measure of the 'state of the art' in EXAFS analysis.

We begin by outlining the most widely used analysis technique, the Fourier transform method, and point out the limitations of this approach when applied to highly disordered systems.

2. COMPARISON OF TECHNIQUES OF ANALYSIS

i) Fourier transform analysis

The advantage of this approach is that it is computationally fast, and provides a graphic representation of the various inter-atomic distances present in a sample. The fine structure obtained from experiment is first expressed as a function $\chi(k)$ of the wave vector of the emitted photoelectron. The integral

$$\int \chi(k) \exp(-i2kr)\, dk$$

is calculated numerically over the range of k values available in the experimental spectrum and the modulus of this integral plotted out as a function of the parameter r. The interpretation then proceeds by correlating the position of the maxima of this function with the various near-neighbour distances in the experimental sample.

This correlation is based upon the most often quoted[3] form of the theoretical EXAFS expression for single scattering :

$$\chi(k) = \sum \frac{N_j}{R_j^2} \sin(2kR_j+2\delta(k)) \exp(-2\sigma_j^2k^2) \exp(-\tau R_j)|f(k)|/k \quad (1)$$

where R_j and N_j are, respectively, the distance and coordination numbers of the jth 'shell' of atoms. τ is related to the mean free path of the emitted photoelectron, and describes the decay of the EXAFS amplitude from various inelastic processes. $|f(k)|$ is the backscattering amplitude from a surrounding atom, and is atomic-type dependent. The factor $\exp(-2\sigma_j^2k^2)$ is a Debye-Waller term that accounts for the vibration of the atoms with respect to their mean positions R_j.

In an ordered system, this Debye-Waller factor represents the thermal vibration of the atoms. In a disordered system, we have, in addition, variations in the mean positions due to bond distortions. In the latter case, the Debye-Waller factor reflects an ensemble average of all the bond length variations in a material. The coefficient is the mean squared relative displacement (MSRD) of all similar atoms with mean separation R_j.

Referring again to the theoretical expression, $\delta(k)$ represents the phase shift of the photoelectron caused by the potential of the absorbing atom. An approximation usually made is that $\delta(k)$ may be expressed as a linear function of k,

$$\delta(k) = \alpha k + \beta$$

and it can be shown that the Fourier integral of the theoretical expression will then peak at $R_j + \alpha$. The value of α is usually obtained by observing the positions of peaks in the Fourier transform for a known material of similar composition. Using this approach, atomic separations may be measured in 'unknown' materials to an accuracy of around 0.02 Angstroms.

The most obvious disadvantage of this approach when applied to highly disordered systems arises when the EXAFS spectrum shows a rapid decline of amplitude with increasing photoelectric k-vector. As will be seen, the smallest particles show very little EXAFS above approximately 8 $Å^{-1}$ and so the range available for the numerical integration to produce the Fourier transform is limited. The peak widths of the resultant are inversely related to the k-space range available, and so greater uncertainty arises in assigning values to atomic separations.

More subtle effects also contribute to the distorting of the accuracy of determinations of interatomic distances. The first is the approximation of the emitting atom phase shift term by a function linear in k. Although this is quite accurate for high values, at lower k, deviation from this form becomes significant, resulting in a further broadening of the resulting peak. In addition, and more seriously, this can result in a shifting of the position of the peak, leading to incorrect determination of values for R_j. Another factor is the difficulty of assigning an absolute value to the photoelectric k-vector. The k=0 point on the EXAFS spectrum is not usually coincident with the absorption threshold, and the correction involved is hard to calculate. Moreover, this correction is dependent upon the chemical nature of the material, and so may vary between samples of similar composition. This also affects the peak position, and in turn, the measured value of R.

Other information about the coordination and disorder of a material may be obtained by measuring the heights of the peaks in the Fourier transform. The maximum height is found to be proportional to $\int N_j \exp(-2\sigma_j^2 k^2)\, dk$. Knowledge of this ratio between 'known' samples and catalyst samples can give information only about the value of the above integral. It is not possible to resolve the separate contributions from N_j and σ_j unless further information is available. Note that increasing N_j and σ_j together can result in the value of the integral being unchanged. This is equivalent to saying that they are highly positively correlated. Although knowledge of coordination and disorder are often secondary considerations in catalyst characterisation, determination of their 'absolute' values can give insight into the nature, not only of the catalyst itself, but also of the bonding to the support.

ii) Ab-initio calculations

A more profitable method of extracting information from EXAFS spectra is to calculate the theoretical expression using knowledge of the supposed structure of a sample, and then comparing the result with experiment. Analysis then proceeds by varying the guessed structural parameters to produce a 'best fit' to the experimental data. Computationally, this is done by defining a function

$$S = \sum_k \left(\chi_{theoretical} - \chi_{experimental}\right)^2$$

the summation being taken over all discrete data points in the experimental spectrum. This function is clearly dependent upon the parameters of the theoretical calculation. The 'best fit' is obtained by minimising the value of S. When this has been done, the values of these parameters give the structural information required.

Unfortunately, calculations of theoretical EXAFS spectra

made using the expression in equation (1) deviate substantially from experiment up to about 250 electron Volts above the absorption threshold. The problem becomes acute in analysis of disordered systems because most of the information in a spectrum is concentrated in this region. The reason for this deviation is that the above expression makes no allowance for the fact that the photoelectron wave function is spherical. The assumption implicit in equation (1) is that the photoelectron wave function is a plane wave. Relaxing this limitation results in a much more complicated expression, but it may be reduced to a computationally convenient form that is analogous to equation (1). The resulting expression is

$$\chi(k') = 2 \sum_j \frac{N_j}{R_j^2} \cos(2k'R + 2\delta(k') + \Phi_j(k')) \exp(-2\sigma_j^2 k'^2) A_j(k') \qquad (2)$$

In this expression the exponential decay term $\exp(-\tau R_j)$ and the backscattering amplitude $|f(k)|$ have been included in the amplitude $A_j(k')$, and an additional phase shift $\Phi(k')$ added.

The equation is formulated in terms of k' where $k'=k-k_o$. k_o accounts for the difference between the absorption threshold, and the true zero of energy for the emitted photoelectron. By making k_o another parameter of the theoretical expression, and minimising the function S, it is possible to eliminate the effect of a chemical shift in the absorption spectrum.

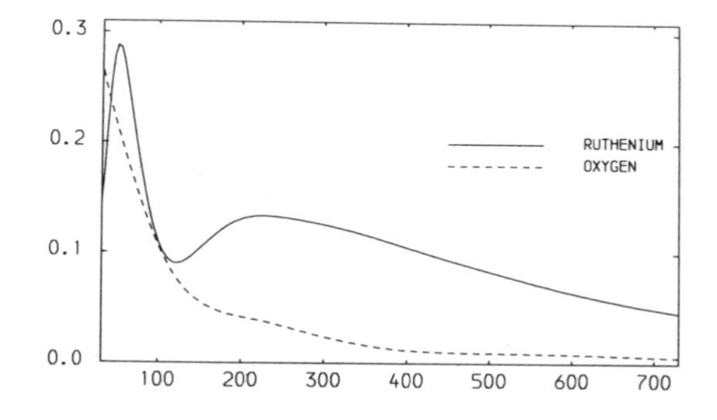

Fig. 1

Amplitude calculations for ruthenium and oxygen.

(x axis: Energy above threshold (eV))

Both the amplitude and phase terms are strongly dependent upon the atomic number of the scattering atoms, and it is possible to exploit this difference to distinguish different kinds of atom. For comparison, Fig. 1 shows the calculations of the amplitudes for oxygen and ruthenium. Oxygen shows a rapid reduction in amplitude with increasing energy above threshold, showing that contributions from this atom will be restricted to the first few hundred electron volts of the spectrum. In contrast, ruthenium shows a more extended structure. Phase-shift calculations of the function Φ (not plotted) show that the contributions to a system from oxygen and ruthenium will be out of phase. These results, presented here as examples for comparison, will be used later in this chapter for the particular case of the dispersed ruthenium catalysts.

3. EXPERIMENTAL PROCEDURE

Four samples of ruthenium catalyst were prepared, differing in both the nature of the silica support, and the compound of ruthenium deposited before reduction. In each case, the ruthenium compound was dissolved in dry toluene, mixed with the support, and the toluene subsequently evaporated slowly in an atmosphere of dry nitrogen. The details of compound and silica are summarised in Table 1. The triruthenium dodecacarbonyl

Table 1

Preparation of Ruthenium Catalysts

Sample Compound	Pore Vol (cc/gm)	Silica Surface Area (m^2/gm)	Dried at
1 $Ru_3(CO)_{12}$	0.4	800*	500°C
2 $Ru_3(CO)_{12}$	0.4	800*	150°C
3 $Ru_3(CO)_{12}$	1.65	300†	500°C
4 $RuCl_3$	1.65	300†	500°C

* Sorbsil Gel, Grade A, Crosfield Chemicals, Warrington.

† Silica Gel, ID/952, W.R. Grace, London.

used was prepared from ruthenium trichloride using the method of Johnson and Lewis[4]. The resultant was then reduced at 350°C with a mixture of 3:1 nitrogen to hydrogen. Catalyst loading on the silica was approximately 2% in every case.

A rough indication of particle size distribution was obtained using an electron microscope. Histograms of particle size are presented in Fig. 2, together with the mean diameters. Small quantities of the catalyst were compressed to form pellets roughly 5 mm thick, and were secured in a cryostat and cooled to liquid nitrogen temperature (77 K) for the absorption spectra to be measured. Measurements were made on the Oxford/Warwick EXAFS apparatus using the 5GeV electron synchrotron 'NINA' at the Daresbury Laboratory of the Science Research Council. The

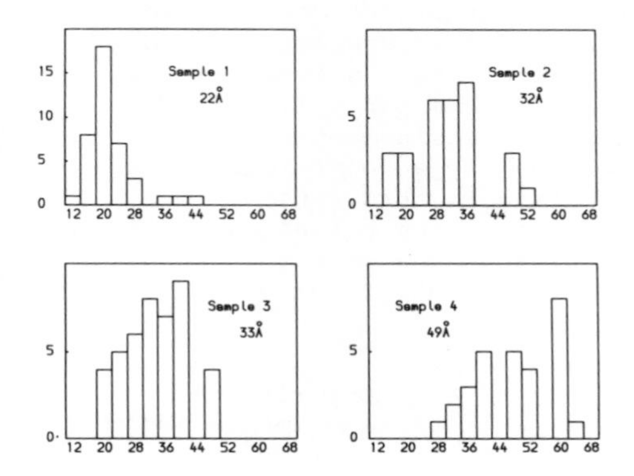

Fig. 2

Particle diameter distribution of ruthenium catalyst samples

apparatus will be described in detail elsewhere[2]. For comparison and calibration purposes, the absorption spectrum of a ruthenium metal foil was obtained on the same apparatus.

4. RESULTS AND DISCUSSION

Fig. 3 shows the fine structure obtained from all five samples. For clarity, the catalyst spectra have been displaced in 5% quanta along the ordinate from the spectrum for the metal foil. The spectra have been arranged in order of increasing particle size to show the difference in the EXAFS spectra more clearly.

As expected, the metal foil spectrum has a complicated fine structure due to the large number of ordered atomic shells within range of the absorbing atom. This complexity is characteristic of a metal. Moving to smaller particle sizes shows a trend of decreasing amplitude, and a faster tail-off of the spectrum with increasing energy above the absorption threshold. This is consistent

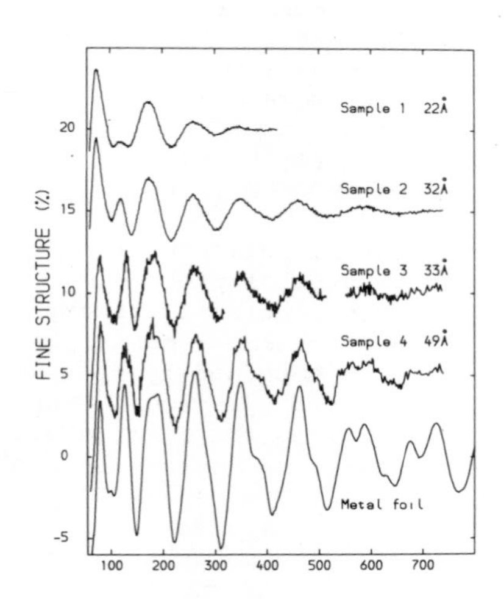

Fig. 3

EXAFS spectra for samples
(x axis: Energy above threshold (eV))
(y axis: Fine structure (%))

with a decreasing coordination number (the ensemble average being lower as a higher proportion of atoms lie on the surface), and a higher disorder coefficient (as atomic sites at various 'depths' within the particle show different nearest-neighbour distances). An interesting feature of these spectra is that the peak at around 120 eV changes shape as the particle size decreases. This trend indicates an increasing contribution from oxygen scattering. As noted before, oxygen contributions are out of phase with ruthenium, resulting in this 'beat'.

These results were quantified using the curved wave fitting approach described above. Fig. 4 shows the experimental data and theoretical best fit obtained for sample 2, which is typical

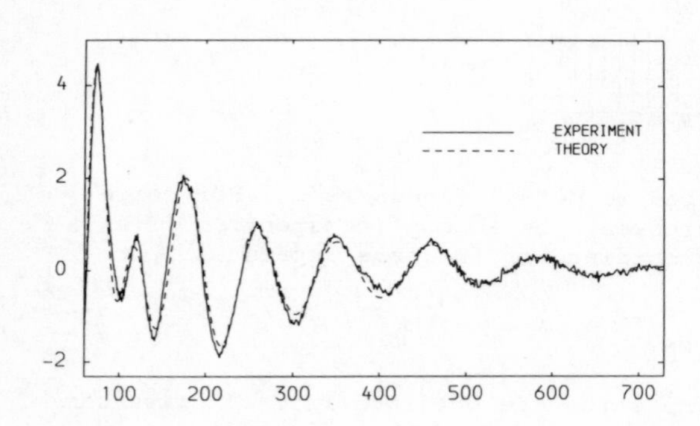

Fig. 4

Comparison of experimental data and theoretical fit for sample 2.
(x axis: Energy above threshold (eV))
(y axis: Fine structure (%))

of the quality of agreement obtained with the other samples. A summary of the parameters extracted from these fits are given in Table 2. These results should be treated as provisional. (In the near future it is intended to publish more accurate results based upon improved phase-shift calculations, and to include an analysis of ruthenium oxide as a further 'model compound'. The results are unlikely to be substantially different from those presented here.)

As noted above, the spectrum for the metal foil contains information from many different scattering paths of photoelectrons within the metal, and so the single scattering theory is demonstrably not accurate enough to model the experimental results. The information from the metal was obtained by fitting the theoretical expression to a Fourier filtered spectrum. The filter was arranged so that information from only the first shell of atoms remained in the spectrum. The additional structure due to multiple scattering was therefore removed. The nearest-neighbour distance obtained was 2.651 Å, which compares with 2.675 Å as obtained from X-ray diffraction. The value obtained for the coordination number of this shell, 4.77, is considerably smaller than the true value of 12 atoms. Both these results require explanation. The discrepancy in the nearest-neighbour distance is most probably due to the inadequate phase-shift calculations. To a first approximation, this will have a

Table 2

Summary of Best-Fit Structural Parameters

	Ruthenium				Oxygen		
	R	σ^2	N	N^*	R	σ^2	N
Sample 1	2.638	8.5	1.58	4.0	1.962	6.4	2.36
Sample 2	2.616	5.1	2.28	5.7	1.941	6.8	1.61
Sample 3	2.643	4.3	2.63	6.6	1.963	1.7	0.16
Sample 4	2.642	1.7	2.80	7.0	1.989	7.6	0.34
Metal	2.651	0.4	4.77	12.0	.	.	.

Values of σ^2 have been multiplied by 10^3 ($Å^2$)

constant effect on the distances extracted from the other samples. The relative changes in interatomic separations (in this case bond length reduction) should be fairly accurate. The discrepancy in the obtained coordination numbers is due to a process known as 'shake up, shake off' (discussed by Lee and Beni[5]). This effect, which also occurs in electron photoemission, has the effect of substantially reducing the amplitude of the EXAFS information, with consequent reduction in the fitted values of N. Unfortunately, the degree of 'shake up, shake off' is not energy independent, and so it does not reduce the amplitude of the EXAFS signal equally throughout the spectrum range. With the approximation that 'shake up, shake off' effects are independent of the photoelectron energy, 'normalised' coordination numbers (N^*) may be calculated. These are also tabulated in Table 2. We shall return to the validity of this assumption later.

Also shown in Table 2 are the parameters corresponding to the presence of oxygen. Samples 3 and 4 give values of coordination number that are very small, and the statistically calculated errors are of the same magnitude, so the other parameters for this atomic 'shell' are unreliable. Samples 1 and 2, however, show the presence of oxygen.

It would have been possible to use these results to obtain information about the bonding of the catalyst samples to the supporting silica. Unfortunately, the samples were exposed to air for some time before the spectra were measured, and so some of the oxygen detected arises, most likely, from adsorption onto the surface. Further observations about the nature of the catalyst-support interaction are therefore precluded.

The ruthenium-ruthenium parameters are more interesting. The reduction of coordination number with particle size suggests a high surface area. This is because surface atoms have a lower coordination than atoms in the bulk of the material. The average coordination will therefore be less. The trend towards higher σ^2 with decreasing size is summarised in Fig. 5. The stars show the calculated radius R and the error-bars extend a distance σ_j (the root mean squared relative displacement) on either side. As can be seen, there is also a small reduction in the first nearest-neighbour distance. These results are consistent with a particle in which the interatomic separations are similar to those of the bulk metal near the 'centre', but become closer nearer to the surface.

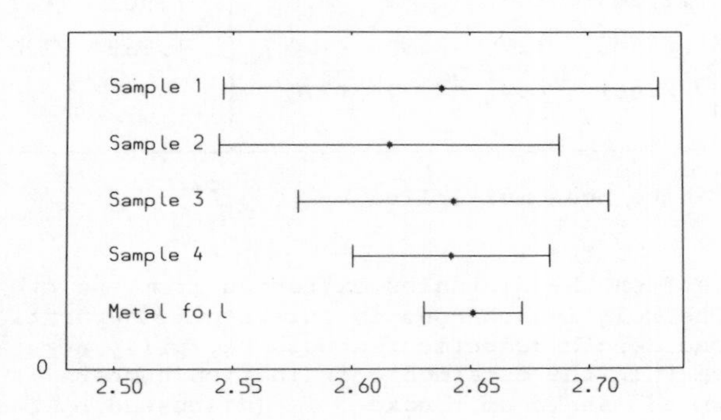

Fig. 5

First nearest-neighbour interatomic separation and root mean squared relative displacement for ruthenium-ruthenium bonds in catalysts and metal foil. (x axis: Interatomic separation (A)).

5. CONCLUSION

The results show that important information may be extracted from highly dispersed catalysts, even when the nature of the samples is such that EXAFS spectra decay rapidly with increasing X-ray energy. Different atomic types may be resolved owing to the dependence of the functional form of $A(k)$ and $\Phi(k)$ upon the atomic number of the scattering atom. Interatomic spacings may be obtained accurately even from rather sparse EXAFS information. Disorder or Debye-Waller coefficients can be obtained that show bond length variations through the bulk of a material, and can give an indication of the nature of the surface. Transmission EXAFS therefore has considerable operational advantages over LEED, surface EXAFS and XPS (ESCA). Coordination numbers are, as yet, more difficult to extract with certainty, owing to the lack of understanding of the phenomena of 'shake up, shake off'. With the availability of a suitable material of similar composition and known structure to use as a model sample, information may be obtained. Results to date on a wide variety of materials suggest that ruthenium is atypical in the magnitude of this effect.

Lee and Beni[5] have concluded, on the basis of photoemission work, that the effect of 'shake up, shake off' above 150 eV is to introduce a constant multiplying factor into the amplitude calculations (this factor being less than unity). This is the basis for the normalisation approach for obtaining coordination information used above. Theoretical calculations of the value of the multiplying factor below 150 eV are clearly very important to the future of the EXAFS technique in disordered systems.

ACKNOWLEDGEMENTS

The author thanks J.H. Beaumont for invaluable help in obtaining the spectra, and R.F. Pettifer and J.B. Pendry for many discussions and encouragement during the course of this work. He is also grateful to R. Murray, A.F. Simpson and R. Whyman of the I.C.I. Corporate Laboratory, Runcorn, Cheshire, for sample preparation and electron microscopy results, and to the Science Research Council for support.

REFERENCES

1. P.A. Lee and J.B. Pendry, Phys. Rev., B11, 2795 (1975).
2. R.F. Pettifer and A.D. Cox, (1980) (in preparation).
3. D.E. Sayers, E.A. Stern and F.W. Lytle, Phys. Rev. Letts, 27, 1204 (1971).
4. B.F.G. Johnson and J. Lewis, Inorganic Synthesis, 13, 92 (1972).
5. P.A. Lee and G. Beni, Phys. Rev., B15, 2862 (1977).

XV

THE PHYSICAL BASIS OF EXTENDED X-RAY ABSORPTION FINE STRUCTURE AND PROBLEMS IN ITS INTERPRETATION

By

R.F. Pettifer

1. INTRODUCTION

Extended X-ray Absorption Fine Structure (EXAFS) is a tool for general structural studies involving all phases of matter. Its main importance lies in its ability to obtain partial pair correlation functions about a particular atomic species in a sample. Further, restrictions on sample types are not great, as gases, solutions, and solids of wide chemical composition are capable of yielding data. Consequently, rapid development has taken place over the last ten years, and the technique is now being applied in situations which hitherto were thought to be inaccessible. Despite its many successes, it is prudent to reconsider the physical basis of the phenomenon and to highlight some of the uncertainties which remain. Gaps in our knowledge place limitations both on the systems which can be studied and upon the accuracies which can be obtained. In this brief article we will firstly consider the equation which describes EXAFS, pointing out its physical origin. The next section describes some of the ways in which structural parameters can be established from the measured data. Finally, conclusions are drawn concerning the best possible techniques of analysis and requirements for further study. Elsewhere in this monograph (Chapters XIII, XIV) specific examples of the application of EXAFS are quoted.

2. THE MATHEMATICAL DESCRIPTION OF THE EXAFS PHENOMENON

A single equation is commonly used as the basis for much of the interpretation of EXAFS information :

$$\chi(k) = -\sum_j \frac{N_j}{R_j^2} \frac{|f_j(\pi)|}{k} \exp\left(-2\sigma_j^2 k^2\right) \exp\left(\frac{-2R_j}{\lambda(E)}\right)$$

$$\mathrm{Sin}\ (2kR_j + \delta_1(k) + \eta_{II,j}(k)) \qquad (1)$$

The basic form of this equation has been derived by Schaich[1], Ashley and Doniach[2], and Lee and Pendry[3], using a scattering formalism developed for low energy electron diffraction (LEED). $\chi(k)$ here is the fine structure function which will be discussed in a later section, and is related to the oscillations observed above the absorption edge. From this equation we can see that the structural parameters we can evaluate are :-

N_j the number of atoms in the jth shell;

R_j the radius of the jth shell of atoms surrounding the absorbing atom; and

σ_j^2 the mean square relative displacement of the emitter from the scatterer.

Equation (1) has been derived under fairly restricted conditions and thus it is wise to review the physical basis for it.

2.1 The Interference Term

The model used is one in which a core electron undergoes photoionisation. The photoelectron is described by its wavefunction which propagates to a surrounding shell of atoms. Part of this wave is then backscattered from the shell and proceeds to interfere constructively or destructively with the original outgoing wave. The matrix element which controls the transition probability, and hence the absorption coefficient, is an overlap integral between the perturbed initial state and the final state. We assume that the perturbation to the initial state takes the form of the dipole operator, as the exciting radiation has a wavelength much larger than the extent of the core state. This implies that the perturbed initial state will be localised in space, and the main contribution to the integral cores from a very small region close to the nucleus. By changing the photoelectron wavevector k we can cause the absorption coefficient to yield maxima and minima. This is precisely the origin of the sine term in equation (1). The argument of the sine term is $(2kR_j + \delta_1(k) + \eta_{II,j}(k))$. The first term in the bracket results from the changing wave vector, and the two remaining terms come from phase-shift changes, (a) as the photoelectron leaves the atom $\delta_1(k)$, and (b) as it is scattered $\eta_{II,j}(k)$. The subscript 1 in $\delta_1(k)$ corresponds to the $\ell = 1$ phase shift which the photoelectron suffers. We include only the $\ell = 1$ term as the dipole selection rules dictate that $\Delta\ell = \pm 1$, and we are emitting from an s state (K shell). Thus equation (1) only applies to photoionisation from states with s-symmetry (e.g. K-edge, and L_I edge). There are, of course, EXAFS oscillations above the L_{II} and L_{III} edge. These edges correspond to excitation of p levels, and thus, using the selection rule, we must expect final states of both s and d symmetry. Equation (1) can thus be modified to take this into account, provided we know the matrix elements[4] for the transitions from $p \rightarrow s$ and $p \rightarrow d$. A further point arises at this stage: we noted that the final state has an angular momentum quantum number $\ell = 1$, and thus the initial outgoing wave is not isotropic but is a p-wave with its axis pointing along the electric field vector. However, equation (1) contains no mention of the orientation of the specimen to the electric field, and was in fact derived for a cubic, polycrystalline or amorphous absorber. Polarisation effects can be, and have been, observed in EXAFS and, indeed, can yield important additional information[5,6,7]. (Heald and Stern (1978), Johansson and Stohr (1979), Cox and Beaumont (1980).) This is particularly relevant with both synchrotron radiation which is polarised and with bremsstrahlung which is partially polarised because the E-vector is always perpendicular to the beam, and thus the preferred orientation needs to be considered if highly anisotropic materials are used.

2.2 The amplitude and phase of backscattering

The third term $\eta_{II,j}(k)$ in the argument of the sine function in equation (1) represents a phase change of the photoelectron wave on scattering. Associated with this phase change is the magnitude of the backscattering amplitude. This can be expressed more economically by noting the backscattering amplitude is a complex quantity which, from scattering theory, is given in terms of partial wave phase shifts by

$$f(\pi) = \frac{1}{k} \sum_{\ell} (2\ell + 1) \sin \delta_1 e^{i\delta_1} (-1)^{\ell} \qquad (2)$$

This expression was originally derived for scattering experiments in which the source and detection of radiation were situated at very large distances from the target compared to the range of the scattering potential. Under these conditions the incident and scattered waves could be assumed to be planar. For an EXAFS experiment, the source-target, target-detection distance is of the order of the interatomic spacing. This, of course, is not very much greater, in extent, than the ion core potential which is responsible for most of the scattering. Thus caution should be exercised in using this asyptotic form for the backscattering amplitude. Lee and Pendry[3] have eliminated this approximation and have shown that the difference between the backscattering given by equation (2) and that taking fully into account the relative size of the ion-core is modest for the case of copper. However, Pettifer, McMillan and Gurman[8] have found that the differences may become very significant when the atomic number of the scatterer is increased. Corrections are needed to take this aberration into account for elements with atomic numbers greater than that of selenium[9].

From equation (2) it can be seen that calculations of backscattering amplitudes require as an input the electron-atom partial wave phase shifts. These can be found by solving Schrodinger's equation. This, in turn, requires the construction of the correct Hartree, exchange and correlation potentials. The phase shifts are extremely sensitive to the potential. This can be demonstrated readily by constructing a muffin-tin potential (see Lourcks[10] for a good discussion of the practical details of this), and varying the muffin-tin radius. Substantial changes in the individual phase shifts are observed when this is done. Similarly, the calculation of the emitter potential phase shift is also sensitive to the model chosen for the excited emitter atom. Further, the tacit assumption of a muffin-tin potential is strictly valid for metallic systems where the inner potential of the solid assumes a constant value between muffin-tin spheres. Clearly this approximation becomes progressively less acceptable with increasing ionicity. We expect, however, that as the kinetic energy of the photoelectron increases, perturbations to the potential from chemical bonding effect will become progressively less important. For example, we may calculate changes of phase shift for the fairly small perturbation of surrounding an arsenic atom with oxygen and tellurium neutral neighbours. Under these conditions the phase $\eta_{II,j}(k)$ shows differences for arsenic between these two environments of 0.6 radians at 50 eV which reduces to a constant 0.2 radian difference above 400 eV. Similarly $\delta_1(k)$ shows differences of 0.15 radian at 50 eV decreasing to 0.1 radian at energies in excess of 150 eV. In total we may well expect difference of phase of at least 0.4 radians if we use environmentally independent phase shifts. Lee and Beni[11] have recognised that inelastic effects within the atom caused by electron-electron correlation give rise not only to a reduction of the emitted wave amplitude but also to perturbation in the phase. These calculations have been tabulated by Teo et al.[12], Lee et al.[13] and Teo and Lee[4]. The phase and amplitude functions in the first two references have been parameterised and consequently the latter set of values should be preferred. It should be noted, however, that these values do not include environmental perturbations and also assume the asyptotic form of the backscattering amplitude, equation (2). Consequently, caution should be exercised when using these tables for highly ionic compounds and heavy element scatterers.

The differences of amplitude and phase for atoms of similar atomic number are small[4,8]. In fact, the differences in $\eta_{II},j(k)$ and $f(\pi)/k$ between elements adjacent in the periodic table are comparable to the environmental effects mentioned above. Consequently, this makes atom discrimination of adjacent elements in the periodic table at present impossible.

2.3 The inelastic loss term

As the photoelectron propagates from the emitter to the scatterer, real excitations of the material occur, resulting in an effective loss of amplitude of the wavefunction. This effect is modelled by the $\exp(2R_j/\lambda(k))$ term in equation (1) where $\lambda(k)$ is the inelastic mean free path. The principal loss mechanism is via coupling to plasmon excitations, and consequently is dependent on electron density. Again an isotropic mean free path is a realistic model for many metals[14], but is of dubious validity for many insulators. It has been argued that the mean free path should be modelled by including in the calculations a constant imaginary part to the self energy of the photo-electron

$$\lambda(k) = 2\frac{E_i}{k}$$

This prediction is in accord with mean free path data from other sources[15] provided the kinetic energy of the photo-electron is greater than 50 eV. Despite this it has been found necessary to treat the imaginary self energy as an adjustable parameter in LEED calculations[16].

We should note here that Lee and Beni[11] have included inelastic loss in their calculations of $f(\pi)$ and thus a term of this form will not occur in equation (1) for the first shell, but it will be required for subsequent shells.

2.4 The Debye-Waller factor

The atoms in the solid are always in motion and consequently the radius of a shell as sampled in the absorption experiment will be an ensemble average over all absorption sites. This effect is included in the calculation via the term $\exp(-2\sigma_j^2k^2)$ where σ_j^2 is the mean square relative displacement of the emitter from the scatterer. We should note here that twice the photo-electron wavevector 2k is equivalent to a momentum transfer Q in X-ray or neutron scattering experiments. This arises in EXAFS because we are backscattering and hence the momentum transfer is 2k. Typical ranges of k in EXAFS spectra are from 4 $Å^{-1}$ to 16 $Å^{-1}$, and making the equivalence with Q we find that our momentum transfer is similar to that found in pulsed neutron time of flight spectroscopy. Consequently we must expect that corrections to the Debye-Waller factor will be necessary to account for the anharmonicity of the interatomic potential. This has been confirmed by Eisenberger and Brown[17]. Consequently we must consider enough parameters in a model pair correlation function to account for this phenomenon. This is particularly important at high temperatures or in loosely bonded materials.

2.5 The origin of the photo-electron wavevector

Thus far in our analysis we have discussed parameters which vary with the photo-electron wavevector k which is defined in atomic units by $2E = k^2$ where E is the energy of the photo-electron with respect to the inner potential of the material. This energy has to be related to the photon energy. It is, of course, located in the vicinity of the absorption edge. If the material is a metal we can expect that the states of appropriate symmetry just above the zero of the inner potential will be occupied, and consequently we can expect the origin of the wavevector to appear below the absorption edge. In contrast we can expect, in some insulators, that excitations to a valence band may well be bound, and in this case the origin of k will lie above the main absorption edge. Unfortunately, this is a basic problem in interpreting EXAFS data. However, in some special cases it is possible to locate a singularity close to the absorption edge[18,19].

2.6 Multielectron Excitation

When theory and experiment are compared it is found that there is a discrepancy in amplitude, with the theory predicting larger values than those measured. Care is necessary in making these comparisons as they are subject to experimental aberrations[9,20,21]. Lee and Beni[11] have argued that multielectron excitation is responsible for a fraction of the photo-electrons losing coherence. The excitations that we are considering are the 'shake-up shake-off' processes familiar in photo-electron spectra. Lee and Beni[11] have argued further, on the basis of rare gas data, that the net effect of this process is to multiply the overall spectrum by a constant which is approximately 0.74. Rehr, Stern, Martin and Davidson[22] have calculated the many-electron overlap integral between the fully relaxed and completely unrelaxed absorbing atom, both containing Z-1 electrons (Z is the atomic number). This overlap integral gives the fraction of the oscillator strength that results in a single channel excitation. Further, these authors also point out that complete incoherence is not achieved in this process. The net result is that the amplitude factor calculated is approximately 0.7 and almost independent of k. This is a comparable figure to that found for bromine by Stern, Heald and Bunker[21] and for As_2Se_3 and As_2S_3, Pettifer[9]. Despite this agreement, other measurements[23] on niobium indicate that the amplitude factor may be much smaller than 0.7 (Pettifer and Cox[23]).

2.7 The fine structure function

The measured experimental spectrum yields the absorption thickness product as a function of photon energy. The fine-structure function $\chi(k)$ is defined, however, by the equation

$$\chi(k) = \frac{\mu(k) - \mu_o(k)}{\mu_o(k)} \qquad (3)$$

where $\mu_o(k)$ is the k shell photoabsorption coefficient of the atoms in isolation and μ is the corresponding function including backscattering. It is clearly inconvenient to measure $\mu_o(k)$ so it is approximated by interpolation. Firstly, the contribution to the absorption from L, M etc excitations is removed by fitting a curve to the region below the absorption edge and extrapolating beneath the edge. Secondly, a smooth curve is fitted through the fine structure oscillations to approximate μ_o. From this $\chi(k)$ can be formed. It is important that this procedure is performed correctly as a false extrapolation can give rise to k dependent amplitude aberrations which may well be confused in particular with the Debye-Waller factor.

3. EXTRACTING STRUCTURAL INFORMATION FROM EXAFS

The particular parameters which we wish to determine are the number of atoms in each shell N_j, the mean radius of the shell R_j, and the pair correlation function about R_j expressed by σ_j^2 (see equation (1)). The methods of extracting these values fall into two groups. Firstly, the non-structural parameters ($|f(\pi)|$, $\lambda(E)$, $\delta_1(k)$, $\eta_{II,j}(k)$, amplitude factor) can be obtained using a known standard material, and the assumption is made then that these parameters are transferable to the unknown material. Comparison is then made between the standard material and the unknown, and relative values of N_j, σ_j^2 and R_j are determined. Secondly, the non-structural parameters can be calculated directly and the unknown fine structure may then be synthesised by variation of the standard parameters. Both techniques may be criticised on similar grounds because they are both subject to the uncertainties listed in Section 2. The comparison techniques are, however, widely recognised as providing more accurate data than those obtained by direct calculation.

3.1 Techniques for Comparison

In applying these techniques we may choose either to compare the fine-structure spectra directly in wavevector space, or alternatively it is possible to localise our structural information in real space by performing a Fourier Transform[24]. Further, hybrid schemes are possible in which the information from scattering from a particular shell is localised to a small region of real space and the information from other shells may then be filtered out. If a back-transform is then performed, the isolated scattering from a single shell is then extracted for further analysis. This technique is necessary when the spectrum contains information from many coordinating spheres. Many examples of this basic approach can be found in the literature[24,25,26]. Unfortunately, the isolation of differing shells may well not be complete, owing to other shells being present nearby. This problem is exacerbated because a Fourier transform is necessarily limited to a finite range in k-space. These limitations are set by a lack of trust in the theory below $k_{min} = 4Å^{-1}$ and a lack of fine structure above $k_{max} = 16Å^{-1}$ (the latter is caused by the decay of $|f(\pi)|$ and the Debye-Waller term $\exp(-2\sigma_j^2 k^2))$. The net result is that the transform is convolved with the Fourier Transform of the window function set by k_{min} and k_{max}. The major limitation is set by k_{min} which creates broad features in real space. Thus limitations are placed on transform-back transform techniques. They are capable of yielding useful information providing (a) suitable standards are found which are chemically similar to the unknown and have shells surrounding the absorber which are well separated radially, (b) caution is exercised in dealing with termination, and (c) consistency is maintained in the choice of the inner potential effects: Hayes and Sen[27] have circumvented the problem (b) by comparing directly in real space. The comparison is performed by synthesising the unknown Fourier Transform from a known transform, and comparing the result over a limited range of real space. The advantage of this approach is that both the radial structure and the appropriate termination effects are synthesised and compared. Owing to the momentum transfer involved in EXAFS ($Q_{min} \simeq 8Å^{-1}$ $Q_{max} \simeq 32Å^{-1}$) some materials only exhibit scattering effects from the first coordination shell. This occurs because the loosely bound second shell has a high mean square relative displacement (σ_j^2) and consequently the $\exp(-2\sigma_j^2 k^2)$ term reduces the contributions to $\chi(k)$ effectively to zero, even at the lowest momentum transfer. This situation is common in studies of metalloenzymes and glasses, and presents a major limitation on the information available. Under these circumstances Fourier filtering is not necessary, and the amplitude

$$\left[\frac{1}{R_j^2} \frac{|f(\pi)|}{k} \exp(-2\sigma_j^2 k^2) \exp\left(- \frac{2R_j}{\lambda(k)}\right)\right] \quad \text{and phase}$$

$(2\delta_1(k) + \eta_{II,j}(k))$ may be parameterised and subsequently used to extract structural information from an unknown material[28].

The basic assumption of this technique is that both amplitude and phase information are transferable between standard and unknown. The validity of these assumptions should be assessed in the light of comments made in Section 2. Clearly the maxim is that the standard should be as closely related chemically to the unknown as possible, and at least a consistent choice of the inner potential is made.

3.2 Direct Calculations

Studies using this approach are generally more inaccurate than those involving standards: however, they do cast important light on the physical basis of the phenomenon. Also direct calculation is the only alternative when standards are not available. Many calculations of this type have been performed[2,3,4,]

[8,9,11,18,29,30,31]. A basic problem in these studies is the choice of the inner potential with respect to the photon energy. Pettifer and Cox[23] have used an iterative fit technique based on Lee and Pendry's[3] theory and have established the correlation coefficients for the fitted parameters. The results show that the choice of inner potential is correlated to the radius of the shell with a correlation coefficient of $\sim$ 0.9. Similarly the parameter σ_j^2 is coupled to N_j again with a correlation coefficient of 0.9. These results suggest that errors in the theoretical phases may be masked by varying the inner potential energy, and this will couple strongly to the shell radii determination. Also, errors in amplitude may be accommodated by variations in N_j and σ_j^2. This is clearly an unhealthy situation and requires further theoretical investigation to fix some of the non-structural parameters; in particular the origin of the inner potential and the many electron excitation amplitude effects.

4. CONCLUSIONS

The attractiveness of EXAFS for the solution of structural problems has encouraged simultaneous activity in both the investigation of the technique and its use in structural studies. Ideally, the problems of the technique should be assessed fully prior to its application. It is hoped that this article will enable the reader to assess for himself the validity of the conclusions presented in the literature. Quantitative values of accuracy have deliberately been avoided as they are highly dependent on the particular circumstances and no global rules can be applied. In general, however, we can state that comparison techniques are capable of giving higher accuracy than direct calculations. Also, the shell radii are often the most reliable parameters to be obtained. In contrast, information from the amplitude of the data, i.e. the number and mean square relative displacements, are still subject to major uncertainties. On the theoretical front, there is still a need for a systematic study of a broad and comprehensive range of compounds to gain information on the validity of the calculated phase shifts and the systematic behaviour of the amplitude dependent terms.

REFERENCES

1. W. Schaich, Phys. Rev. B., 8, p.4028 (1973).
2. C.A. Ashley and S. Doniach, Phys. Rev. B., 11, p.1279 (1975).
3. P.A. Lee and J.B. Pendry, Phys. Rev. B., 11, p.2795 (1975).
4. B.K. Teo and P.A. Lee, J. Am. Chem. Soc., 101, 2815-2832 (1979).
5. S.M. Heald and E.A. Stern, Phys. Rev. B., 17, 4069-81 (1978).

6. L.I. Johansson and J. Stöhr, Phys. Rev. Letts, 43, 1882-1885 (1979).
7. A.D. Cox and J. Beaumont, Phil. Mag. (in press).
8. R.F. Pettifer, P.W. McMillan and S.J. Gurman, The Structure of Non-Crystalline Materials (Ed. P. Gaskell) p.63, Taylor and Francis (1979).
9. R.F. Pettifer, 4th E.P.S. Gen.Conf., Chapter 7, p.522, in "Trends in Physics", Hilger (1979).
10. T. Lourcks, Augmented Plane Wave Method, p.47, Benjamin (1967).
11. P.A. Lee and G. Beni, Phys. Rev. B., 15, 2962 (1977).
12. B.K. Teo, P.A. Lee, A.L. Simmons, P. Eisenberger and B.M. Kincaid, J. Am. Chem. Soc., 99, 3854 (1977).
13. P.A. Lee, B.K. Teo and A.L. Simmons, J. Am. Chem. Soc., 99, 3856-3859 (1977).
14. B.S. Ing and J.B. Pendry, J. Phys. C., 8, 1087 (1975).
15. I. Lindau and W.E. Spicer, J. Electron Spect. and Related Phenom., 3, 409-413 (1974).
16. J.B. Pendry, Low Energy Electron Diffraction, Academic Press (1974).
17. P. Eisenberger and G.S. Brown, Solid State Comm., 29, 481 (1979).
18. B.M. Kincaid and P. Eisenberger, Phys. Rev. Lett., 34, 1361 (1975).
19. B. Holland, J.B. Pendry, R.F. Pettifer and J. Bordas, J.P.C., 11, 633 (1978).
20. R.A. Van Nordstrand, Handbook of X-rays for Diffraction Emission Absorption and Microscopy (Ed. E.F. Kaeble) Chapter 43, McGraw-Hill (1967).
21. E.A. Stern, S.M. Heald and B. Bunker, Phys. Rev. Lett., 42, 1372 (1979).
22. J.J. Rehr, E.A. Stern, R.L. Martin and E.R. Davidson, Phys. Rev., B17, 560 (1978).
23. R.F. Pettifer and A.D. Cox, to be submitted (1980).
24. E.A. Stern, D. Sayers and F.W. Lytle, Phys. Rev. B11, 4836 (1975).
25. P. Eisenberger, R.G. Shulman, B.M. Kincaid, G.S. Brown and Ogawa, Nature, 274, 30 (1978).
26. A.J. Bourdillon, R.F. Pettifer and E.A. Marseglia, J.Phys. C., 12, 3889 (1979).
27. T.M. Hayes and P.N. Sen, Phys. Rev. Lett., 34, 956 (1975).
28. S.P. Cramer, Stanford Synchrotron Radiation Laboratory Report 78/07 (1978) and references therein.
29. P. Lagarde, Phys. Rev., B14, 741 (1976).
30. S.J. Gurman and J.B. Pendry, Solid State Comm., 20, 287 (1976).
31. S.J. Gurman and R.F. Pettifer, Phil. Mag. B., 40, 345 (1979).

XVI

CONCLUDING REMARKS

COMMENTS FROM A CHEMICAL ENGINEER

(Dr. C.N. Kenney)

At first sight the world of chemical engineering seems remote from that of surface science where distances are measured to a fraction of an Ångström at pressures of 10^{-10} Torr, since in industrial catalysis pressures are commonly above 20 bar and chemical reactor diameters range from centimetres to several metres. However, the understanding of surface processes in the laboratory is a crucial preliminary step to their extrapolation to actual reactor operating conditions. Nonetheless, since this involves some dozen orders of magnitude in pressure, it will continue to provide many challenges to catalyst theory. The usefulness of an industrial catalyst, in addition to its cost and mechanical strength, depend on :

1) adequate activity,
2) high selectivity,
3) long life and resistance to sintering and/or poisoning.

All these facets involve rate processes, and so it is perhaps appropriate to pose the question "Whatever happened to catalyst kinetics?" and try to indicate briefly why, in industrial catalysis, there is now a need for more effort to understand the dynamics of surface processes to complement the impressive research effort on the static studies of surface structure. The measured range of surface reaction rates like homogeneous reaction kinetics is considerable.

If rates are expressed in turnover numbers, a typical value is one molecule reacting per site per second, although reactions a hundred times as fast or as slow as this are known. Such figures presuppose an accurate knowledge of surface atom site density, usually taken as 10^{15} sites/cm^2, and make no allowance for inactive surface or a spectrum of sites of different activity. They nonetheless provide a useful basis for comparisons of activity[1]. More widespread reporting of reaction rates obtained from surface science studies would be helpful, since it is often not possible to compare data obtained at low pressures with rate measurements at atmospheric pressure and above (for reactor calculations) because papers frequently omit crucial details about catalyst surface area, pumping speed, or reactor gas volume.

Chemical reactor design is still rather empirical in its use of kinetic equations and relies heavily on mechanisms of the Langmuir-Hinshelwood form to model reactions, employing well-recognised but often incorrect assumptions about sites of uniform activity and a slow surface reaction constituting the rate-limiting step. Power law equations can also be used to relate the rate of reaction to the product of the concentrations (partial pressures) of reactants, each raised to a power found from experimentally observed rates. The intention is then to relate the order of reaction, frequency factor and the activation energy to the thermodynamic properties of the reactants and simple activated intermediates. Theoretical reaction mechanisms reflect the simplest description of the surface process accounting for the formation of observed products in terms of elementary steps. However, the rate-limiting step can be one of a number of processes:- diffusion of reactants to the catalyst, adsorption of reactants on catalyst, reaction of catalyst involving one or more adsorbed reactants, desorption of products from catalyst, or transport of products away from catalyst. The mechanism depends on the nature of the intermediates on the surface, the nature of the interaction, and relative concentrations. The order of reaction only applies to the rate-limiting step and gives no information on further steps. Kinetic measurements alone cannot lead to an unambiguous statement of mechanism, but their central role in reactor performance has attracted considerable research effort in the last two decades[2,3] to

a) elucidate the way in which diffusion and heat and mass transfer can modify the observed behaviour of a catalyst,
b) collect and correlate rate data over a range of experimental conditions.

Catalyst preparation itself, it will be noted in passing, involves further rate and diffusion phenomena associated with pellet impregnation precipitation, calcining and sintering. The vast majority of catalytic reactors are fixed beds, i.e. vessels through which gas (or liquid) flows contacting the catalyst pellets. Very small beds of particles (< 1 mm) lead to high pressure drops with significant pumping costs. If larger particles are used, and the reaction is fast, the reactants cannot diffuse quickly enough into the pellet and the reaction may be confined to a skin of catalyst near the outer surface of the pellet. This means that the 'effectiveness' with which the catalytic metal (often costing thousands of pounds or more) is utilised is significantly less than 100%, and so an understanding of how reactants (and heat) are transferred from the moving fluid to the pellet interstices is important.

The basic ideas may be illustrated simply. For a first-order reaction with velocity constant k (sec^{-1}), the reactant concentration will decay with a time constant of ca $1/k$. The time a molecule takes, on average, to diffuse a distance 'L', say the catalyst particle radius, is given by the Einstein diffusion equation, i.e. L^2/D where D is the molecular diffusion coefficient which may be bulk, Knudsen or a surface diffusion process depending on pressure, temperature and pellet structure in addition to the chemical nature of diffusing species. Clearly if the reactants are to have access in the pellet to the catalyst $L^2/D \approx 1/k$ or $L^2k/D \approx 1$. The dimensionless group L k/D is called the Thiele modulus and plays a central role in determining whether or not diffusional processes affect catalyst behaviour. Three important consequences of a strong diffusional resistance are :-

1) the apparent reaction order is falsified. Kinetic experiments give an apparent order of $n + \frac{1}{2}$ for a reaction of order 'n';

2) the apparent activation energy is Eact/2 where Eact is the 'true' activation energy;

3) the selectivity for intermediates in a consecutive reaction chain is reduced.

These phenomena can often be observed for a given catalyst reaction system from the familiar Arrhenius plot of rate of reaction against the reciprocal of temperature. At low temperatures, the slope corresponds to the activation energy of the heterogeneous reaction, but as the temperature is increased, the slope falls because of significant pore diffusion effects and approaches a value corresponding to half the true activation energy. Further increase in the temperature results in a fast reaction confined to the outer surface of the pellet with the transfer of reactants through the gas film controlling the rate, and the slope tends to a small value since gas phase diffusion processes have negligible activation energies. At very high temperatures, the slope changes again and is associated with a homogeneous gas phase activation energy, since reactants are removed ultimately by gas phase reactions, with the catalytic reaction making only a small contribution to the removal of reactant.

Of comparable importance in industrial reactors is the role of heat. In experimental reactors it is difficult to measure operating catalyst pellet temperature which in exothermic reactions may be over a hundred degrees hotter than that of the surrounding gas. Of greater practical importance is the problem of removing heat of reaction. Reversible exothermic reactions are intrinsically safe, because the net rate falls to zero through self-heating. Sulphur dioxide oxidation is an example and adiabatic reactors of several metres diameter are used. With organic oxidation processes, such as the oxidation of ethylene to ethylene oxide, destructive oxidation, possibly explosive, is a constant hazard. Reactors consequently consist of hundreds of thousands of narrow tubes, a few centimetres in diameter mounted in parallel and surrounded by a coolant. In this way the axial temperature profile can be regulated and the radial temperatures can be controlled.

There appear to be several important gaps in the knowledge of the interaction of diffusion and heat transfer on the microscopic scale. Dr. Acres pointed out that hydrogen absorption rates on catalyst particles of different sizes prepared in different ways can differ markedly. The growth of filaments on platinum used for ammonia oxidation are indicative of the high local temperatures which can occur. At lower temperatures less dramatic but important interactions might be expected if the heat produced through reaction on a metal crystallite is not conducted away rapidly through the ceramic matrix of the catalyst support. A number of examples of hysteresis and oscillatory effects have now been described which may have their origin in transport phenomena[4,5]. It is highly unlikely that such dynamic effects can be incorporated in the scope of Langmuir-Hinshelwood theory with its central assumption of a steady-state on the catalyst surface.

An interesting recent example of the effects of diffusion through the gas film to the outer surface of a catalyst pellet is shown in the oxidation of ammonia to nitric oxide, for nitric acid, over a cobalt oxide catalyst. An unwanted side reaction is oxidation to nitrogen. Nitric oxide formation is first order but oxidation to nitrogen is second order. Both reactions are gas-diffusion-limited and so simply increasing the catalyst surface area, without altering its intrinsic activity, will reduce the ammonia pressure at the catalyst surface and so favour formation of nitric oxide relative to that of nitrogen. Catalyst experiments which failed to consider the role of diffusion could result in unnecessary effort to develop a more selective catalyst.

The sintering of metal crystallite particles is another well-recognised form of catalyst deactivation[6,7]. Other factors contributing to catalyst particle behaviour are the deposition of coke or poisons at a pore mouth at the pellet surface which penetrate slowly into the particle. Alternatively the whole pellet may be permeated by the deactivating species leading to a uniform decay in activity. The economic consequences of having poison-resistant catalysts are considerable, and in many processes some activity would readily be sacrificed if higher poison resistance were available. Thus a cursory examination of the flow sheet of an ammonia plant shows many steps in the process designed to remove oxygen, sulphur, carbon monoxide and carbon dioxide before the stoichiometric nitrogen-hydrogen mixture is allowed to contact the iron based synthesis catalyst.

Catalyst testing is another topic where there have been significant developments in recent years[8]. The batch reactor of the elementary physical chemistry text gives concentration-time data which can be fitted directly to integrated forms of simple rate equations. For complex reactions and exothermic systems, where temperature control is difficult even at low conversions, the treatment of the data may be difficult. Because of this, catalyst evaluation commonly employs continuous reactors, and several versions have been developed. The differential reactor consists of a small plug of catalyst and the conversion per pass is kept to a few percent so that the rate can be obtained directly. However, a serious drawback is the associated measurement error and its influence on the derived rate value. The integral reactor overcomes the analytical problems because large conversions are used. However, for highly exothermic reactions temperature control is difficult and severe axial as well as radial temperature profiles may exist within the reactor[3]. In addition, each experiment only produces a single data point at a given throughput, and so to obtain the rate, a comprehensive experimental programme is necessary. Often the data needs to be numerically differentiated to obtain the rate.

A spinning basket or internal recycle reactor is often used to measure rates of reaction. It is necessary to know the expected main products and also those intermediates which, although present, could easily be ignored in the analysis, especially if their concentrations are low but which might act as catalyst poisons. Integration of differential data may not give good predictions if the chemistry of intermediate formation and reaction is complex and ill-understood. One method of checking this is to pass the reagent-product composition from a pre-reaction through a spinning basket reactor and determine whether the conversion changes when the reactor is fed with an ostensibly similar feed made up from pure components. In these recycle reactors heat and mass transfer conditions can be achieved that are similar to commercial units, in contrast to once-through reactors, and still maintain the simplicity of mathematical analysis associated with the differential reactor. If the recycle is provided by an external circulating pump, the role of mass transfer to catalyst pellets can also be studied independently and quantitatively by varying the gas velocity.

These developments have been accompanied by progress in the design of experiments and data analysis. The need for this is apparent when it is remembered that the conventional 'one variable at a time' approach would necessitate several hundred experiments to explore the effects of temperature, several concentrations and total pressures, gas flow rates in addition to several catalyst formulations on particles of different sizes. The aim is usually to 'fit' hyperbolic models of the form that are widely used to describe rate phenomena in heterogeneous and homogeneous catalysis as well as biochemistry and homogeneous reactions: chemical equilibria are often important in these complex reaction networks. The Langmuir-Hinshelwood equations are a typical example, and the approach is to find a functional form, which describes with the postulated mechanism, and best fits the data. A further step is to determine the parameters of this best model, such as velocity constant, adsorption equilibrium constants, activation energies and heats of adsorption. The crudity of this approach does not need stressing since the probability of a postulated mechanism being correct should be tested, at the very least, by the direct determination of adsorption characteristics of reactants and products. However, it does seem that hyperbolic models are more effective in correlating experimental rates in heterogeneous catalysts than mathematically more flexible power series expressions. The literature on this subject is extensive and excellent review articles exist[9,10].

This brief survey has attempted to indicate the importance of the interaction of heat and mass transfer as an important topic influencing observed catalyst behaviour. In addition, attention has been drawn to improvements in the laboratory reactors used for testing catalysts and the computational methods now available for analysing collected data. Those who toil in the vineyard of catalysis do above all need to be versatile !

REFERENCES

1. M. Boudart, A.I. Ch.E.J., 18, 465 (1972).
2. J.M. Thomas and W.J. Thomas, Introduction to the principles of heterogeneous catalysis, Academic Press (1967).
3. J.J. Carberry, Chemical and catalytic engineering, McGraw Hill (1976).
4. M. Scheintuch and R.A. Schmitz, Cat. Rev. Sci. and Eng. 15, 107 (1977).
5. M.B. Cuttip and C.N. Kenney, ACS, Sump. ser.65, 39 (1978).
6. J.B. Butt, Catalyst Deactivation, Advan. Chem., 109, 259 (1972).
7. S.E. Wanke and P.C. Flyn, Cat. Rev. Sci. and Eng., 12, 93 (1975).
8. V.W. Weekman, A.I. Ch.E.J., 20, 833 (1974).
9. J.R. Kittrell, Advances in Chemical Engineering, 8, 98 (1970).
10. P.M. Reilly and G.E. Blau, Can. J. Chem. Eng., 52, 289 (1974).

XVII

CONCLUDING REMARKS

COMMENTS FROM A PHYSICAL CHEMIST: PROSPECTS AND PROGNOSES

(R.M. Lambert)

Among the criteria which could be considered in discussing and assessing methods of catalyst characterisation are the following :

1. The use of centralised *versus* local laboratory facilities.
2. Speed.
3. Post-mortem or *in situ* examination of the catalyst.
4. Destructive *versus* non-destructive methods.
5. How universally applicable (sensitivity to all chemical elements; works with supported and unsupported catalysts, insulators and conductors).
6. The ability to provide useful information in the presence of a large amount of contamination which may result from operating the catalyst.
7. Cost.
8. Sampling depth of the technique.

While the usefulness of centralised facilities is not in doubt, it must be borne in mind that for many industrial applications speed is of the essence. The testing and development of catalysts often involves screening a large number of different catalyst preparations, each of which might be operated under a variety of different conditions. Such work, although basically of a routine nature, will continue to play a crucially important part in much of the exploratory and developmental research which is carried out in industrial laboratories. In this area the greatest emphasis will therefore be placed on speed, reliability and simplicity. Very often the requirement is for a "fingerprinting" technique which will enable the investigator to recognise a particular preparation as good or bad on the basis of empirically determined correlations between the measured property and subsequent catalytic performance. If the property in question can conveniently be converted into an electrical signal, and, more importantly, if the instrumentation can suitably be interfaced with a microprocessor or minicomputer, then the opportunity exists for automatic data acquisition, processing and storage. It seems likely that such developments in analytical techniques will play an increasingly important part in catalyst research laboratories. A case in point is provided by the work of McNicol (Chapter IX) and the results reported by Acres (Chapter IV). Both temperature programmed reduction and cyclic voltammetry are quick and convenient techniques which are well suited to developments of the kind suggested above. Although electrochemical methods of characterisation are open to the criticism that the surface and its properties may be radically altered by the presence of the electrolyte[1], McNicol has shown that a very good correlation can exist between the properties of alloy catalysts at the gas-solid interface and their behaviour in cyclic voltammetry. While some of the particle beam methods available at central facilities are capable of producing results on a large number of samples very quickly (Cairns, Chapter XI), the EXAFS technique generates information on a very different time-scale (Cox, Chapter XIV and Joyner, Chapter XIII). Although EXAFS seems to offer considerable promise as a means of elucidating the atomic structure of real catalysts, it is likely that the constraints imposed by the availability of machine-time and the non-trivial computational tasks involved will limit its use to basic studies of specific problems, at least in the early phases of its application. As Pettifer points out (Chapter XV), a number of theoretical issues remain to be clarified before the usefulness of the technique to catalyst studies can be fully assessed.

Most of the methods described in this book deal with post-mortem rather than in situ characterisation. The contributions of Acres, Thomson and Wright show how obvious and very considerable advantages can be gained by examining real-time changes in the structure and composition of a working catalyst under realistic conditions. Such studies can be particularly valuable in the investigation of problems of catalyst start-up and poisoning (Sampson, Chapter V). As illustrated by Howie (Chapter VI) and Murray (Chapter VII) it seems clear that electron microscopy will continue to play a central rôle in catalyst characterisation, thanks to the wealth of detailed information which the technique can provide about the most complex of materials. The increasing use of hot stages and environmental cells is particularly to be welcomed. Developments along these lines are being pursued in other areas[2], although progress is inevitably made difficult whenever the technique involves working with charged particles incident upon or ejected by the specimen.

The question of sampling depth is obviously an important one. In catalyst characterisation, one's interest is often directed towards that portion of the solid which is the seat of its catalytic activity, i.e. the immediate surface region, and, in particular, the first few atomic layers. This by no means implies that methods which probe on a much greater scale of depth are without value. Murray's contribution shows how the depth distribution of an active component can critically affect the performance of a catalyst. However, a technique such as conventional emission spectroscopy (sampling depth $\leq$ 1.0 mm) is unlikely to provide information which is of direct relevance to the surface properties of a solid, although it does provide an excellent way of determining bulk levels of contamination. Even here, developments such as laser vaporisation offer the prospect of analysing surfaces with a probing depth of $\sim$ 10μ and a spatial resolution of the same order. X-ray diffraction can provide structural information on a scale of 0.5 to 1.0 μ, which has made it valuable for the study of polycrystalline films and the detection of intermetallic phases in the device and semiconductor industry, as well as in the field of catalysis.

Many of the newly-evolved physical methods of surface analysis have been developed from the well-established techniques of electron microscopy, X-ray fluorescence and particle beam scattering. While it would not be appropriate to attempt a critical review of these newer techniques at this point, it is pertinent to ask what their impact has been and is likely to be in the development of catalytic science. Current practice is to combine a number of complementary probes in a single instrument, and from the point of view of catalyst characterisation some combination of the following techniques would seem to be quite attractive: X-ray photoelectron spectroscopy, secondary ion mass spectrometry, scanning Auger spectroscopy and scanning electron microscopy. Of course the use of such energetic probes always involves some risk of electron induced damage, and, as Edmonds shows in his contribution (Chapter III), this can include relatively subtle changes such as the alteration in oxidation state of a metal ion. Continual improvements in methods of signal detection[3] offer real hope that such effects can be minimised by working at substantially reduced levels of dosage.

From a more theoretical standpoint, one might question the extent to which these newer methods (a number of which are particularly suited to single crystal studies) have increased our understanding of catalysis and catalyst behaviour. Thomson (Chapter XII) makes the point that in many cases the catalytic chemistry takes place on top of a strongly bound "primary layer", so that characterisation of this layer becomes a matter of some importance. Physical methods based on electron diffraction and spectroscopy can be very effective in studying such primary layers which may be generated by the cracking of reactants or products or by the adsorption of deliberately introduced 'promoter' or 'moderator' foreign atoms. Two recent examples of such an approach are (i) a study of the way in which alkalis and halogens modify the oxygen chemistry of silver surfaces in relation to ethylene epoxidation[4,5], (ii) the crystallographic face specificity of CO cracking and carbide formation on cobalt in relation to Fischer Tropsch synthesis[6,7]. A striking example of the effectiveness of basic studies with single crystals is provided by a comparison of the W-NH_3 and Fe-NH_3 catalytic systems. Of the two, far more work has been done on W than on Fe, but there is almost a complete lack of agreement between different groups of investigators regarding even the most basic experimental facts of the W-NH_3 interaction[8,9]. The work on W has involved very little use of single crystal/electron spectroscopic methods. In contrast to this, a relatively small number of single crystal studies on the Fe-NH_3 problem have served to produce a coherent picture of the essential surface chemical features of the system.

In this work, extensive use was made of photoemission, LEED and related techniques[10,11,12]. One might conclude, therefore, that although measurements of this kind are never likely to lead to the development of a new catalyst or a new process in any direct way, they still have a significant and perhaps important rôle to play. By elucidating the nature of elementary surface reactions, intermediate species, and the structure, stoichiometry and stability of surface phases, they provide a guide to thinking in the formulation of concepts and in suggesting new lines of investigation in catalytic research.

REFERENCES

1. Characterisation of Solid Surfaces, Eds P.F. Kane and G.B. Larrabee, Plenum (1974) London, p.190.
2. R.W. Joyner and M.W. Roberts, Chem. Phys. Lett., 60, 459 (1979).
3. Characterisation of Solid Surfaces, Eds P.F. Kane and G.B. Larrabee, Plenum (1974) London, p.318.
4. D. Briggs, R.A. Marbrow and R.M. Lambert, Surface Sci., 65, 314 (1977).
5. P.J. Goddard and R.M. Lambert, Surface Sci., 67, 180 (1977).
6. M.E. Bridge, C.M. Comrie and R.M. Lambert, Surface Sci., 67, 393 (1977).
7. K.A. Prior, K. Schwaha and R.M. Lambert, Surface Sci., 77, 193 (1978).
8. Y.K. Peng and P.T. Dawson, Canad. J. Chem., 52, 1147 (1974).
9. K. Matsushita and R.S. Hansen, J. Chem. Phys., 54, 2278 (1971).
10. M. Grunze, F. Bozso, G. Ertl and M. Weiss, Appl. Surface Sci., 1, 241 (1978).
11. M. Drechsier, H. Hoinkes, H. Kaarmann, H. Wilson, G. Ertl and M. Weiss, Appl. Surface Sci., 3, 217 (1979).
12. M. Weiss, G. Ertl and F. Nitschke, Appl. Surface Sci., 2, 614 (1979).